# Algorithms and Combinatorics 1

Study and Research Texts

W0042865

Karl Heinz Borgwardt

The Simplex Method

A Probabilistic Analysis

With 42 Figures in 115 Separate Illustrations

Springer-Verlag Berlin Heidelberg New York
London Paris Tokyo

Prof. Dr. Karl Heinz Borgwardt
Institute of Mathematics
University of Augsburg
Memminger Str. 6
D-8900 Augsburg, West Germany

Mathematics Subject Classification (1980): 68 C 25, 90 C 05

Library of Congress Cataloging-in-Publication Data
Borgwardt, Karl Heinz, 1949-
The simplex method.
(Algorithms and combinatorics : study and research texts; 1)
Bibliography: p.
Includes index.
1. Linear programming. I. Title. II. Title: Simplex method. III. Series. Algorithms and
combinatorics ; 1.
T57.76.B67 1987 004'.015'1972 86-25995
ISBN-13: 978-3-540-17096-9   e-ISBN-13: 978-3-642-61578-8

DOI: 10.1007/978-3-642-61578-8

© Springer-Verlag Berlin Heidelberg 1987

2141/3140-543210

This book is dedicated

to

KARSTEN and STEFFEN

DORIS

MOTHER and FATHER

I had very little time for them while I was writing it.

PREFACE

For more than 35 years now, George B. Dantzig's Simplex-Method has been the most efficient mathematical tool for solving linear programming problems. It is probably that mathematical algorithm for which the most computation time on computers is spent. This fact explains the great interest of experts and of the public to understand the method and its efficiency. But there are linear programming problems which will not be solved by a given variant of the Simplex-Method in an acceptable time. The discrepancy between this (negative) theoretical result and the good practical behaviour of the method has caused a great fascination for many years. While the "worst-case analysis" of some variants of the method shows that this is not a "good" algorithm in the usual sense of complexity theory, it seems to be useful to apply other criteria for a judgement concerning the quality of the algorithm.

One of these criteria is the average computation time, which amounts to an analysis of the average number of elementary arithmetic computations and of the number of pivot steps. A rigid analysis of the average behaviour may be very helpful for the decision which algorithm and which variant shall be used in practical applications.

The subject and purpose of this book is to explain the great efficiency in practice by assuming certain distributions on the "real-world"-problems. Other stochastic models are realistic as well and so this analysis should be considered as one of many possibilities.

This book was written to collect and to summarize the ideas and results of several papers. I began with the analysis of the average complexity of the Simplex-Method in my dissertation under the advice of Professor H. Brakhage. I want to thank him for directing my mathematical interest towards this fruitful field of research and for many valuable discussions during that time.

My research on this subject has two aspects:

 - the search for a theoretical approach

- the evaluation and estimation of rather difficult expectation values given as integrals.

The theoretical aspect consists of two parts

- finding a Phase II-algorithm which is appropriate for such an analysis (done in the dissertation 1977)

- finding a Phase I-algorithm which meets the necessary stochastic assumptions (done in 1981).

The evaluation turned out to be the greater problem. Only step by step could I obtain the desired results

- asymptotic bounds under special distributions (dissertation 1977), which have been improved significantly in this book

- upper and lower asymptotic bounds under rather general stochastic assumptions (1978, 1979 and 1984)

- polynomial upper bounds under general assumptions for Phase II and for the complete method (1981, improved 1984)

- polynomial upper bounds for the problem type with nonnegativity constraints (1984).

Some of the considerations and calculations are very lengthy, technical and complicated. For that reason I tried to explain in detail and to illustrate what I mean in a great number of figures.

The Introduction gives a survey over the most important developments in this field of research. It consists of four parts. In a first part the formulation of the problem and basic notation are introduced. After that we give a rather informal survey over the main developments in the analysis of the algorithm in the past. Part 3 deals with the question which stochastic model seems to be appropriate. Part 4 summarizes the following chapters, the methods, the results and the conclusions. Here the improvements in the results (compared with their original version) become apparent. This chapter may be interesting even for readers who are not interested in details. In Chapter I the Simplex-Method and the special variant used for our analysis are explained. Here I use an approach which differs from the usual terminology using "basic" and "nonbasic" variables. I hope that this part will be instructive even for people who are not familiar with that algorithm. Chapter II describes the stochastic model and requires elementary probability theory. Rather technical and lengthy are Chapters III and V. Here it is shown that the average number of steps is polynomial. For the proof some elementary techniques of integration in $\mathbb{R}^n$ are necessary. Chapter V shows that the results of Chapter III can be saved even when the assumption of rotational symmetry is weakened to a certain degree. In Chapter IV various methods for the analysis of the asymptotic behaviour are demonstrated. And the Appendix gives some formulae and estimations which are frequently used.

I want to thank Prof. B. Korte, Prof. L. Lovász and Prof. M. Grötschel for many valuable hints and Mrs. Th. Konnerth for the excellent typesetting.

Finally, I want to make two remarks. I have used the "we"-form in the book in order to include the reader into the considerations and to let him participate. And, of course, my English is not perfect. Please do not mind!

TABLE OF CONTENTS

Chapter 0

INTRODUCTION

FORMULATION OF THE PROBLEM
AND BASIC NOTATION

0.1 THE PROBLEM

This book deals with the computational effort required for solving linear programming problems of the following type

	Maximize	$v^T x$
(0.1.1)	subject to	$a_1^T x \le b^1, \ldots, a_m^T x \le b^m$ ($x \ge 0$ optional)
	where	$v, x, a_1, \ldots, a_m \in \mathbb{R}^n$, $b \in \mathbb{R}^m$.

Remark.

Problems containing constraints in equation-form can be transformed into problems of our type by appropriate coordinate transformations (or by writing the constraints as two inequalities).

All these problems have in common that a linear objective function $v^T x$ is to be maximized on a polyhedron X, given as the intersection of m (resp. $m + n$) halfspaces in $\mathbb{R}^n$.

Figure 0.1

The feasible region X

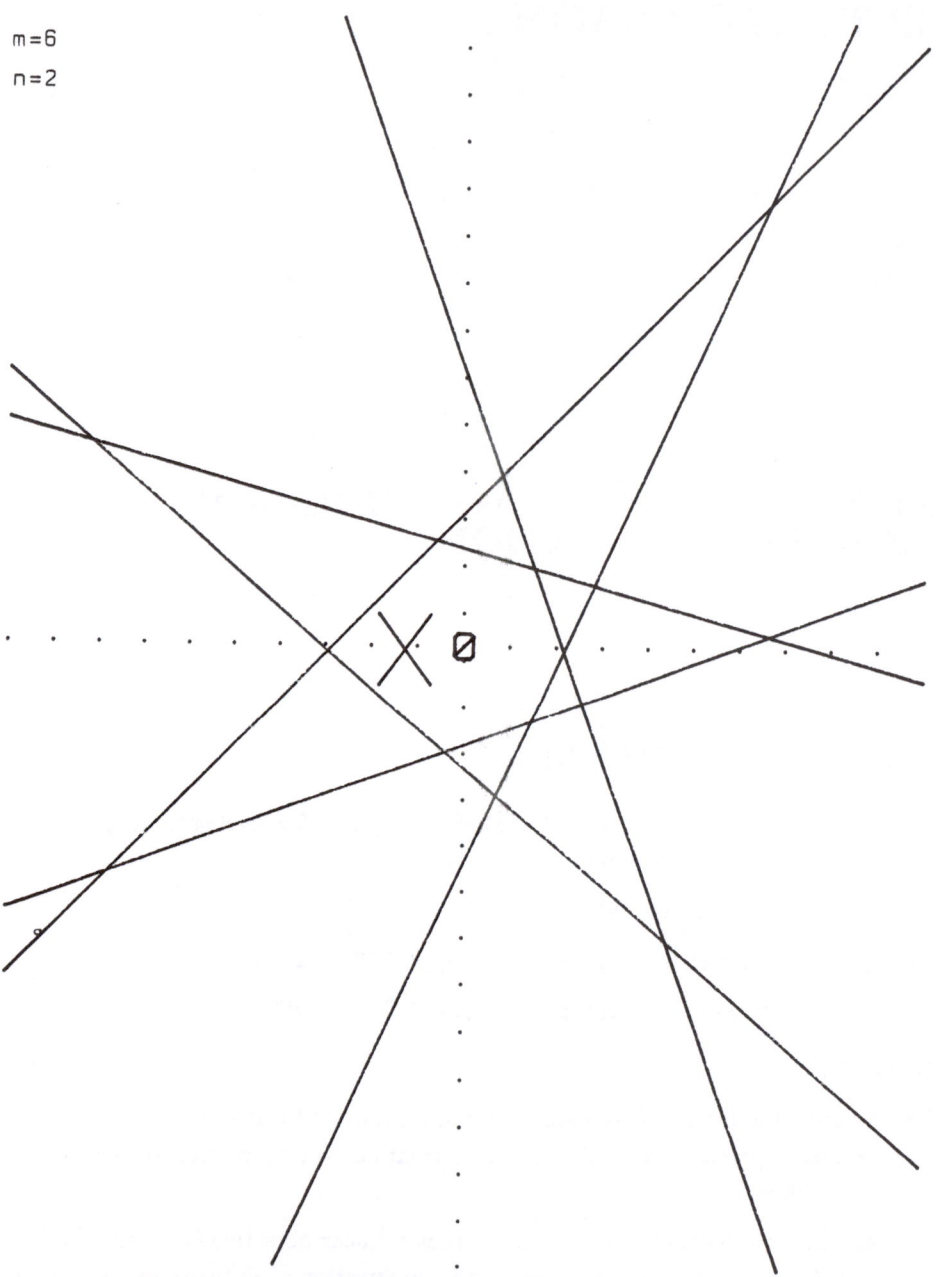

m=6
n=2

Each restriction excludes a halfspace.
X is the intersection of all non-excluded halfspaces.

$$X = \{x \mid a_1^T x \le b^1\} \cap \ldots \cap \{x \mid a_m^T x \le b^m\} \qquad (\cap\{x \mid x \ge 0\})$$

is bounded by some (or all) of the "restriction-hyperplanes" $\{x \mid a_i^T x = b^i\}$.

Clearly the sign constraints can be formulated in the normal way by

$$-e_1^T x \le 0, \ldots, -e_n^T x \le 0,$$

where e_i denotes the i-th unit vector in $\mathbb{R}^n$.

Let us list some basic definitions and notation.

Throughout the book vectors are regarded as **column-vectors**. Row-vectors are written in the form x^T.

Different vectors will be distinguished by lower indices as $a_1, \ldots, a_m$; whereas upper indices $a^1, \ldots, a^n$ denote the **different components** of the same vector a.

A set X is called **convex**, if for all $y, z \in X$ and for all $\lambda \in [0, 1]$ the points $\lambda y + (1 - \lambda)z$ belong to X.

The **convex hull** of a set M is denoted by $\mathrm{CH}(M)$ and defined as the smallest convex set containing M.

The **linear hull** of a set M is denoted by $\mathrm{span}(M)$ and defined as the smallest linear space containing M.

The **convex cone** of a set M (with vertex 0) is denoted by $\mathrm{CC}(M)$ and defined as the smallest set containing all nonnegative linear combinations of the elements of M and the origin.

A restriction $a_i^T x \le b^i$ is called **active in a point** $x_0 \in X$, if $a_i^T x_0 = b^i$.

The restriction is called **redundant**, if $a_i^T x < b^i$ for all $x \in X$.

A **supporting hyperplane** for X is a hyperplane $H = \{x \mid a^T x = b\}$ such that $H \cap X \ne \emptyset$ and $a^T x \le b$ for all $x \in X$.

The intersection of a supporting hyperplane H and X is called a **face** of X if X is not contained in H.

If x_0 is a point of X and if $\{x_0\}$ is a face of X, then x_0 is called a **vertex** of X (a face of dimension 0).

An **edge** of X is a face of dimension 1.

A **facet** of X is a face of dimension $\dim(X) - 1$.

A set of k points $\{a_1, \ldots, a_k\}$ is in **general position** if $\dim(\mathrm{CH}(a_1, \ldots, a_k)) = k - 1$.

Such a polyhedron X (also called the feasible set or region of the inequality system) has some very useful properties, which can be be employed to solve linear programming problems.

1) X is a convex set.

2) The vertices of X are intersection points of (at least) n restriction-hyperplanes. The edges of X are intersection sets of (at least) $n - 1$ restriction-hyperplanes.

3) Every vertex of X is adjacent to (at least) n edges.

4) If X contains a vertex and a nonempty solution set (of points with maximal value of $v^T x$), then this subset contains (at least) one vertex.

5) Every nonoptimal vertex of X is adjacent to an edge, where the objective function $v^T x$ is improved.

A vertex which has maximal value with respect to $v^T x$ (an optimal vertex) will often be called "maximal" vertex.

The terms "at least" can be dropped or replaced by "exactly", if the given problem is not degenerate.

We define **nondegeneracy** *in the following way*

$$(0.1.2) \qquad \textit{All the submatrices of} \quad \begin{bmatrix} a_1^T & \cdots & b^1 \\ \vdots & & \vdots \\ a_m^T & \cdots & b^m \end{bmatrix} \quad \textit{and of} \quad \begin{bmatrix} a_1^T \\ \vdots \\ a_m^T \\ v^T \end{bmatrix}$$

are of full rank.

Note that our definition concerns the complete formulation of the problem, not only the feasible polyhedron X.

Now the problem of finding a solution for (0.1.1) can be reduced to

a) Find a maximal vertex of X

or equivalently

b) Find a set of n linearly independent restriction-hyperplanes such that their intersection point is feasible and maximal.

The mostly used and — as far as we know today — most efficient solution algorithm for such problems is the so-called Simplex-Method, which had been introduced by GEORGE B. DANTZIG in 1947/48. Alternative methods have turned out to be less practicable and much slower.

Figure 0.2a

Bounded feasible region X and bounded objective $v^T x$

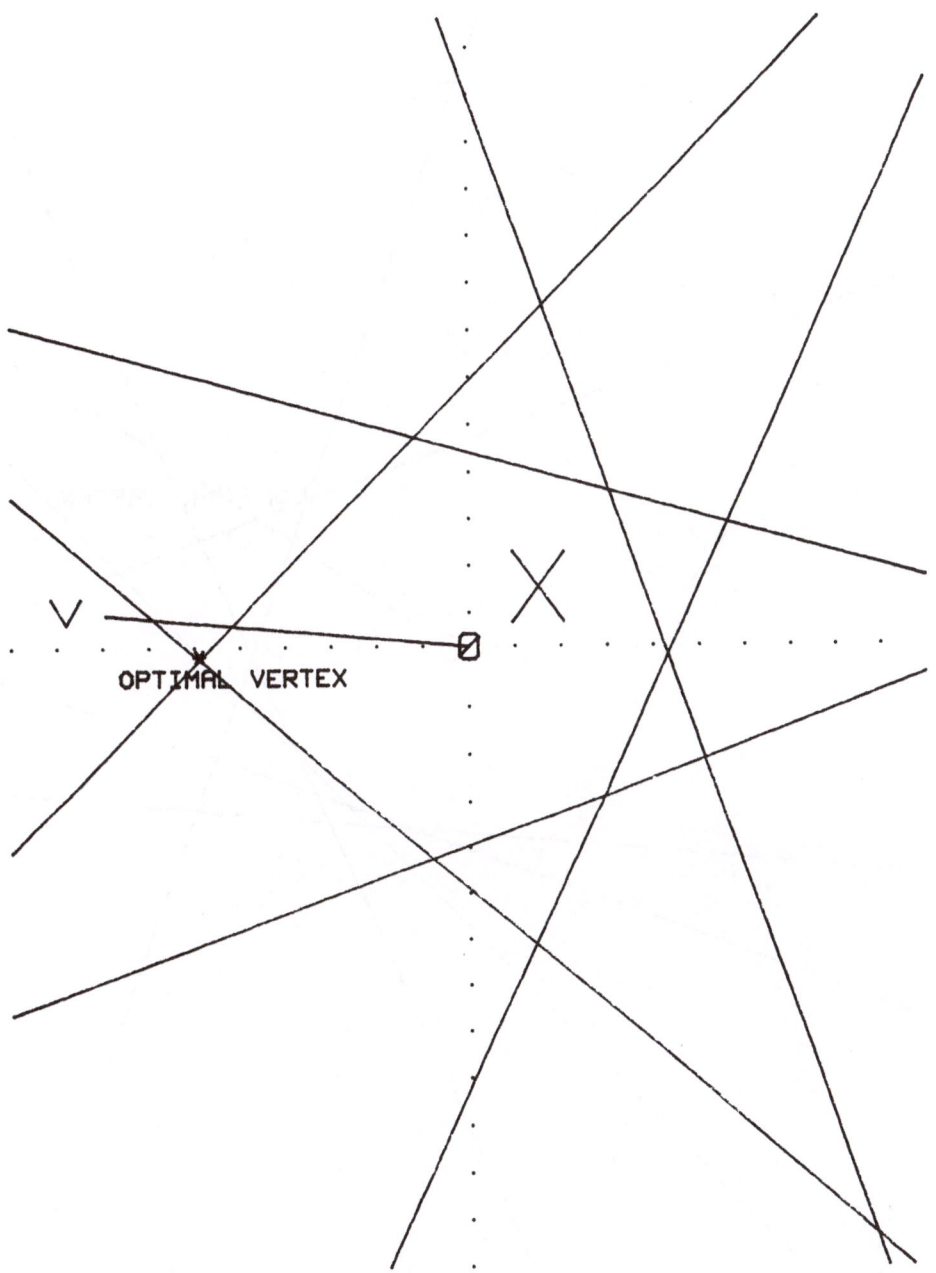

Figure 0.2b

Unbounded feasible region with bounded objective

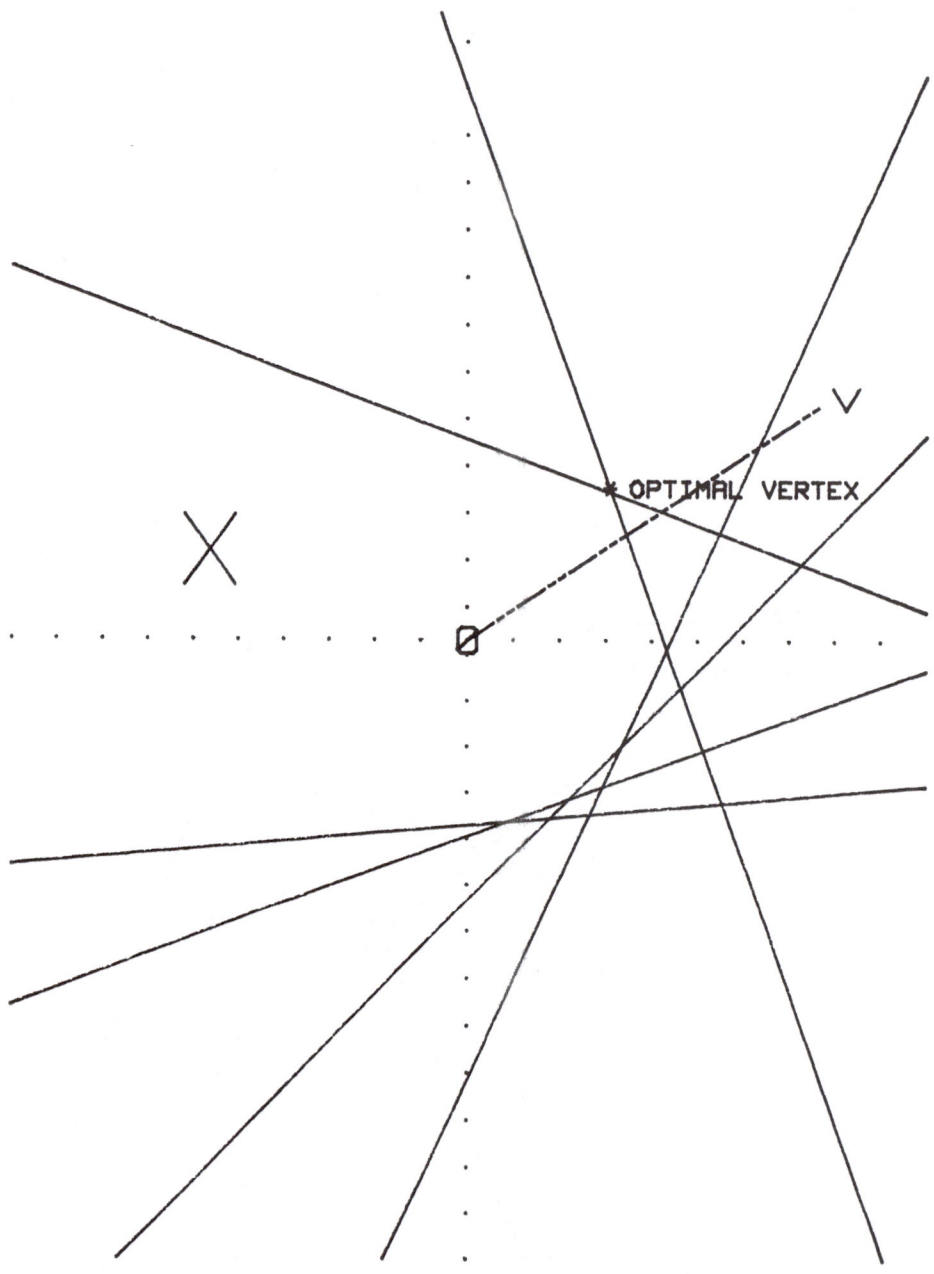

Figure 0.2c

Unbounded feasible region with unbounded objective

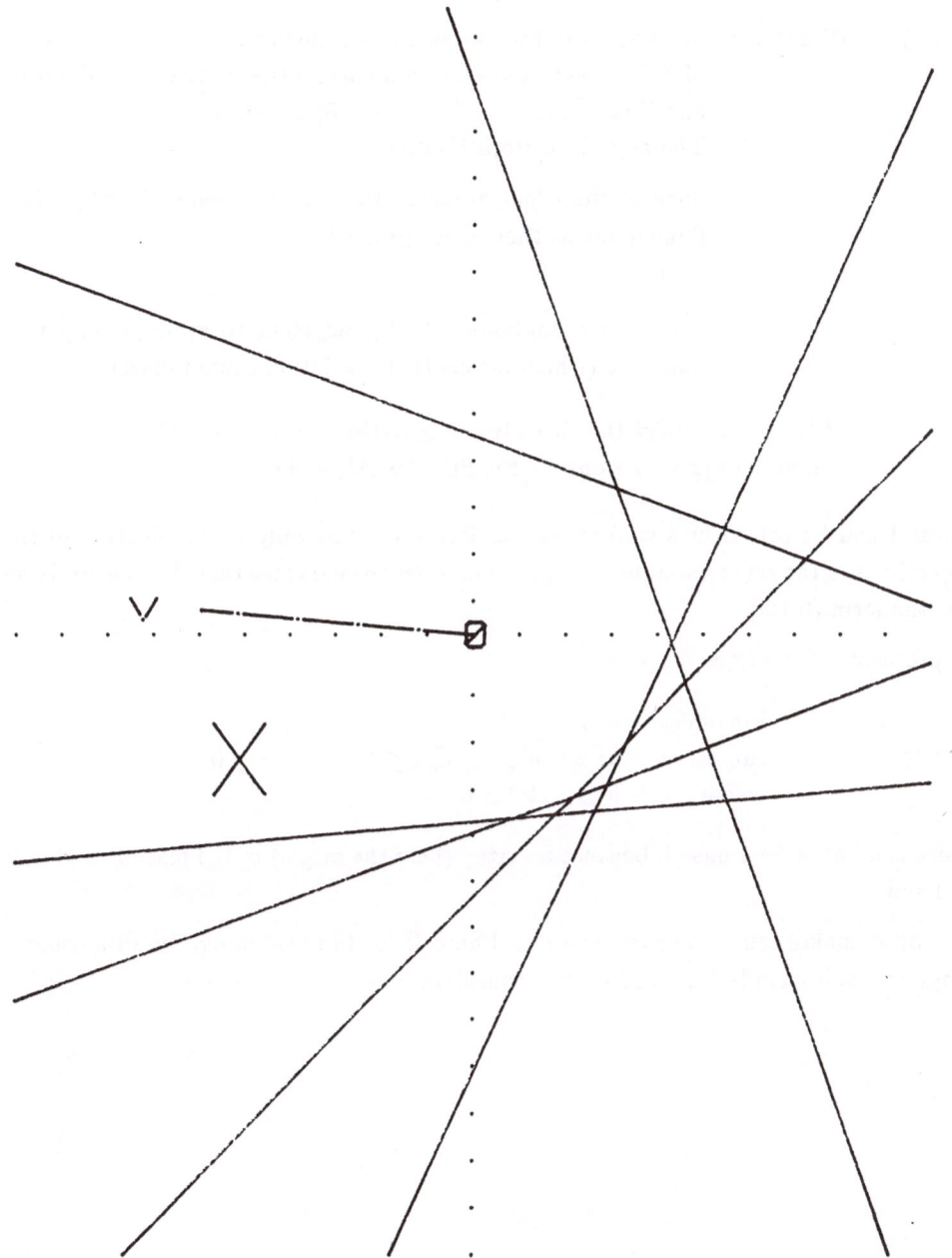

In Dantzig's algorithm (compare DANTZIG 1963) the properties 1) through 5) are exploited to a high degree. The method runs in two phases.

(0.1.3)   **Phase I:**   A feasible vertex x_0 of X is determined and calculated. The algorithm stops if this is not possible.

(0.1.4)   **Phase II:**   Starting from the vertex x_0, a sequence $x_0, \ldots, x_s$ of vertices of X is constructed such that consecutive vertices are adjacent and that $v^T x_{i+1} > v^T x_i$ for $i = 0, \ldots, s - 1$.
The algorithm stops if either

none of the edges adjacent to x_s is increasing the objective (which means that x_s is optimal)

or if

one finds an unbounded edge adjacent to x_s improving the objective (which means that a solution cannot exist).

Phase II is called the **Simplex-Algorithm.** *s is called the number of (pivot) steps for the Simplex-Algorithm.*

Phase I can be solved in a similar way to Phase II. Here only a modification of the objective and the set of variables is required in order to guarantee that the new problem has the form (0.1.5).

In problems of the type

(0.1.5)
$$
\begin{array}{ll}
\text{Maximize} & v^T x \\
\text{subject to} & a_1^T x \le b^1, \ldots, a_m^T x \le b^m \text{ and } x \ge 0 \\
\text{where} & b^1, \ldots, b^m > 0
\end{array}
$$

there is no need for Phase I, because a vertex (here the origin) of the feasible region is at hand.

So it makes sense to concentrate on Phase II for the beginning. In an advanced stage we shall include Phase I into our considerations.

Each of the s pivot steps in Phase II induces a walk from one vertex to an adjacent vertex. At the same time one element of the set of the n active restrictions is replaced by a restriction which had not been active before. Now the new vertex is determined by the system of equations given by the new active restriction set. The knowledge of the old vertex — which is the solution of the old system — is very valuable for the calculation of the new, because both systems have $n - 1$ equations in common. So, much computing time can be saved. The calculation of the new solution is done by using a so-called Simplex-Tableau (see Chapter 1, Section 3). Every transformation of that tableau (as described before) requires a computational effort of at most

$$(0.1.6) \qquad \begin{array}{l} 0(mn) \quad \text{additions/subtractions} \\ 0(mn) \quad \text{multiplications/divisions.} \end{array}$$

Also the effort for finding the restriction which is to be replaced has size $0(mn)$, when the usual rules are applied. This holds particularly for the rule used in our analysis.

Whereas the number of calculations for each of the single pivot steps is easily analyzable, information on the number s is extremely rare. But it is clear that this number is the crucial quantity determining the computational effort.

Information on s is very important for several reasons. So it is advantageous to have an a-priori-estimation for the expenses (costs) of the run of a linear programming job. A second reason results from the mistakes caused by round-off-errors. Since they cumulate during the computation, the results get more and more doubtful. It is clear that the size of possible errors highly depends on the number of pivot steps. In practice one tries to avoid that danger by repeated "reinversions" of the tableau data from the original input data after a certain number of pivot steps. But here our argument remains valid, because a high number of pivot steps requires a great number of reinversions. And every reinversion causes significant additional effort. A third reason is our hope to solve greater problems when the capacity of computers will be greater. We cannot test such greater problems today as we can do with problems of moderate size empirically. And all our good experience with small problems could hide a little exponential growth of s with the dimensions.

Hence we regard $s = s(m, n)$ as a function of the dimensions or parameters m and n.

Until now we have not defined Phase II completely. We still need a rule for choosing the successor vertex for the case that more than one $v^T x$-improving edges are adjacent to the current vertex. Such a rule defines the variant of the algorithm.

Figure 0.3a

<u>Variants</u>

Different variants cause different Simplex-Paths for the
same problem. The length of the path depends upon the
variant.
The example is given for m=12 (4 redundant constraints)
and n=3.

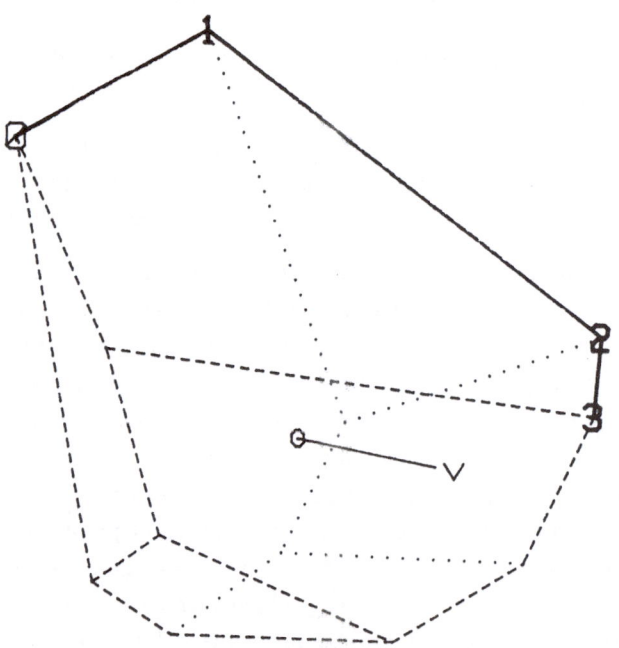

The sequence of numbers shows the sequence of vertices on
the Simplex-Path for maximization of $v^T x$.
Here s=3.
The lengths of the possible paths range from 2 to 6
(compare the following figures).

Figure 0.3 b

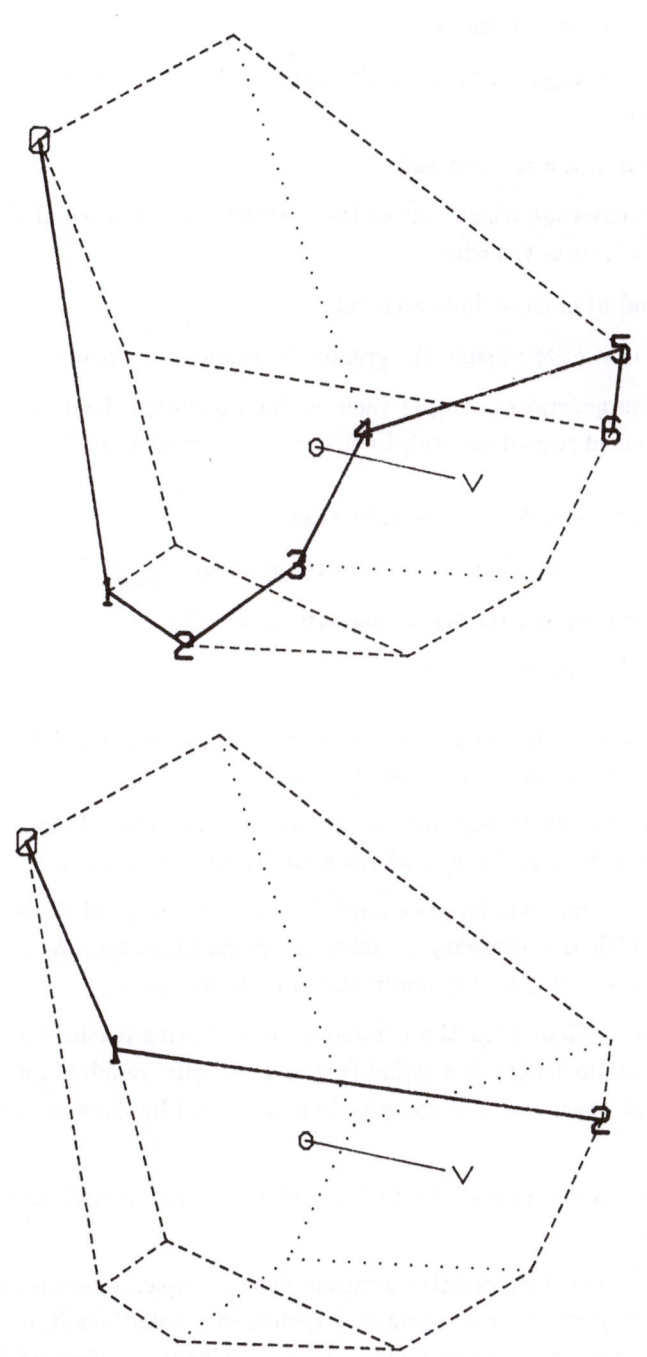

Examples for such variants are (compare LAU 1981, KLEE and MINTY 1972, JEROSLOW 1973, GOLDFARB and SIT 1979)

   – the first (or random) edge rule

       take that edge which is (randomly) first recognized to be improving the objective

   – the method of the steepest edge

       take that edge which causes the greatest improvement of the objective per length unit of the edge

   – the method of greatest improvement

       take that edge causing the greatest absolute improvement of the objective.

There is no general ranking of such variants (compare LAU 1981), although the variants mentioned second and third usually behave very well.

> *So s depends on three influences*
>
> *– the dimensions m and n of the problem*
>
> *– the input data $a_1, \ldots, a_m, b, v$*
>
> *– the variant.*

For every investigation on s one has to define in advance which class of problems shall be solved, and which variant shall be used.

Then one can try to find results on the maximal possible number of steps for (m, n)-problems, denoted by $\bar{s}(m, n)$ (m restrictions, n variables).

Note that our measure for the complexity is the "number of elementary arithmetic operations" which are necessary to solve the given problems. Also we measure the "size" of a problem only by the parameters m and n.

This model differs from the commonly used Turing-machine-model, where the size is defined as the length of a string (number of digits) required for encoding all the input data, and where the time-complexity is measured by the number of moves of the machine.

But in the special case of the Simplex-Method, both models are compatible and closely related.

The reason is that every entry occuring in the Simplex-tableaus during the run of the algorithm is part of the solution of a system of n equations in n variables, whose coefficients are the original input data (compare Chapter 1, Section 3). Application of Cramer's Rule then shows that the length of every such entry is bounded by a polynomial in the length of the total original input. So every elementary operation can be done within a polynomial number of moves of the Turing-machine.

This observation justifies to concentrate on the number of arithmetic operations and even (see 0.1.6) on the number of pivot steps.

For worst case analysis, we are looking for a function $f(m, n)$ such that for every (m, n)-problem the number of pivot steps is bounded from above by $f(m, n)$.

If we try to find out the "average complexity", we look for a function $g(m, n)$, such that the expected number of pivot steps for an (m, n)-problem is bounded from above by $g(m, n)$. Here the expected number is defined by averaging over all (m, n)-problems belonging to a certain probability-space.

The task of **this** book is to obtain information on the **average** or **expected** number of pivot steps.

For that purpose it will be necessary to make a third decision (choice), namely the definition of a stochastic model.

But before diving into probability theory, we want to give a survey over the development of the research on the computational complexity of linear programming and of the Simplex-Method.

A HISTORICAL OVERVIEW

0.2 THE GAP BETWEEN WORST CASE AND PRACTICAL EXPERIENCE

Since the invention of the Simplex-Method in the late forties, most of the users have been very content with its computational speed. So they were optimistic concerning the worst case behaviour, too. It seemed that it would take only little time until an upper bound for $\bar{s}(m, n)$ could be derived theoretically.

Recall that $\bar{s}(m, n)$ is the maximal possible number of pivot steps when a given variant is applied to (m, n)-problems.

Such a bound was expected to be extremely lower than the only bound known at that time

$$(0.2.1) \qquad\qquad \bar{s}(m, n) \leq \binom{m}{n}.$$

This bound is obvious, because there are only $\binom{m}{n}$ sets of n restrictions out of m. At most $\binom{m}{n}$ intersection points have a chance to be a vertex of X.

The number of pivot steps observed in practical applications was very much lower. DANTZIG (1963) wrote:

> "For an m-equation problem with m different variables in the final basic set, the number of iterations may run anywhere from m as a minimum to 2m and rarely to 3m. The number is usually less than $\frac{3}{2}m$ when there are less than 50 equations and 200 variables" (to judge from empirical observations).

This empirical result does even include the effort for Phase I. He continues

> "It has been conjectured that, by proper choice of the variables to enter the basic set, it is possible to pass from any basic feasible solution to any other in m or less pivot steps, where each basic solution generated along the way must be feasible."

This is a quotation of the so-called Hirsch-conjecture given in 1957 by W. M. Hirsch (see DANTZIG 1963)

> "in a convex region in $n - m$ dimensional space defined by
> n halfspaces, is m an upper bound for the minimum length
> chain of vertices joining two given vertices?"

Note that the Hirsch conjecture does **not** refer to a fixed or given variant.

Translated into our type of problems these claims, judgements and conjectures would mean that

- $3(m - n)$ pivot steps should be enough to solve a linear programming problem
- $\frac{3}{2}(m - n)$ is about the average number of steps for such problems
- for every problem it is possible to find a variant of the Simplex-Algorithm which solves the problem in $m - n$ pivot steps.

The background of the Hirsch conjecture is the hope that it might be possible to prevent any restriction which is active at the initial (or current) vertex and nonactive at the final vertex from becoming active again after it once has lost this property.

These statements were so fascinating that a lot of papers were written on this subject. Here the work of Victor Klee should be mentioned. Except for $m - n \leq 5$ and for $n = 3$ the Hirsch conjecture is still open (compare KLEE and KLEINSCHMIDT 1985).

Also it was Victor Klee, who in 1965 (KLEE 1965c) constructed a sequence of linear programming problems with m linear equations and n variables ($m \leq n$), where a pair of vertices can be found which is connectable (by a certain variant) in not less than $m(n - m - 1) + 1$ pivot steps.

In our notation this means that there are (m, n)-problems requiring $(m - n)(n - 1) + 1$ pivot steps.

The surprise was so great that Klee immediately conjectured that this is the absolute maximum. After all David Gale wrote in 1969 (GALE 1969)

> "Thus there is a large and embarrassing gap between what
> has been observed and what has been proved. This gap has
> stood as a challenge to workers in the field for twenty years
> now and remains in my opinion, the principal open question
> in the theory of linear computation."

The great disappointment came in 1971, when it became clear that for many of the usual variants $\bar{s}(m, n)$ is not bounded by a polynomial in m and n.

The first landmark proof in this direction is due to Klee and Minty (KLEE & MINTY 1972). In their simplest example they constructed a sequence of polyhedra for $n = 2, 3, 4, \ldots$ with $m = 2n$ restrictions $a_i^T \leq b^i$.

These polyhedra enable the first edge rule to run through all of its 2^n vertices on a walk from the minimal to the maximal vertex.

So we obtain $2^n - 1$ pivot steps for every problem of the class

Maximize $e_n^T x$

subject to

$$
\begin{aligned}
x^1 &\geq 0 & x^1 &\leq 1 \\
x^2 &\geq \varepsilon x^1 & x^2 &\leq 1 - \varepsilon x^1 \\
x^3 &\geq \varepsilon x^2 & x^3 &\leq 1 - \varepsilon x^2 \\
&\;\vdots & &\;\vdots \\
x^n &\geq \varepsilon x^{n-1} & x^n &\leq 1 - \varepsilon x^{n-1}
\end{aligned}
$$

where ε is arbitrary out of $(0, \frac{1}{2})$.

So, for problems of dimension $(2n, n)$ the number of steps is exponential in the parameters.

To illustrate that behaviour one starts from $x = \begin{bmatrix} 0 \\ \vdots \\ 0 \end{bmatrix}$. On our walk to the solution vertex it is possible to run through all the vertices. This path satisfies all conditions on the Simplex-Algorithm (see attached figure).

Some more sophisticated ideas show that there are positive values α_n and β_n (depending on n) such that

(0.2.2) $\alpha_n m^{\lfloor n/2 \rfloor} < \bar{s}(m, n) < \beta_n m^{\lfloor n/2 \rfloor}$ for all $m > n$, where
$\alpha_n > 0$

and where $\lfloor \cdot \rfloor$ denotes the greatest integer less or equal the argument..

Such a result was derived for a second variant, Dantzig's pivot rule, too. Here the most negative element in the objective row of the Simplextableau determines the pivot column.

Still there remained a little hope that "faster variants" would not fall into the trap of such artificially constructed polytopes. But 1973 Jeroslow showed that for the greatest improvement rule (probably one of the fastest in usual cases) a similar result could be proven. Also for the method of steepest edge a proof of nonpolynomiality was published by Goldfarb and Sit (GOLDFARB & SIT 1979).

Figure 0.4

Klee-Minty examples

Polyhedron and Simplex-Path under first edge rule

n=2
m=4
s=3

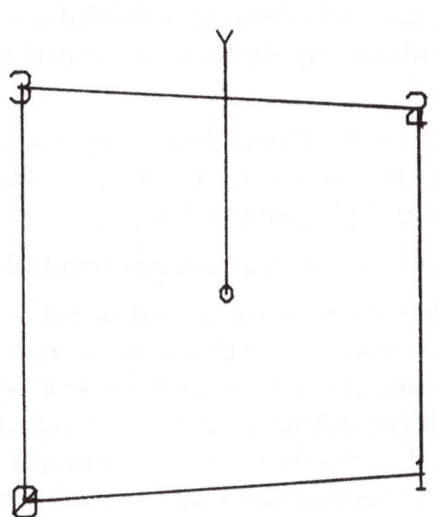

n=3
m=6
s=7

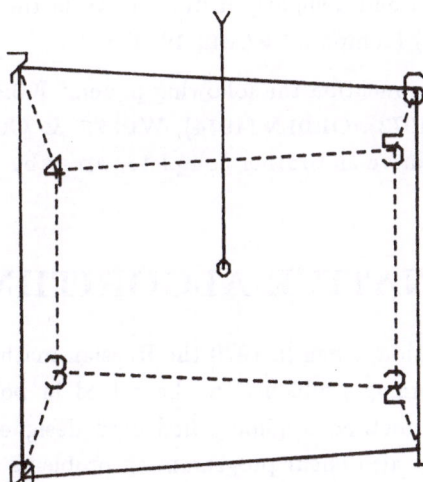

During the following years more variants were analyzed. People were inspired by the hope that still there could be a polynomial variant. But mostly the outcoming was the nonpolynomiality for one more variant (e. g. Bland's Pivot Rule).

The current research is trying to construct new variants which avoid certain disadvantages of the known nonpolynomial ones (see ZADEH 1981).

For the variant used in this book, the shadow-vertex algorithm, the question of polynomiality has been settled in 1983, when Goldfarb gave a proof of nonpolynomiality using polyhedra which are very similar to the original examples of Klee and Minty (GOLDFARB 1983).

The same result can be deduced from a paper of Murty, who showed that the Parametric Simplex-Algorithm is not polynomial, by a slight transformation and modification of the given examples (MURTY 1980).

These theoretical results mean a tremendous contradiction to practical experience.

Very early, before the worst case behaviour had turned out to be so bad, the Simplex-Method was "tested" in numerical experiments. Some of these results were published in the scientific literature. Since the stochastic models and the variants differ to a high degree, it is very difficult to come to a precise conclusion which could include all the observations. And, as a matter of fact, it is not possible to describe and evaluate the complexity of an algorithm finally by numerical examples and tests of examples having bounded size.

But there is one overall impression. Most of the experiments show that there is a slow growth of s depending on the greater parameter (in our case m), which seems even to be sublinear, and a slightly faster growth in the lower of the two dimensions (slightly superlinear) (compare SHAMIR 1984).

Here we want to mention the following papers: KUHN & QUANDT (1963), LAU (1981), LIEBLING (1972), ORDEN (1974), WOLFE & CUTLER (1963), KELLY (1981) and the very informative and rather complete survey by SHAMIR (1984).

0.3 ALTERNATIVE ALGORITHMS

It was a great sensation, when in 1979 the Russian mathematician Khachiyan proved that linear programming problems can be solved in polynomial time (KHACHIYAN 1979). His ellipsoid-method originally had been designed for solving systems of linear inequalities. But also linear programming problems can be solved by use of that method, either by reformulating them as an inequality-system or by applying the "sliding objective function method", which is a slight modification of the original method. For a short description we follow the paper of BLAND, GOLDFARB & TODD (1981).

Suppose that we want to solve the system

(0.3.1) $a_i^T x \le b^i$ for $i = 1, \ldots, m$

                                where $a_i \in \mathbb{R}^n$, $b \in \mathbb{R}^m$.

Let the entries of the input vectors be integers. The length of the input string required for binary encoding of all the input data is

(0.3.2) $L = \sum_{i,j}(\lg_2 |a_i^j|) + \sum_i(\lg_2 |b^i|) + \lfloor \lg_2 n \rfloor + \lfloor \lg_2 m \rfloor + 2mn + 2m + 4.$

The above mentioned restriction on the input data means no loss of generality for usual complexity theory. It can be shown that (0.3.1) has a solution if and only if the following system has a solution

(0.3.3) $a_i^T x \le b^i + 2^{-L}$ for $i = 1, \ldots, m$.

A solution of (0.3.1) can easily be deduced from a solution of (0.3.3) and also the transformation can be done in polynomial time. The main reason why (0.3.3) is solved, is the following: If (0.3.3) has a solution, then the solution set contains even a ball of dimension n with radius 2^{-2L}. Now a sequence of points $x_0, x_1, \ldots \in \mathbb{R}^n$ and a corresponding sequence of symmetric, positive definite matrices $A_0, A_1, \ldots \in \mathbb{R}^{n \times n}$ can be constructed recursively.

The ellipsoid method starts with

(0.3.4) $x_0 = 0$ and $A_0 = 2^{-L} I$, where I is the identity matrix.

On the $(k + 1)$st iteration, the algorithm checks whether x_k satisfies (0.3.3). If yes, we have found a solution and we are ready. If no, we take one of the violated inequalities, where $a_i^T x_k > b^i$ and calculate a new pair (x_{k+1}, A_{k+1}) by

(0.3.5) $x_{k+1} = x_k - \dfrac{1}{n+1} \dfrac{A_k a_i}{\sqrt{a_i^T A_k a_i}}$

                       $A_{k+1} = \dfrac{n^2}{n^2 - 1}\left(A_k - \dfrac{2}{n+1}\dfrac{(A_k a_i)(A_k a_i)^T}{a_i^T A_k a_k}\right)$

Geometrically, x_k and A_k define an ellipsoid E_k by

$$(0.3.6) \qquad E_k := \{x \mid (x - x_k)^T A_k^{-1} (x - x_k) \leq 1\}$$

whose center is x_k. As a result of the construction, E_k contains all feasible points belonging to E_{k-1}. If the solution set of (0.3.3) is nonempty, then E_0 contains a feasible ball with radius 2^{-2L}. Then this ball belongs to every E_k of the iteration sequence. And the construction of the E_k's guarantees that the volume of the ellipsoids shrinks by a factor of less than $e^{-1/(2(n+1))}$ in every step. So it is clear that — after a certain number of iterations — the volume of E_k would be less than the volume of the ball contained in the solution set and in E_k. Hence there is a $C \in \mathbb{R}$, such that whenever a solution exists, the algorithm terminates within the first

$$(0.3.7) \qquad C\,n^2\,L \text{ iterations.}$$

If it does not terminate during that time, then we know that there is no solution. Consequently, the algorithm is polynomial with respect to the number of iterations.

Because computers work with finite precision, it will not be possible to run the algorithm exactly in the way described above on a computer. Some modifications will be necessary, some results of calculations have to be rounded (0.3.5), and one has to find a compromise of the following kind: In order to guarantee that the final result is true, one has to calculate with rather high precision. On the other side the precision must not be too high, because then the polynomiality of the single steps could be lost. And in fact, it is possible to find such a compromise-precision which satisfies both wishes.

The effort for the single step (calculation of a new pair) is greater than for the Simplex-Algorithm, but low-polynomial.

The "method of sliding objective functions", a variation of the method described above, solves linear programming problems of the type

$$(0.3.8) \qquad \begin{array}{ll} \text{Maximize} & v^T x \\ \text{subject to} & a_1^T x \leq b^1, \ldots, a_m^T x \leq b^m\,. \end{array}$$

In the first stage of this method we determine a feasible solution of the inequality-set in (0.3.3) as described above. Let this feasible point be $\bar{x}$.

Then we add the inequality $v^T x \geq v^T \bar{x}$ to the restrictions and continue the iteration. (It is possible to do the iteration step even when the restriction in question is active — not only when it is violated).

Whenever a feasible iterate x_k satisfies $v^T x_k > v^T \bar{x}$, we set $\bar{x} := x_k$ and continue as above.

It can be shown that this method solves the problem (0.3.8) in a polynomial number of steps.

This was not exactly the kind of polynomiality which had been expected and desired by OR-experts all the time. They had wanted the step-number to be independent of L.

In addition, the storage requirements and the necessary precision of the calculations turned out to be so high, that Khachiyan's algorithm did not affect the use of the Simplex-Method. Now it is not regarded as a competitor in practical applications.

In 1982 Megiddo (MEGIDDO 1982) could prove that problems in n variables and with m restrictions of the type

(0.3.9)
$$\begin{array}{ll} \text{Minimize} & c^T x \\ \text{subject to} & Ax \geq b \\ \text{where} & c, x \in \mathbb{R}^n, \ A \in \mathbb{R}^{(m \times n)}, \ b \in \mathbb{R}^m \end{array}$$

can be solved in $0(m)$ steps when n is hold fixed.

The algorithm used in his considerations is recursive and applies the Simplex-Algorithm as a subroutine.

In each of the successive stages of the algorithm the set of restrictions having a chance to be active in the final solution is decreased by a fixed factor.

This can be done by a test which includes the solution of three linear programs of dimension $(n-1)$ with m restrictions. After the test a problem of dimension n with $(1 - \beta)m$ restrictions $(0 < \beta < 1)$ remains. The total effort for that algorithm is

(0.3.10) $C(n)m$ calculation-units,

where $C(n)$ is growing faster than 2^{2^n}.

So, Megiddo shows a moderate growth in m, but the behaviour in n is still unsatisfactory.

In late 1984, a paper of KARMARKAR (1984) was disseminated. He develops a new poynomial time algorithm for solving linear programming problems and shows that its worst-case running time is $0(n^{3.5}L^2)$, which is better than that of the ellipsoid algorithm.

Karmarkar exploits similar complexity-theoretical facts as Khachiyan, but his geometric concept and the approximation method are different. The essential part of Karmarkar's paper is a fast algorithm for solving a special type of problems.

(0.3.11)
$$\begin{array}{lll} \text{Maximize} & v^T x & x, v \in \mathbb{R}^n \\ \text{subject to} & x \in U \cap S & \\ \text{where} & U = \{x \in \mathbb{R}^n \mid Ax = 0\} & \text{with } A \in \mathbb{R}^{(m,n)} \\ \text{and} & S = \{x \in \mathbb{R}^n \mid x \geq 0, \ \sum_{i=1}^{n} x^i = 1\}. & \end{array}$$

Karmarkar's fast algorithm works under four additional assumptions

(0.3.12)   a) U is defined by homogeneous equations.

              b)   The minimal value of (0.3.11) is 0.

              c)   The center of S, the point $\frac{1}{n}e$, is feasible

              d)   A parameter $q \in \mathbb{N}$ is known such that we can stop as
soon as $\frac{v^T x^{(k)}}{v^T x^{(0)}} \leq 2^{-q}$.

The algorithm constructs a sequence of iterates $x^{(0)}, x^{(1)}, x^{(2)}, \ldots$ starting from $x^{(0)} = \frac{1}{n}e$. In every iteration step, a so-called potential function

(0.3.13)
$$f(x) = \sum_{i=1}^{n} \ln \frac{v^T x}{x_i}$$

is diminished by a constant value.

This can be achieved by applying a projective transformation from S onto S which maps $x^{(k)}$ into $\frac{1}{n}e$ of the image space. In this image space one has to minimize a certain linear objective (different in different steps) on the ball with center $\frac{1}{n}e$ and with radius $\frac{1}{4}\frac{1}{\sqrt{n(n-1)}}$ intersected with the image of U. The resulting point belongs to S again and we can map it back into the original space by inverting the projective transformation. There it delivers the iterate $x^{(k+1)}$.

The proof for the (at least) constant improvement of the potential function exploits the fact that a ball around $\frac{1}{n}e$ belongs to S if it has radius $\frac{1}{\sqrt{n(n-1)}}$ and contains S if it has radius $\frac{\sqrt{n-1}}{\sqrt{n}}$.

The constant decrease of f guarantees that after

(0.3.14) $O(n(q + \ln n))$ iteration steps the criterion in d) is satisfied.

So it can be shown that the complete algorithm delivers the solution (after necessary roundings) within a running time of

(0.3.15) $O(n^{3.5}L^2)$

In addition, Karmarkar shows how to reformulate a given problem in general form such that the conditions a) – d) are met.

0.4 RESULTS OF STOCHASTIC GEOMETRY

Observing that the gap between practical experience and worst-case step-number had such an enormous size, scientists became very afraid about the "normal" number of steps. They feared that only a selection of harmless problems out of a large set of very complex and difficult problems had been solved so far. In addition, they were anxious that the dimensions of practical examples had been too small to demonstrate the very bad behaviour of the Simplex-Method. So the average number of pivot steps got into the focus of interest. Only theoretical bounds on the average behaviour for arbitrary m and n could settle these questions.

At that time (about 1972/73) one could find a lot of results of stochastic geometry which turned out to be rather useful for our question. Some probability-theory experts had already studied the number of faces (of arbitrary dimension), the volume, the number of vertices, or edges of polyhedra, which had been generated as the convex hull of m random points in $\mathbb{R}^n$.

Renyi and Sulanke investigated such expected numbers for convex polyhedra generated as the convex hull of m random points in $\mathbb{R}^2$ (particularly for $m \to \infty$). When these points are distributed according to the Gaussian distribution over $\mathbb{R}^2$, then the average number of edges of such polyhedra grows like $\sqrt{\ln m}$ for $m \to \infty$ (see RENYI & SULANKE 1963).

In addition, these papers give valuable information on the methods how line-measures in $\mathbb{R}^2$ can be defined and evaluated.

Similar considerations were made by Carnal in 1970 (CARNAL 1970). He postulated that the random points had to be distributed symmetrically under rotations. In addition, they should be distributed independently and identically. His main interest was directed towards the asymptotic behaviour ($m \to \infty, n = 2$) of the average number of vertices of such polyhedra.

He showed that the order of growth is higher when the distribution of the random points is concentrated close to the boundary of a bounded support in $\mathbb{R}^2$.

In 1965 Bradley Efron developed an integral formula and the according integration-technique for the calculation of expected values of random variables such as the number of vertices, faces, the volume etc. for arbitrary dimension n (EFRON 1965).

In a paper of Hervé Raynaud of 1970 (announced in 1965 RAYNAUD), the author gave two important asymptotic ($m \to \infty$, n fixed) results concerning the expected number of $n - 1$-dimensional faces of such random convex hulls (RAYNAUD 1970).

1) *If $a_1, \ldots, a_m$ are distributed according to the uniform distribution on the (full) unit ball of $\mathbb{R}^n$, then the average number of $n - 1$-faces satisfies*

$$(0.4.1) \qquad E_{m,n}(V) = C(n) \, m^{(n-1)/(n+1)} \, (1 + \gamma(m,n))$$

with $\gamma(m,n) \to 0$ for $m \to \infty$ and fixed n and

$$C(n) = 2 \frac{\lambda_{n^2-1}(\Omega_{n^2-1})}{\lambda_{n^2}(\Omega_{n^2})} \frac{\Gamma(\frac{n^2+1}{n+1})}{(n+1)!} (n+1) \cdot \left[\frac{\lambda_n(\Omega_n)}{\lambda_{n-1}(\Omega_{n-1})} \right]^{\frac{(n^2+1)}{(n+1)}} .$$

Here $\lambda_k(\Omega_k)$ denotes the k-dimensional Lebesque-measure of the k-dimensional unit ball.

(For the size of these quantities and estimations see Appendix Section 1 and 2.)

2) *When these points are distributed according to Gaussian distribution, then*

$$(0.4.2) \qquad E_{m,n}(V) = \frac{2^n}{\sqrt{n}} \, (\ln m)^{\frac{1}{2}(n-1)} \, (1 + \gamma(m,n))$$

with $\gamma(m,n) \to 0$ for $m \to \infty$ and fixed n.

3) The parallel proof for *uniform distribution on the unit sphere* was published by Kelly and Tolle 1979 (KELLY & TOLLE 1979). They showed that there are constants $\alpha > 0$, $\beta > 0$ with

$$(0.4.3) \qquad \alpha^n \, n^{\frac{(n-6)}{2}} \, m \, (1 - \gamma(m,n)) \leq E_{m,n}(V) \leq$$
$$\leq \beta^n \, n^{\frac{(n-5)}{2}} \, m \, (1 - \gamma(m,n))$$

where again $\gamma(m,n) \to 0$ for $m \to \infty$, n fixed.

These results are very important for our purpose, because $E_{m,n}(V)$ can equivalently be interpreted as the expected number of vertices of a polyhedron X, which is generated randomly as the feasible region satisfying $a_1^T x \leq 1, \ldots, a_m^T x \leq 1$, when $a_1, \ldots, a_m$ are distributed as mentioned above.

In addition we want to mention W. Schmidt and Lindberg. They proved that for certain distributions on $\mathbb{R}^n$, which satisfy symmetry under rotations, independence and identity, the number $E_{m,n}(V)$ converges to a **constant** for $m \to \infty$ and fixed n (depending on n only) (see SCHMIDT 1968, LINDBERG 1981, BORGWARDT 1980).

It was the work of Thomas Liebling to find the direct connection between those papers and the average complexity of the Simplex-Method (LIEBLING 1972).

Liebling considered problems

$$
\begin{array}{ll}
\text{Maximize} & v^T x \\
\text{subject to} & a_1^T x \le 1, \ldots, a_m^T x \le 1 \\
\text{where} & v, a_1, \ldots, a_m \in \mathbb{R}^n \ \text{ and } \ m \ge n.
\end{array}
$$

(0.4.4)

Equivalent to his "primal" polyhedron

(0.4.5)
$$
X = \{x \mid a_1^T x \le 1, \ldots, a_m^T x \le 1\} \subseteq \mathbb{R}^n
$$

he introduces the "dual" polyhedron

(0.4.6) $\quad Y = \{y^T x \le 1 \ \text{ for all } \ x \in X\}(\subseteq \mathbb{R}^n) = \mathrm{CH}(0, a_1, \ldots, a_m)$

(CH stands for convex hull.)

He showed that there is a one-to-one correspondence (under nondegeneration) between the vertices of X and the $(n-1)$-dimensional faces of Y not containing the origin.

In the same manner the problem of finding the maximal vertex on X with respect to $v^T x$ can be translated into the problem of finding that facet of Y which is intersected by the ray $\mathbb{R}^+ v \quad (\mathbb{R}^+ := (0, \infty))$.

The sequence of X-vertices touched by the Simplex-Algorithm corresponds to a sequence of such $n-1$-dimensional facets.

Liebling's stochastic model postulates that the vectors a_i are distributed independently and identically and that degeneracy has probability 0. So he is able to translate many known results about $E_{m,n}(V)$ based on various distribution-assumptions.

He does not succeed in making the step from V (the number of vertices in X) to s (the number of pivot steps). The reason is that there is no simple characterization of those vertices lying on the Simplex-Path available. This characterization has to be very simple in order to make stochastic considerations, calculations and the evaluation of mean values in integral form possible.

Figure 0.5

The dual polyhedron Y and the primal polyhedron X

n=3, ·m=12 (4 redundant constraints)

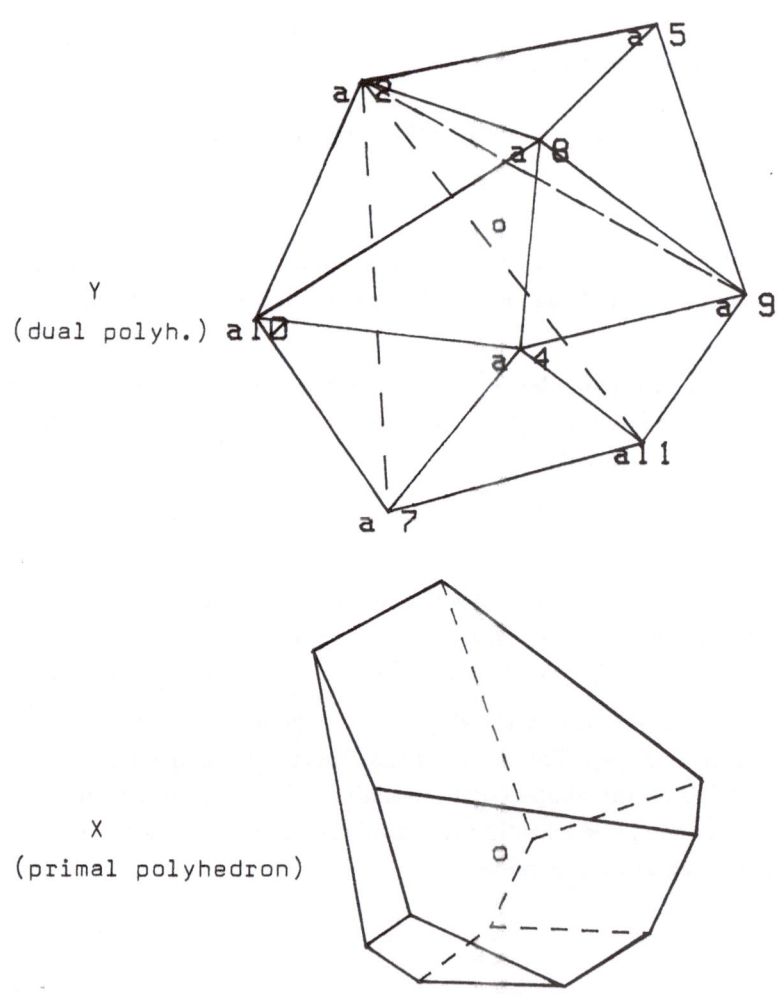

0.5 THE RESULTS OF THE AUTHOR

Such a characterization was given in the work of the author. I used a so-called shadow-vertex algorithm for Phase II. The type of problems to be solved is

$$
\begin{array}{lll}
& \text{Maximize} & v^T x \\
(0.5.1) & \text{subject to} & a_1^T x \leq 1, \ldots, a_m^T x \leq 1, \\
& \text{where} & v, x, a_1, \ldots, a_m \in \mathbb{R}^n \text{ and } m \geq n.
\end{array}
$$

The stochastic assumptions concerning the input data are

(0.5.2) $\qquad a_1, \ldots, a_m, v$ are distributed on $\mathbb{R}^n \setminus \{0\}$

- **independently**
- **identically**
- **symmetrically under rotations.**

(These assumptions give **degeneracy the probability 0**).

The solution variant (for Phase II) is the

(0.5.3) **shadow-vertex algorithm.**

Some reasons for these choices will be given in Sections 9 and 10. Let us — in a few words — describe the variant . Since we concentrate on Phase II for the moment, the vertex x_0 is supposed to be given in advance.

Let u be a vector of $\mathbb{R}^n$ such that $u^T x$ is maximized in x_0. Then the shadow-vertex algorithm constructs a sequence of so-called shadow-vertices $x_0, x_1, \ldots, x_s$ with respect to the two-dimensional plane $\text{span}(u, v)$.

A vertex $\bar{x}$ of X is called shadow-vertex if it keeps its vertex-property even when $\mathbb{R}^n$ and X are projected onto $\text{span}(u, v)$.

Readers who are familiar with parametric or multiobjective programming know that shadow-vertices are those which are efficient with respect to one of the objective-pairs $(u^T x, v^T x)$, $(-u^T x, v^T x)$, $(u^T x, -v^T x)$, $(-u^T x, -v^T x)$.

In 1955 Gass and Saaty (GASS & SAATY 1955) had described a variant generating all the efficient vertices of the parametric problem $(u^T x, v^T x)$, which proceeds in the same way as my variant.

The main progress and advantage of the method is the fact that one can develop a dual description of the variant, which turns out to be very useful for theoretical considerations. This is because it relies directly on our input data $a_1, \ldots, a_m, v$.

Figure 0.6

<u>Shadow-vertices</u>

X

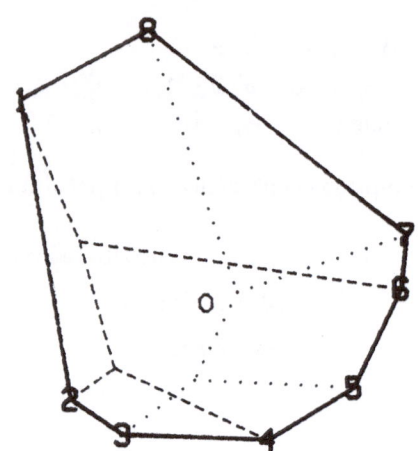

$\Gamma(X)$

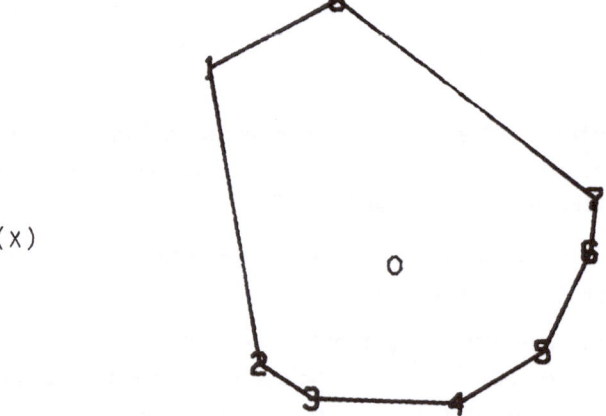

A vertex x of X is called shadow-vertex iff $\Gamma(x)$ is a
vertex of $\Gamma(X)$, where Γ is the projection on the two-
dimensional plane span(u,v).

Since all vertices on the path are shadow-vertices, the number
of pivot steps is bounded from above by S, the number of
(u, v)-shadow-vertices.

This method enabled me to derive integral formulae for $E_{m,n}(S)$ and for $E_{m,n}(s)$, where $\frac{1}{4}E_{m,n}(S) \leq E_{m,n}(s) \leq E_{m,n}(S)$. The evaluation of these formulae yields a lot of interesting results:

First (in the dissertation 1977) I obtained assymptotic ($m \to \infty$, n fixed) bounds for $E_{m,n}(s)$ under special assumptions on the distribution of the a_i's.

Theorem. (Borgwardt, 1977)

There is a function $\varepsilon(m, n)$ with $\varepsilon(m, n) \to 0$ for $m \to \infty$ and fixed n such that

1) for Gaussian distribution

$$(0.5.4) \qquad E_{m,n}(S) \leq 2\sqrt{\pi}\ n^{3/2}\ \sqrt{\ln m}\ (1 + \varepsilon(m, n))$$

2) for uniform distribution on the unit ball of $\mathbb{R}^n$

$$(0.5.5) \qquad E_{m,n}(S) \leq \sqrt{2\pi}\ n^2\ m^{1/(n+1)}\ (1 + \varepsilon(m, n))$$

$$(0.5.6) \qquad E_{m,n}(S) \geq C_2\ m^{1/(n+1)}\ (1 - \varepsilon(m, n))$$

3) for uniform distribution on the unit sphere of $\mathbb{R}$

$$(0.5.7) \qquad E_{m,n}(S) \leq \sqrt{2\pi}\ n^2\ m^{1/(n-1)}\ (1 + \varepsilon(m, n))$$

$$(0.5.8) \qquad E_{m,n}(S) \geq C_3\ m^{1/(n-1)}\ (1 - \varepsilon(m, n))$$

where $C_2, C_3 > 0$.

In 1978 and 1979 I could generalize these results.

Theorem. (Borgwardt, 1979, 1980, 1982a)

1) For all distributions according to our stochastic model (0.5.2) we know that

$$(0.5.9) \qquad E_{m,n}(S) = 0(m^{1/(n-1)})\ \text{ as function of } m \text{ while } n \text{ is fixed.}$$

2) For each such distribution $\tilde{F}$ with bounded support there is a function $\varepsilon_{\tilde{F}}(m, n)$ with $\varepsilon_{\tilde{F}}(m, n) \to 0$ for fixed n and $m \to \infty$ (depending on the distribution) such that

$$(0.5.10) \qquad E_{m,n}(S) \le \sqrt{2\pi} \; n^2 \; m^{1/(n-1)} \; (1 + \varepsilon_{\tilde{F}}(m, n))$$

3) Under the conditions of 2) we have even for problems with nonnegativity constraints

$$(0.5.11) \qquad E_{m,n}(S) = 0(m^{1/(n-1)}) \quad \text{as function of } m \text{ while } n \text{ is fixed.}$$

In the following year I obtained some information on the minimal possible order of growth for $E_{m,n}(S)$ as a function of m.

Theorem. (Borgwardt 1980, 1982a)

1) For all distributions according to our stochastic model, where $P(\|a\| \ge r)^{-1}$ is of polynomial order in r for $r \to \infty$, we have

$$(0.5.12) \qquad E_{m,n}(S) = 0(1) \quad \text{as function of } m \text{ while } n \text{ is fixed.}$$

2) No distribution in our model with bounded support yields $E_{m,n}(S) = 0(1)$, but for every $\delta > 0$ there is a distribution with bounded support such that

$$(0.5.13) \qquad E_{m,n}(S) = 0(m^\delta).$$

$\square$

Since 1977 I had been trying to get rid of the disturbing asymptotic correction term $\varepsilon(m, n)$. This was necessary to get information on the behaviour of $E_{m,n}(S)$ when both m **and** n grow.

Finally in 1981 I reached my long-desired aim.

Theorem. (Borgwardt 1981, 1982b)

For all distributions according to our stochastic model we know that

$$(0.5.14) \qquad E_{m,n}(S) \le e\pi(\frac{\pi}{2} + \frac{1}{e}) \; n^3 \; m^{1/(n-1)}.$$

This was the proof of polynomiality for $E_{m,n}(S)$ in m and n.

But still I had to show polynomiality of the complete method. For this purpose I developed a method for solving the total problem (including Phase I). It works in $n-1$ stages with growing dimension $k = 2, 3, \ldots, n$. In every stage the shadow-vertex algorithm is applied.

This method enabled me to obtain results on the number of pivot steps required for the complete solution and not only for Phase II. The crucial feature of the complete method is that in every stage all the stochastic conditions of our stochastic model are met. This is not the case for usual Phase I-methods. However, the usual methods may do their task in a faster and more effective way.

For the analysis of that complete method I could apply the result of the Theorem above (Phase II) on any of the $n-1$ stages and prove

Theorem. (Borgwardt 1981, 1982b)

For all distributions according to our stochastic model our method for solving the complete problem does not require more than

$$(0.5.15) \qquad \frac{e\pi}{4}\left(\frac{\pi}{2} + \frac{1}{e}\right) n^2(n+1)^2 \, m \qquad \text{pivot steps on the average.}$$

This was the proof of average polynomiality of the complete method.

Many of these results have been improved in the meantime, some others have been added. See Section 11!

0.6 THE WORK OF SMALE

In 1980/1981 George B. Dantzig (DANTZIG 1980) gave an interesting plausibility-proof for polynomiality of the expected number of pivot steps.

Far the greatest public attention gained the work of Steve Smale in 1982 (SMALE 1982, 1983). His papers and results gave this field of research a strong push forward, because

- he treated the problem from a more general view

- he involved a new solution method which unifies the work of Phase I and Phase II

- he obtained interesting asymptotic results

- many researchers became familiar with this field.

We want to describe Smale's considerations briefly. To remain consistent, we try to translate them into our notation.

Smale deals with problems of the type

(0.6.1)
$$\text{Maximize} \quad v^T x$$
$$\text{subject to} \quad a_1^T x \le b^1, \ldots, a_m^T x \le b^m \text{ and } x \ge 0$$
$$\text{where} \quad v, x, a_1, \ldots, a_m \in \mathbb{R}^n, b \in \mathbb{R}^m.$$

This linear programming problem is imbedded into a more general problem, the Linear Complementarity Problem (LCP) (see COTTLE & DANTZIG 1968).

When such an LCP is solved, then the solution of (0.6.1) and of the corresponding dual problem (0.6.2) are at hand.

(0.6.2)
$$\text{Minimize} \quad b^T y$$
$$\text{subject to} \quad -A^T y \ge v \text{ and } y \ge 0$$
$$\text{where} \quad A = \begin{bmatrix} a_1^T \\ \vdots \\ a_m^T \end{bmatrix} \in \mathbb{R}^{m \times n}, \ b \in \mathbb{R}^m, \ y \in \mathbb{R}^m, \ v \in \mathbb{R}^n.$$

The LCP in general form is as follows

(0.6.3)   For a given matrix M ($n + m$ rows, $n + m$ columns) and

a given vector $q \in \mathbb{R}^{n+m}$, find vectors $w \in \mathbb{R}^{n+m}, z \in \mathbb{R}^{n+m}$,

$w \ge 0$, $z \ge 0$, such that $w - Mz = q$ and $w^T z = 0$.

For the special case of linear programming M must have the form

(0.6.4)
$$M = \begin{bmatrix} 0 & -A \\ A^T & 0 \end{bmatrix} \quad \text{and} \quad q = \begin{bmatrix} b \\ -v \end{bmatrix}.$$

These problems are solved by application of the self-dual algorithm of Lemke. One tries to represent vectors $q_\lambda = \lambda q + (1 - \lambda)e$ in the form $w - Mz$ for growing λ ($\lambda \varepsilon (0, 1)$), where $e = (1, \ldots, 1)^T \in \mathbb{R}^{n+m}$.

The initial representation (for $\lambda = 0$) is immediately at hand with $w = (1, \ldots, 1)^T$ and $z = 0$.

As soon as we have a representation for $\lambda = 1$, then we have solved the complete LCP.

Starting with $\lambda = 0$, we observe that for growing λ some positive entries in w must become 0. And it will not be necessary for them to stay at 0 during the whole process. Every such change in the set of positive entries in w is equivalent to a pivot step.

There is a one-to-one correspondence between the index-set of positive entries in w and a matrix Φ_M which is defined as follows:

> If $w^j = 0$ then the j-th column of Φ_M coincides with e_j, else
> it coincides with the j-th column of $-M$.

Hence the respective vector q_λ is a positive combination of the columns of Φ_M.

Smale tries to derive an upper bound for the maximal possible number of such matrices Φ_M, which generate vectors out of $[e, q]$ as positive combinations. For that purpose he exploits the so-called dominance relations between rows of Φ_M: where all entries of one row are greater or equal than the corresponding entries of another row.

On the other hand the structure of q determines a complete "dominance relation" between the entries of all the vectors out of $[q, e]$.

So a positive combination is only possible, if the entries of q satisfy all the actual dominance relations of the rows of Φ_M.

Smale estimated the probability for these events in two papers (SMALE 1982, SMALE 1983) under different stochastic models

(0.6.5) 1) The distribution of the random variables (A, b, v) is absolutely continuous.

  2) The random variables A, b, v are idependent and the columns are distributed independently.

  3) The probability measures of the random variables A, b, v are invariant under rotations of the respective spaces $(\mathbb{R}^{mn}, \mathbb{R}^m, \mathbb{R}^n)$.

In the proof in (SMALE 1983) assumption 3 was not used in this strong form, so SMALE (1982) could weaken this assumption in the following model

(0.6.6) 1) as above

  2) as above

  3) The probability measure of A and b is invariant under columnwise-independent permutations of coordinates in columns of (A, b).

Dominance relations between rows of Φ_M appear more frequently if Φ_M contains a lot of zeros. This is the case for matrices M resulting from linear programming problems (see 0.6.4). That effect is intensified when m is very much greater than n. Then dominance relations between the first m rows are more likely and the intersection-conditions between $[e, q]$ and the cone spanned by the columns of Φ_m become stronger. This effect compensates the growth of the number of cones for increasing m.

Using some further estimations concerning spherical measures and combinatorial sums, Smale obtains an upper bound for the average number of pivot steps required by Lemke's algorithm.

$$(0.6.7) \qquad E_{m,n}(s^L) \leq C(n)\, (1 + \ln{(m+1)})^{n(n+1)}$$

$$\text{for fixed } n \text{ and } m \text{ tending to infinity (asymptotically).}$$

This is a remarkable weak growth in m, but not polynomiality.

Note that Smale demands that the entries of the vectors a_i are distributed idependently. There are only few distributions in our model which satisfy this condition (e. g. Gaussian distribution).

In 1983 Charles Blair proved similar bounds for the expected number of vertices (without regard to the number of pivot steps and to the variant) (BLAIR 1983). He deals with problems of the type

$$(0.6.8) \qquad \begin{aligned} &\text{Maximize} \quad v^T x \\ &\text{subject to} \quad a_1^T x \leq b^1, \dots, a_m^T x \leq b^m, \ x \geq 0. \end{aligned}$$

His results hold for all variants which avoid redundant restriction hyperplanes. The stochastic assumptions are similar to those of Smale

$$(0.6.9) \qquad 1) \quad \text{The columns of } A = \begin{bmatrix} a_1^T \\ \vdots \\ a_m^T \end{bmatrix} \text{ are independent.}$$

        2)   Columnwise-independent permutations of coordinates in columns of (A, b) are without any effect on the probability measure.

A slight weakening of Smale's assumptions (no continuity of the distribution needed) makes it possible to generalize Blair's results even to discrete distributions, which is very important for the analysis of degenerate problems. Blair comes to the estimation

$$(0.6.10) \qquad E_{m,n}(V) \leq C(n)\, (\ln m)^{n(n+1)\ln{(n+1)}+n}.$$

Again, we observe a very moderate growth in m.

After that, MEGIDDO (1983) could "improve" the estimation method of Smale (only under Smale's original model (0.6.5)) so far that the dependence upon m disappeared completely.

(0.6.11) $$E_{m,n}(s) \leq C(n).$$

But still there was no explicit information about $C(n)$, which could still be exponential in n.

0.7 THE PAPER OF HAIMOVICH

Also in 1983, Mordecai Haimovich applied a very elegant and astonishing simple idea to a stochastic model introduced by May and Smith (HAIMOVICH 1983).

His solution method for Phase II (he does not deal with Phase I) is the shadow-vertex algorithm in its primal form (first described by GASS & SAATY 1955).

He considers the length of cooptimal simplex paths. These are simplex sequences consisting of all solutions for problems

(0.7.1)
$$\begin{array}{ll} \text{Maximize} & (u + \rho v)^T x \\ \text{subject to} & x \in X, \\ \text{where} & \rho \in \mathbb{R} \text{ runs from } -\infty \text{ to } +\infty. \end{array}$$

Now the sequence is generated by the continuing growth of ρ.

Instead of postulating certain assumptions on the distribution of the vectors a_i, he demands that the distribution is symmetrical under sign inversions

(0.7.2)   of rows in (A, b) for the type without sign constraints.

This means that the inequalities $a_i^T \leq b^i$ can (independently from each other) be flipped into $a_i^T x \geq b^i$ with probability $\frac{1}{2}$. Such a flip will make the opposite halfspace feasible. This is exactly the stochastic model proposed by MAY & SMITH (1982).

The considerations and results of Haimovich can easily be transferred to Phase II of problem types as

(0.7.3)   Maximize $v^T x$

               subject to $a_1^T x = b^1, \ldots, a_m^T x = b^m$, $x \geq 0$

        or

(0.7.4)   Maximize $v^T x$

               subject to $a_1^T x \leq b^1, \ldots, a_m^T x \leq b^m$, $x \geq 0$,

where the sign constraints (one for each component) are allowed to be "flipped", too.

No additional stochastic assumptions are required. Only nondegeneracy is necessary. Now every choice of $a_1, \ldots, a_m$ and b generates a special instance class of 2^m (for our type) inequality combinations. So 2^m different problems are generated. The corresponding feasible regions will be called cells.

Normally, a lot of the 2^m cells are empty (see section 9). Imagine that every candidate for being a vertex, which is the intersection point of n restricting hyperplanes, has to satisfy $m - n$ additional constraints. This can occur only in 2^n of the 2^m cells. So only in one of 2^{m-n} problems our candidate will become a vertex.

Simultaneoulsy, the model creates a lot of problems with great share of redundant restrictions. So it becomes obvious that the problems created have a small average number of vertices and of pivot steps. Haimovich now restricts his considerations to those cells where the set of cooptimal vertices is nonempty and develops a formula for the conditional expected number.

> A point y is called cooptimal, if $y \in X$ and if there is no $z \in \mathbb{R}^n$ with $v^T y = v^T z$ but $u^T z > u^T y$.

Theorem. (Haimovich 1983)

The expectation value of the numbers of pivot steps for (m, n)-problems (without sign constraints) in the sign-invariance model and under the condition that a cooptimal path exists, is not greater than

(0.7.5) $$n \, \frac{m - n + 2}{m + 1}.$$

Figure 0.7

The sign-invariance model for n=2 and m= 6

The model generates 2^6=64 problems of identical likelihood.
Only 22 of them have feasible points, only 10 bounded X.

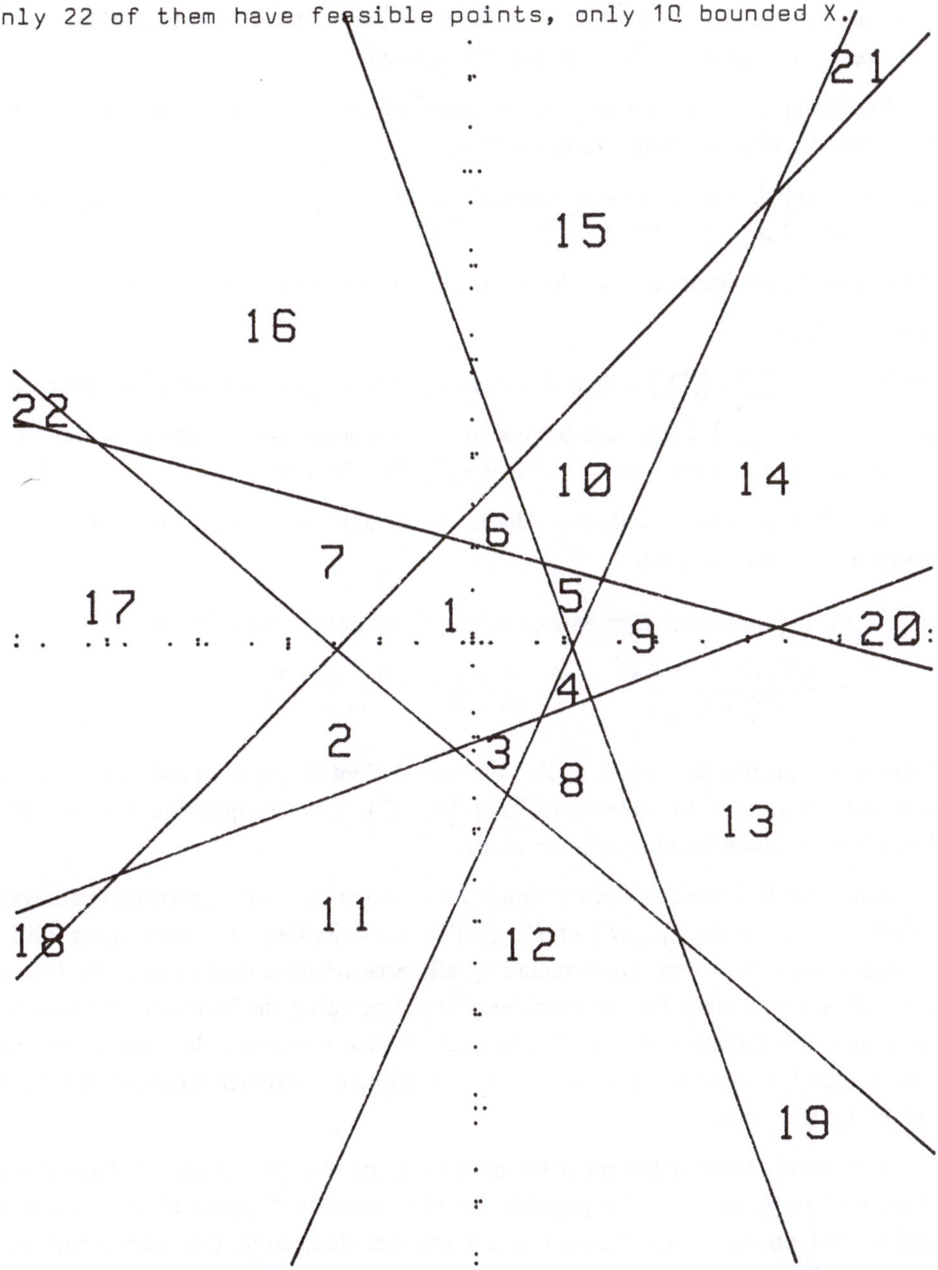

The average number of vertices and of facets is rather low
(even after conditioning on the feasible problems). Note
that cell 1 would be a typical X for the rotation-symmetric
model.

Proof. (Translated into our terminology)   There are 2^m cells and $\binom{m}{n}$ intersection points. The intersection sets of $n - 1$ restricting hyperplanes will be called lines.

Every such line is itself intersected by $m - n + 1$ restriction hyperplanes. So we obtain $m - n + 2$ segments on each line. The two outside segments are rays, the $m - n$ inner ones are bounded. Note that each of the segments belongs to a different cell. And every segment is cooptimal in exactly one cell.

Exploiting the usual nondegeneracy assumptions, Haimovich observes the following properties of an arbitrary instance class:

1) Every cell has at most one optimal vertex (relative to $v^T x$) and at most one cooptimal path (relative to $v^T x$ and $u^T x$).

2) Each intersection point of n hyperplanes is optimal in exactly one cell.

So he concludes

3) There are $\binom{m}{n} + \binom{m}{n-1}$ cells with a nonempty cooptimal path for a fixed pair (u, v).

4) There are $\binom{m}{n-1}$ lines, each having $m - n + 2$ segments. So the total number of line segments of the instance class is $\binom{m}{n-1}(m - n + 2)$.

Now it is possible to calculate the expected number of segments contained in a **nonempty** cooptimal path.

$$E_{m,n} \text{ (number of segments on a cooptimal path / path nonempty) } =$$
$$= \frac{\binom{m}{n-1}(m - n + 2)}{\binom{m}{n} + \binom{m}{n-1}} = \frac{m - n + 2}{\frac{m-n+1}{n} + 1} = n \frac{m - n + 2}{m + 1}.$$

Before one transfers this result to the number of pivot steps, it must be clear how the unbounded rays shall be counted (as 0 or $\frac{1}{2}$ or 1?). But this question does not affect the extremely small number of pivot steps.

Note that Haimovich's expectation value (though the same algorithm is analyzed) is quite different from $E_{m,n}(S)$ or $E_{m,n}(s)$ in our definition. We are dealing with an unconditional expectation value including all cases where a cooptimal path does not exist. This is necessary for our considerations concerning the inclusion of Phase I. In addition, our estimation of $E_{m,n}(S)$ is done in such a manner that the given upper bounds are also upper bounds for the corresponding conditional expectation value (as defined by Haimovich).

The size of Haimovich's upper bound is so small, that even linearity of the average number of steps seems to be possible for the complete Simplex-Method under his model. But unfortunately Phase I is not yet included, while this part seems to be quite troublesome. So even polynomiality is not yet proven in this paper (HAIMOVICH 1983).

Such a generalization to the complete method and the derivation of polynomiality has been given by Haimovich (verbatim) and by TODD (1983), ADLER & MEGIDDO (1983) and ADLER, KARP & SHAMIR (1983b) in written papers (see the next section).

Remark.

I was told that similar results have been derived by Ilan Adler independently and for the same stochastic model.

0.8 QUADRATIC EXPECTED NUMBER OF STEPS FOR SIGN-INVARIANCE MODEL

In late 1983 a remarkable and significant progress has been made by several authors. Encouraged by the astonishing good results for Phase II (Haimovich and Adler), they derived algorithms for Phase I and Phase II which require less than $0(\min(m^2, n^2))$ steps on the average under the sign-invariance model.

Michael Todd (November 1983) and Ilan Adler/Nimrod Megiddo (December 1983) independently analyzed the so-called lexicographic Lemke-algorithm. This algorithm is applied to linear complementarity problems which arise from linear programming problems of the type

(0.8.1)

$$\begin{array}{ll} \text{Maximize} & v^T x \\ \text{subject to} & a_1^T x \leq b^1, \ldots, a_m^T x \leq b^m, \ x \geq 0 \\ \text{where} & v, x, a_1, \ldots, a_m \in \mathbb{R}^n, \ b^i \in \mathbb{R}. \end{array}$$

So far there is a direct analogy to Smale's paper. But instead of starting with the vector $q_0 = e \in \mathbb{R}^{m+n}$, these authors use an artificial vector $q_0 = (\delta^n, \ldots, \delta^2, \delta^1)^T$ for initiating the algorithm. Here δ should be regarded as an arbitrary small positive value. It does not need to be determined in advance, because it is only required to formalize the lexicographic selection process analytically. When we proceed on the line segment $[q_0, q]$ with arbitrary q, then the first little move will guarantee that $\text{sign}((q_0 + \lambda(q - q_0))^1) = \text{sign}((q)^1)$, since δ^n is extremely small. Afterwards, the first component keeps its sign all the time. Now the second component obtains the final and correct sign etc.

Finally, the $m + n$-th component shall cross the value 0, if this is still necessary. Translated into the language of linear programming this means that a first move confirms the satisfaction of the restriction $a_1^T x \leq b^1$. Then keeping $a_1^T x \leq b^1$, the value of $a_2^T x$ is improved (if necessary), until it satisfies $a_2^T x \leq b^2$ etc.

When we are ready with the first m components, then all concentration is directed towards the correction of the v-part of $q_0 + (q - q_0)$. During certain iterations the first components of q_0 can be taken as essentially zero. This fact simplifies the estimations of the probability that basic cones are intersected by $[q_0, q]$. So these authors obtain the following astonishing small upper bounds under the sign-invariance model and regularity assumptions.

Theorem. (Todd 1983)

When the probability distribution of (A, b, v) satisfies regularity and sign-invariance, then the expected number of steps in the lexicographic Lemke-algorithm for problems of type (0.8.1) is at most

$$(0.8.2) \qquad \min\{(m^2 + 5m + 11)/2, \, (2n^2 + 5n + 5)/2\}.$$

□

Sign invariance means that

$$(0.8.3) \qquad \text{the distributions of } (A, b, v) \text{ and of } (S_1 A S_2, S_1 b, S_2 v)$$
$$\text{are identical for all sign matrices } S_1 \text{ and } S_2.$$

Sign matrices are diagonal matrices with ± 1 in the entries s_{ii}. This condition is equivalent to the definition of the "flipping model" in section 7.

Adler's and Megiddo's result can be summarized as follows.

Theorem. (Adler/Megiddo 1983)

The average number of steps for the lexicographic Lemke algorithm under regularity and sign-invariance conditions as above is bounded by a quadratic function

$$(0.8.4) \qquad E_{m,n}(s^L) \leq c_2(\min(m, n))^2.$$

In addition, they derive a lower bound under somehow stronger probability assumptions on

$$A^* = \begin{bmatrix} A & b \\ v^T & 0 \end{bmatrix}.$$

Theorem. (Adler/Megiddo 1983)

Under regularity conditions and if the entries of A^* (except for the 0 in the right hand corner) are independent, identically distributed random variables, whose individual distribution is symmetric about the origin, the average number of pivot steps in the lexicographic Lemke algorithm is bounded from below by a function

$$(0.8.5) \qquad c_1(\min(m, n))^2.$$

Closer related to the solution method analyzed in this book is the argumentation in the paper of ADLER, KARP & SHAMIR (1983b).

The authors deal with the same type of problem

(0.8.6)
$$\text{Maximize} \quad v^T x$$
$$\text{subject to} \quad a_1^T x \le b^1, \ldots, a_m^T x \le b^m, \ x \ge 0.$$

Their stochastic assumptions are the same as described above. But the analyzed algorithm is a so-called Constraint-By-Constraint-algorithm (CBC), which proceeds in $\mathbb{R}^n$, not in $\mathbb{R}^{n+m}$ as Lemke's algorithm. Here the algorithmic procedure is as follows. Let

(0.8.7)
$$X^{(k)} = \{x \in \mathbb{R}^n \mid x \ge 0, a_1^T x \le b^1, \ldots, a_k^T x \le b^k\}$$

We run through $m + 1$ stages.

(0.8.8) **Stage 0:** Determine the unique vertex $\bar{x} = 0$ of $X^{(0)}$ and choose a u such that $u^T x$ is maximized at $\bar{x}$ on $X^{(0)}$ (e. g. $u = -e$).

Go to stage 1.

Stage k: ($1 \le k \le m$)
Starting at $\bar{x}$, which maximizes $u^T x$ on $X^{(k-1)}$, use the parametric objective algorithm (= shadow-vertex-algorithm) with span$(u, -a_k)$ as projection plane. If $\bar{x} \in X^{(k)}$ go to $k+1$. Else Stop as soon as $a_k^T x \le b^k$ is achieved. Then the last traversed edge contains a point x such that $a_k^T x = b^k$. Since x lies on the efficient path, we know that x maximizes $u^T x$ on $X^{(k)}$. If it is impossible to achieve $a_k^T x \le b^k$, then the original problem's feasible region is empty. Then we can Stop. Else we set $\bar{x} = x$ and go to stage $k + 1$.

Stage $m + 1$: Again we start at $\bar{x}$, maximizing $u^T x$ on $X^{(m)} = X$. We apply the shadow-vertex-algorithm with span(u, v) as the projection plane and get to the solution or we end up on an unbounded ray, demonstrating that there is no solution.

Now we sum up over all 2^{n+k} instances appearing in stage $k + 1$ and over all basic cones and over the 2^n possible directions of the objective when the signs are flipped. The results of Haimovich and Adler tell us that a fixed intersection point of

n hyperplanes is cooptimal in exactly $n + 1$ of the 2^{n+k} instances, if the objective is fixed, too. So we obtain

$$E_{m,n}(s) \leq \sum_{k=1}^{m} \binom{k+n}{n} 2^{-k}(n+1).$$

Evaluation of the right side leads to the upper bound $2^{n+1}(n+1)$. But this result can be improved significantly, if we choose u as $(-\delta^n, \ldots, -\delta)$ with $\delta > 0$ sufficiently small.

Then it can be shown that an intersection point of the first r nonnegativity constraint-hyperplanes and with $x^{r+1} > 0$ is — for a fixed instance — efficient for none or for at least 2^{n-r-1} of the 2^n possible objectives. This observation enables the authors to prove that

(0.8.9) $$E_{m,n}(s^{\mathrm{CBC}}) \leq 2(n+1)^2.$$

In case of $m < n$, the authors switch to the dual problem and prove that $E_{m,n}(s) \leq 2(m+1)^2$.

If the CBC-algorithm is regarded as a special kind of Lemke's algorithm and if the problem is imbedded into an LCP, one observes that the sequences of cones which are intersected, are actually the same as in the algorithms of Todd resp. Adler/Megiddo (compare MEGIDDO 1984). Whereas the usual algorithm proceeds by letting μ grow in $q_0 + \mu(q - q_0)$, the CBC-method starts with q_0, too. But then it makes a little move in the direction of $e_1 \in \mathbb{R}^{m+n}$, then in direction $e_2 \in \mathbb{R}^{m+n}$ etc., just until all the according constraints are satisfied after the move in direction $e_m \in \mathbb{R}^{m+n}$.

Since δ can be chosen sufficiently small, these detours can be kept so small that the set of intersected LCP-cones remains the same. This results from the fact that every intersected cone is entered in the interior of one of its side-cones because of nondegeneracy.

So the subject under consideration (problem type, probability model, algorithmic sequence) is quite the same in all three papers.

DISCUSSION OF DIFFERENT STOCHASTIC MODELS

0.9 WHAT IS THE "REAL WORLD MODEL"?

This is a philosophical question and nobody can answer it satisfactorily. But one should discuss the ideas, conjectures and experiences of practical and theoretical experts of linear programming.

Motivation and inspiration for dealing with the average number of steps mostly come from the bad Klee-Minty examples and the fear that such a bad behaviour could occur frequently. We do not share that concentration on that small set of artificial problems, because we believe that the information on the average step number itself is very important, valuable and useful. However, it is clear, that models excluding the Klee-Minty examples cannot give serious results.

A critical feature and a real point of weakness in all stochastic models used for investigating average step numbers is the **exclusion of degeneracy** (or probability 0 for degeneracy). Many practical problems generate matrices A^*, which contain a lot of zeros and/or have many entries with the same value. So such problems tend to be degenerate. May be that Blair's ideas or similar considerations could help here. However, we do not believe that degenerate problems usually behave worse than the nondegenerate, when parametric algorithms are appplied. The trouble comes from ambiguity in certain situations, which can cause detours or even loops. But parametric algorithms seem to manage most of these critical situations perhaps better than the commonly used algorithms. This holds particularly for the comparison with variants relying on the value of the objective after the next step. These questions seem to be a fruitful field of future research.

Another point to be discussed is the use of the terms "weak" and "strong" in connection with stochastic models. Here these terms have quite different meanings from those in the context of assumptions of conditions in mathematical claims.

Example

If we calculate an upper bound for the mean value of a nonnegative random variable over a set M_1 and afterwards over a set $M_0 \supset M_1$, then the result for M_0 is not better or more useful or more important. Imagine that the random variable could be 0 all over $M_0 \setminus M_1$! So we do not obtain an improvement until the behaviour on M_1 or on $M_0 \setminus M_1$ is estimated or analyzed, too. The same argumentation holds if $M_0 = M_1$, but the distributions over the random space are different.

Now we want to compare the sign-invariance model and our rotation-invariance model. Let us demonstrate the differences between both models at the problem type with sign constraints, because that type is more symmetrical under dualization. So we consider the type

(0.9.1)
$$\begin{aligned}
\text{Maximize} \quad & v^T x \\
\text{subject to} \quad & A^T x \leq b, \; x \geq 0 \\
\text{where} \quad & v, x \in \mathbb{R}^n, \; b \in \mathbb{R}^m, \; A \in \mathbb{R}^{m \times n}.
\end{aligned}$$

(0.9.2)   – Our rotation-invariance model requires that $b = e \in \mathbb{R}^m$ and that the rows of A are distributed on $\mathbb{R}^n \setminus \{0\}$ identically, independently and and symmetrically under rotations.

(0.9.3)   – The sign-invariance model requires that the probability measure is invariant under inverting the sign of complete rows and/or of complete columns in the matrix

$$A^* = \begin{bmatrix} A & b \\ v^T & 0 \end{bmatrix} \in \mathbb{R}^{(m+1) \times (n+1)}.$$

Row-multiplications in A^* with -1 can be interpreted as flipping the inequality direction in $a_i^T x \leq b^i$, whereas column-sign-inversions mean an exchange of the variable x^i by $-x^i$ or equivalently flipping the constraint $x^i \geq 0$.

The motivation for the sign-invariance model comes from

(0.9.4)   – mathematical elegance

          – great tractability with proofs using only combinatorical arguments

          – preservation of the stochastic model under dualization of the problems

          – inclusion of many types of problems

          – requirement of only a finite number of symmetries.

The motivation for the rotation-invariance model and the inherent preference for problems with a given feasible point results from the following considerations:

(0.9.5)   – Even Phase I can be done by solving a linear programming problem with a given feasible point.

          – Any problem satisfying Slater's constraint qualification (X contains an inner point) can be changed into our problem type by a coordinate-transformation.

– In most practical problems an initial feasible point is known
(a vertex or an inner point). Experience shows that most applications are done in order to improve a known solution of an
inequality system, not to find one. Mostly practical intuition
has already found such a solution, but it is not optimal.
– It seems to be plausible that relative to the given feasible point
the directions of the restrictions are distributed symmetrically
under rotations.
– Rotational invariance still allows a great freedom in the choice
of the complete distribution through the variation of the radial
distribution.

The last point seems to be very important, because it enables us to study models
with completely different redundancy rates. The redundancy rate seems to have a
tremendous influence on the behaviour of the Simplex-Method. The higher the average
redundancy rate — the better the average behaviour. Our model allows redundancy
rate 0 (for uniform distribution over the unit sphere) as well as redundancy rates close
to 1 (see the W. Schmidt example and the extremely good results in IV, 4). Using the
sign-invariance model, we do not have such freedom.

The concentration on ≤ 1 inequalities in our model has a great effect on the results
of the analysis. Whereas we obtain only feasible problems (0 is feasible in any case), the
chance for emtpyness of X in the sign-invariance model is very large. So, the effort for
Phase II is very low. Consider the number of feasible problems in the sign-invariance
model (here and in the following compare MAY and SMITH 1982)

$$(0.9.6) \qquad \binom{m+n}{0} + \binom{m+n}{1} + \ldots + \binom{m+n}{n}$$

while the number of generated problems is 2^{m+n}.

So for $m \gg n$ the quotient

$$\frac{\text{number of feasible problems}}{\text{number of generated problems}}$$

tends to 0. Also, the expected number of vertices per problem is rather low. Here we
have

$$(0.9.7) \qquad E_{m,n}(V) = \frac{2^n \binom{m+n}{n}}{2^{m+n}} \qquad \text{in the sign-invariance model.}$$

Even conditioning on nonempty cells does not change very much. Here

$$(0.9.8) \qquad E_{m,n}(V/ \text{ cell nonempty}) \; = \; \frac{2^n \binom{m+n}{n}}{\binom{m+n}{0} + \ldots + \binom{m+n}{n}}$$

which tends to 2^n for $m \to \infty$, n fixed.

This is not much for an expected number of vertices in feasible problems (all counted problems are feasible). Compare with the size in the results of RAYNAUD 1970 and of KELLY & TOLLE 1979 (0.4. 1–3).

Another interesting aspect is the average redundancy rate. Consider that the introduction of one additional restriction (no. $m+n$) increases the number of nonempty cells from $\binom{m+n-1}{0} + \ldots + \binom{m+n-1}{n}$ to $\binom{m+n}{0} + \ldots + \binom{m+n}{n}$. That means that the additional restriction divides $\binom{m+n-1}{0} + \ldots + \binom{m+n-1}{n-1}$ into two. Consequently we have nonredundancy of that restriction in $2\{\binom{m+n-1}{0} + \ldots + \binom{m+n-1}{n-1}\}$ of 2^{m+n} generated resp. $\binom{m+n}{0} + \ldots + \binom{m+n}{n}$ feasible problems. In $\binom{m+n-1}{n}$ of the new feasible cells the new restriction $(m+n)$ is redundant.

It is clear that both nonredundancy rates (the conditional and the unconditional one) tend to 0 for $m \to \infty$, n fixed.

$$(0.9.9) \qquad \frac{2\{\binom{m+n-1}{0} + \ldots + \binom{m+n-1}{n-1}\}}{2^{m+n}} \qquad \text{(unconditional nonredundancy rate)}$$

and

$$(0.9.10) \qquad \frac{2\{\binom{m+n-1}{0} + \ldots + \binom{m+n-1}{n-1}\}}{\binom{m+n}{0} + \ldots + \binom{m+n}{n}} \qquad \text{(conditional nonredundancy rate)}.$$

The corresponding expectation values of nonredundant constraints per problem (resp. per cell) is calculated by multiplication of $(m+n)$ with the nonredundancy rate. The unconditional expectation value tends to 0 asymptotically, whereas the conditional one is

$$(0.9.11) \qquad \frac{(m+n)2\{\binom{m+n-1}{0} + \ldots + \binom{m+n-1}{n-1}\}}{\binom{m+n}{0} + \ldots + \binom{m+n}{n}} \qquad \longrightarrow \qquad 2n \text{ for } m \to \infty, n \text{ fixed}.$$

In the opposite case, namely $m \ll n$, the sign-invariance model yields a high probability of unbounded problems (conditional probability). This conditional probability is

$$(0.9.12) \qquad \frac{\text{number of unbounded problems}}{\text{number of nonempty cells}} = \frac{\binom{m+n}{0} + \ldots + \binom{m+n}{n-1}}{\binom{m+n}{0} + \ldots + \binom{m+n}{n}} \geq \frac{n}{m+n+1},$$

which can become rather large (close to 1). So there is a great chance of recognizing unboundedness very soon. And in these cases the algorithm will stop quickly.

These observations suggest, that in the sign-invariance model a great number of extremely harmless problems overcompensates the effect of the bad examples.

So we expect that the size of $E_{m,n}(s)$ differs extremely between both models. This holds particularly for the cases $m \gg n$ and $m \ll n$. Only for $m \simeq n$ the models seem to be somehow compatible. In fact, the Klee-Minty examples mostly are of this type and both models are able to show that these bad examples are very seldom.

It is interesting that ADLER, KARP & SHAMIR (1983b) and ADLER & MEGIDDO (1983) remark that in the sign-invariance model the conditional expectation value (under feasibility) of $E_{m,n}(s)$ is about $0(n^{5/2}) = 0(m^{5/2})$ if $m = n$. Note the similarity to our results for Phase II !

Another interesting question concerns the possible advantage of dualization of the problem in our model when $m \gg n$. Will such a transformation lead to a shorter calculation time as it does under the sign-invariance model ?

At the first glance, there is great similarity between the distribution of our dual problems and the sign-invariance model. Sign-invariance is preserved under dualization of the problem

$$
(0.9.13) \qquad \text{I} \quad
\begin{array}{ll}
\text{Maximize} & v^T x \\
\text{subject to} & Ax \le b, \ x \ge 0 \\
\text{where} & v, x \in \mathbb{R}^n, \ b \in \mathbb{R}^m, \ A \in \mathbb{R}^{m \times n}
\end{array}
$$

into the problem

$$
(0.9.14) \qquad \text{I'} \quad
\begin{array}{ll}
\text{Minimize} & b^T y \\
\text{subject to} & A^T y \ge v, \ y \ge 0 \\
\text{where} & y \in \mathbb{R}^m, \ v, b, A \text{ as above.}
\end{array}
$$

In our rotation-invariance model the matrix (A^T, v) is invariant under multiplications of rows and/or columns with -1. This is a kind of inequality sign-invariance. But still we are far from sign-invariance, because in $A^{*T} = \begin{bmatrix} A^T & v \\ b^T & 0 \end{bmatrix}$ the vector $b = e$ is fixed. The only difference to the situation in the primal problem is that the crucial part (showing that sign-invariance is not satisfied) has moved from the right side of the restrictions to the objective vector. In order to compare the two models here, consider a fixed realization of A and v with $b = e$, which is $\begin{bmatrix} A^T & v \\ b^T & 0 \end{bmatrix}$. In the sign-invariance model the probability for such a realization is as high as for any realization $\begin{bmatrix} A^T S & v \\ b^T S & 0 \end{bmatrix}$, where S is a diagonal matrix with ± 1 in its diagonal entries. But in most of the problems

$$
(0.9.15) \qquad
\begin{array}{ll}
\text{Minimize} & (b^T S)y \\
\text{subject to} & (A^T S)y \ge v, \ y \ge 0
\end{array}
$$

the objective will be unbounded, since the number of rows in $(A^T S)$ is smaller than the number of columns. So the Simplex-Method can be expected to stop quickly.

In our model our fixed realization is stochastically equivalent to the 2^m problems

$$(0.9.16) \qquad \begin{aligned} \text{Minimize} \quad & b^T y = e^T y \\ \text{subject to} \quad & (A^T S)y \geq v, \; y \geq 0 \end{aligned}$$

The probability for infeasibility is not very high since $m \ll n$. And for all feasible problems solvability is guaranteed because $b^T 0 = e^T 0 = 0$ is a lower bound for the objective.

(All these considerations could also and more directly be derived from the duality theorems, but there they may be less plausible). So our model will be much more inconvenient even after dualization of the problem.

As a consequence, it is not yet clear that dualization will improve the calculation speed in our model.

OUTLINE OF CHAPTERS 1 – 5

0.10 THE BASIC IDEAS AND
THE METHODS OF THIS BOOK

When I began to write this book, I planned to summarize the results, and to unify the notation. In the meantime I could, in addition, simplify some of the proofs and improve some of the results. It was not my aim to demonstrate the considerations very briefly and concisely, because I have the seldom chance to motivate and to make ideas plausible. Simultaneously, I want to demonstrate the proofs explicitly, leaving little effort to the reader.

As already mentioned in section 5, we deal with problems of the type

$$(0.10.1) \quad \begin{array}{ll} \text{Maximize} & v^T x \\ \text{subject to} & a_1^T x \leq 1, \ldots, a_m^T x \leq 1 \\ \text{where} & v, x, a_1, \ldots, a_m \in \mathbb{R}^n, \quad \text{and } m \geq n. \end{array}$$

The solution variant is the

$$(0.10.2) \quad \text{shadow-vertex algorithm.}$$

And my stochastic assumptions are

$$(0.10.3) \quad a_1, \ldots, a_m, v \text{ are distributed on } \mathbb{R}^n \setminus \{0\}$$

  – independently

  – identically

  – symmetrically under rotations.

One of the reasons for choosing this type of problems (no sign constraints) was the wish to make the considerations as simple as possible. Here we have the advantage that only one type of constraints occurs. So, complicated case-studies are avoided.

By demanding positive right sides in the inequalities, we concentrate on a subclass where the optimization part is rather hard (the Phase I-part is easier than for other types). I conjecture that in general the optimization is at least as hard as the process of finding a solution for an inequality-system. Now $x = 0$ is feasible in any problem. And feasible problems seem to be preferred in practical applications. And by using our type we try to avoid too comfortable and too convenient problems.

> As long as the restrictions are of the form $\tilde{a}_i^T x \leq b^i$ with
> $b^i > 0$, we are allowed to normalize without loss of generality.
> We replace $\frac{1}{b^i}\tilde{a}_i$ by a_i and obtain the restrictions $a_i^T x \leq 1$.
>
> Note that when the vectors $\tilde{a}_i$ are distributed according to
> (0.10.3) and when the b^i's are somehow distributed on the
> interval $(0, \infty)$ then the resulting (normalized) vectors a_i are
> again distributed according to (0.10.3). The new distribution
> may differ from the old, but it is symmetrical under rotations.

Other types of problems can be reduced to our type by coordinate transformations
and — if necessary — by reduction of the dimension. Particularly the following type
can be reduced:

$$
\begin{aligned}
&\text{Maximize} \quad v^T x \\
&\text{subject to} \quad a_1^T x \leq a_1^T x_0 + b^1, \ldots, a_m^T x \leq a_m^T x_0 + b^m \\
&\text{where} \quad v, x, a_1, \ldots, a_m, x_0 \in \mathbb{R}^n, \text{ and } b \in \mathbb{R}^m, b > 0, \\
&\text{and where} \quad x_0 \text{ is known,}
\end{aligned}
$$

(0.10.4)

with

(0.10.5)
$$
\begin{aligned}
& a_1, \ldots, a_m, v \text{ distributed according to (0.10.3)} \\
& \text{and } b^1, \ldots, b^m \text{ distributed over } (0, \infty) \\
& \quad \text{– independently} \\
& \quad \text{– identically.}
\end{aligned}
$$

> Here x_0 is feasible in any case. If we set $x' = x - x_0$, then
> x_0 is mapped into the origin. After that we can normalize as
> described above.
>
> Finally, we have a problem of our original type (0.10.1) sat-
> isfying the stochastic assumptions (0.10.3). Hence our theory
> and our results hold for problems of that kind (0.10.4) **and**
> (0.10.5), too. Note that it is essential to know the point x_0
> in advance, because our Phase I-algorithm has to start from
> a given feasible point!

As already discussed in Section 9, nobody knows the real world stochastic model.
Hence there are divergent requirements on the stochastic model. Some authors emphasize the wish to keep the different components of the restriction vectors independent
(example: Gaussian distribution). Others give great importance to the tractability of

discrete distributions (a lot of degenerate problems as transportation problems). We want to keep our considerations free and independent of a special choice of a coordinate system. So we regard rotational symmetry as rather natural and typical for most of the practical problems.

And, last but not least, rotational symmetry is tractable by means of "hard analysis" much better than many other assumptions on the distribution. It assigns probability 0 to the set of degenerate problems. Rotational symmetry is very appropriate as long as we deal with problems without nonnegativity constraints. In Chapter V we shall generalize our methods and our main results to the problem type with such restrictions. In Chapter I we are going to explain our solution variant in two different ways. There is a "primal" and a "dual" interpretation. The primal deals with the feasible set X and the dual is concerned with properties of the dual polyhedron Y (compare section 4 Liebling's work).

Whereas the primal interpretation seems to be very popular and to be known by most of the readers, the dual will be much more convenient and appropriate for our stochastic considerations.

First we concentrate on Phase II and suppose that the start vertex x_0 is given. For the theoretical determination of x_0 we use a vector $u \in \mathbb{R}^n \setminus \{0\}$, such that

$$(0.10.6) \qquad x_0 \text{ is maximal on } S \text{ with respect to } u^T x.$$

The following assumption of nondegeneracy enables us to avoid complicated case studies.

(0.10.7)　　Every n-element subset of $\{a_1, \ldots, a_m, u, v\}$ is linearly independent and each $(n+1)$-element subset of $\{a_1, \ldots, a_m\}$ is in general position.

Here it is useful that the degenerate cases form a nullset. So they do not influence our results on the expected number of pivot steps. The term "shadow-vertex" means the following.

Consider the n-dimensional polyhedron X and its image $\Gamma(X)$ under orthogonal projection Γ on the two dimensional plane $span(u, v)$.

A vertex of X will be called shadow-vertex iff its image is a vertex of $\Gamma(X)$. Now it is important that each pair of shadow-vertices is connected by a Simplex-Path visiting only shadow-vertices. As a result of the projection on $span(u, v)$, x_0 and x_s are shadow-vertices.

Our shadow-vertex-algorithm just realizes such a path from x_0 to x_s touching only shadow-vertices.

Figure 0.8a

Simplex-Paths generated by the shadow-vertex-algorithm

We start at x_0 and move to the optimal vertex with respect to $v^T x$.

Here n=3, m=12 and span(u,v) is the two-dimensional observation-plane.

s=4
S=8

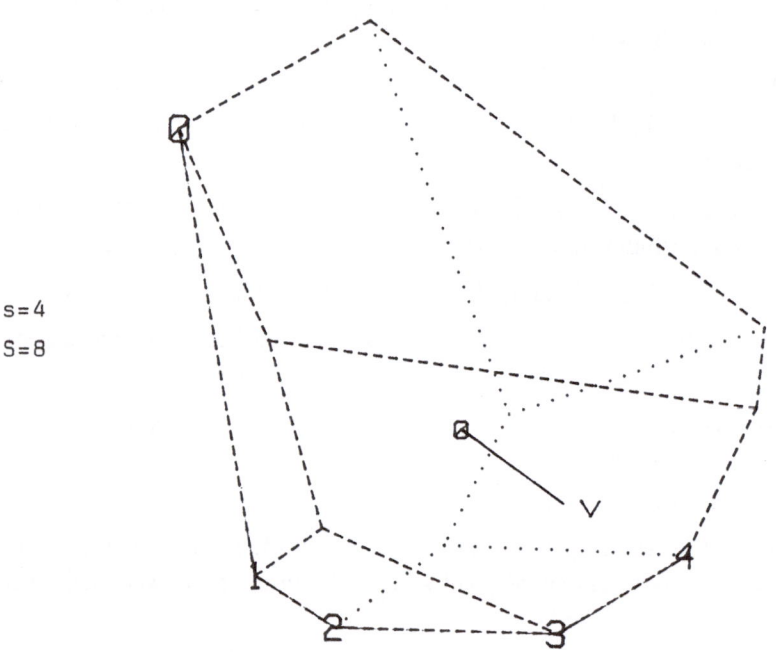

In this special example, the shadow-vertex-algorithm could also follow another path (see figure 0.8b).

An explicit choice of u would make the path unique.

Figure 0.8b

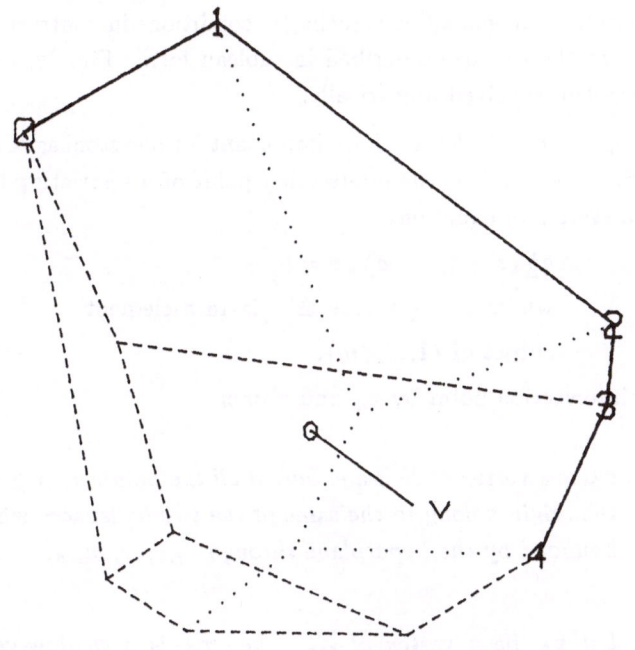

Hence it is clear, that S, the number of shadow-vertices, is an upper bound for s, the number of pivot steps in Phase II.

We can even show that

(0.10.8)
$$\frac{1}{4} E_{m,n}(S) = E_{m,n}(s).$$

So we are allowed to analyze the value of S rather than the value of s.

We explain how linear programming problems can be solved by our algorithm using Simplex-tableaus. Further we develop a method for solving the complete problem, including Phase I. It is based on the shadow-vertex algorithm and applies it for increasing dimension. During the process, it provides us with the solution for some "projected problems" of dimension lower than n. The method looks sophisticated, but is chosen because it meets all our stochastic conditions in contrast to usual Phase I-methods. This method is also described in tableau form. Finally, we demonstrate, how a special problem is solved numerically.

The following results of Chapter I are important for our stochastic considerations. Recall that every vertex of X is the intersection point of n restricting hyperplanes, or the solution of a system of equations

(0.10.9)
$$a_{\Delta^1}^T x = 1, \ldots, a_{\Delta^n}^T x = 1,$$

where $\Delta = \{\Delta^1, \ldots, \Delta^n\}$ is an n-element

subset of $\{1, \ldots, m\}$.

We denote this intersection point by x_Δ and obtain

Lemma.

(0.10.10)
x_Δ *is a vertex of X if and only if all the points a_i ($i \notin \Delta$) and the origin belong to the same of the two halfspaces which are bounded by the hyperplane through $a_{\Delta^1}, \ldots, a_{\Delta^n}$.*

Lemma.

(0.10.11)
Let x_Δ be a vertex of X. Then x_Δ is a shadow-vertex if and only if $\text{span}(u, v)$ intersects $CC(a_{\Delta^1}, \ldots, a_{\Delta^n})$, the convex cone generated by $a_{\Delta^1}, \ldots, a_{\Delta^n}$.

Hence S, the number of shadow-vertices, is equal to the number of Δ's where the corresponding point x_Δ satisfies the conditions of both lemmas.

In Chapter II we define a probability space of all possible (m, n)-problems of our type. The probability in this space is induced by the distribution of the vectors $a_1, \ldots, a_m, v$. They are distributed according to (0.10.3). Hence their distributions are uniquely characterized by "radial-distribution functions"

$$F : [0, \infty) \to [0, 1] \quad \text{with} \quad F := P(\|x\| \le r, x \in \mathbb{R}^n).$$

Denoting the original distribution function over $\mathbb{R}^n$ by $\tilde{F}$ we get to the following result.

Theorem.

$$E_{m,n}(S) = \binom{m}{n} \frac{n}{2} \int_{\mathbb{R}^n} \cdots \int_{\mathbb{R}^n} P(a_{n+1}, \ldots, a_m \text{ lie "below" the}$$

(0.10.12)

$$\text{hyperplane through } a_1 \ldots, a_n).$$

$$\cdot P(\text{span}(u, v) \text{ intersects } \text{CH}(a_1, \ldots, a_{n-1})) \, d\tilde{F}(a_1) \ldots d\tilde{F}(a_n)$$

Now we exploit our assumption of rotational symmetry. So the integral formula can be simplified to

Theorem.

$$E_{m,n}(S) = \binom{m}{n} n \int_{\mathbb{R}^n} \cdots \int_{\mathbb{R}^n} G(h(a_1, \ldots, a_n))^{m-n}.$$

(0.10.13)

$$\cdot W(a_1, \ldots, a_{n-1}) \, d\tilde{F}(a_1) \ldots d\tilde{F}(a_n)$$

Here $h(a_1, \ldots, a_n)$ is the distance of the origin to the hyperplane through $a_1, \ldots, a_n$. $G(h)$ is the probability that $x^n \leq h$ for a random vector x. And $W(a_1, \ldots, a_{n-1})$ is the normalized spherical measure of the cone generated by $a_1, \ldots, a_{n-1}$.

A certain coordinate transformation simplifies the integral formula in (0.10.13) once again. Now it remains to evaluate these integral formulae. This is the task of the Chapters III-V.

0.11 THE RESULTS OF THIS BOOK

The evaluation of these integral formulae turns out to be very difficult. For that reason we compare the expectation value of S with the expectation value of a second random variable Z. Z is the number of optimal vertices with respect to $v^T x$ in X.

We make use of the facts that

1) $E_{m,n}(Z) \leq 1$, because Z cannot be greater 1 in nondegenerate cases,

2) the integral formula for $E_{m,n}(Z)$ is quite similar to that for $E_{m,n}(S)$.

So we are able to evaluate and to estimate the quotient $\frac{E_{m,n}(S)}{E_{m,n}(Z)}$. By this way we derive the following main results.

The term "according to our model" or "acc. to the rotation-invariance model" appearing in the claims means that $a_1, \ldots, a_m, v$ are distributed on $\mathbb{R}^n \setminus \{0\}$

     - identically

     - independently

     - symmetrically under rotations.

Theorem.

For all distributions according to the rotation-invariance model

$$(0.11.1) \qquad E_{m,n}(S) \leq m^{1/(n-1)} \, n^3 \, \pi(1 + \frac{e\pi}{2}).$$

The following result concerns the number of steps required by our method for solving the complete problem.

Theorem.

For all distributions according to the rotations-invariance model our method for solving the complete problem requires not more than

$$(0.11.2) \qquad m^{1/(n-1)} \, (n+1)^4 \, \frac{2}{5} \, \pi(1 + \frac{e\pi}{2})$$

pivot steps on the average.

These two results hold for every pair (m, n) with $m \geq n$. Notice the improvement in the exponents of m and n in comparison with the results of section 5.

The integral formula (0.10.11) can be evaluated much easier in the asymptotic case (i. e. $m \to \infty$ and n fixed). Note that the main problem arising during the evaluation is the calculation or estimation of the spherical measures of $CC(a_1, \ldots, a_{n-1})$ under the condition that $CH(a_1, \ldots, a_n)$ is a boundary simplex.

If the angles $\text{arc}(a_i, a_j)$ with $i, j \in \{1, \ldots, n-1\}$ are **all** very small and if $h(a_1, \ldots, a_n)$ is close to 1, then $\dfrac{\lambda_{n-1}(CH(a_1, \ldots, a_{n-1}))}{\lambda_{n-1}(\Omega_{n-1})}$ is a very good approximation for the desired spherical measure. And for that quantity there are quite convenient formulae. But the asymptotic case emphasizes exactly the probability of these events.

So it is not surprising that we obtain somehow better results for the asymptotic case. And here it becomes possible to study the influence of special distributions or of special classes of distributions. For the asymptotic case we restrict our considerations to the estimation of $E_{m,n}(S)$. The first results in Chapter IV deal with distributions on a bounded support.

Theorem.

Let F be a radial distribution function of a distribution with bounded support according to the rotation-invariance model. Then there is a function $\varepsilon_F(m, n)$ depending on F and tending to 0 for $m \to \infty$ while n is fixed, such that

$$(0.11.3) \qquad E_{m,n}(S) \leq m^{1/(n-1)} \, n^2 \, \sqrt{2\pi} \, (1 + \varepsilon_F(m, n)).$$

Also we have results concerning the minimal possible growth for $m \to \infty$, n fixed.

Figure 0.9

The simplex $CH(a_1, a_2, a_3)$

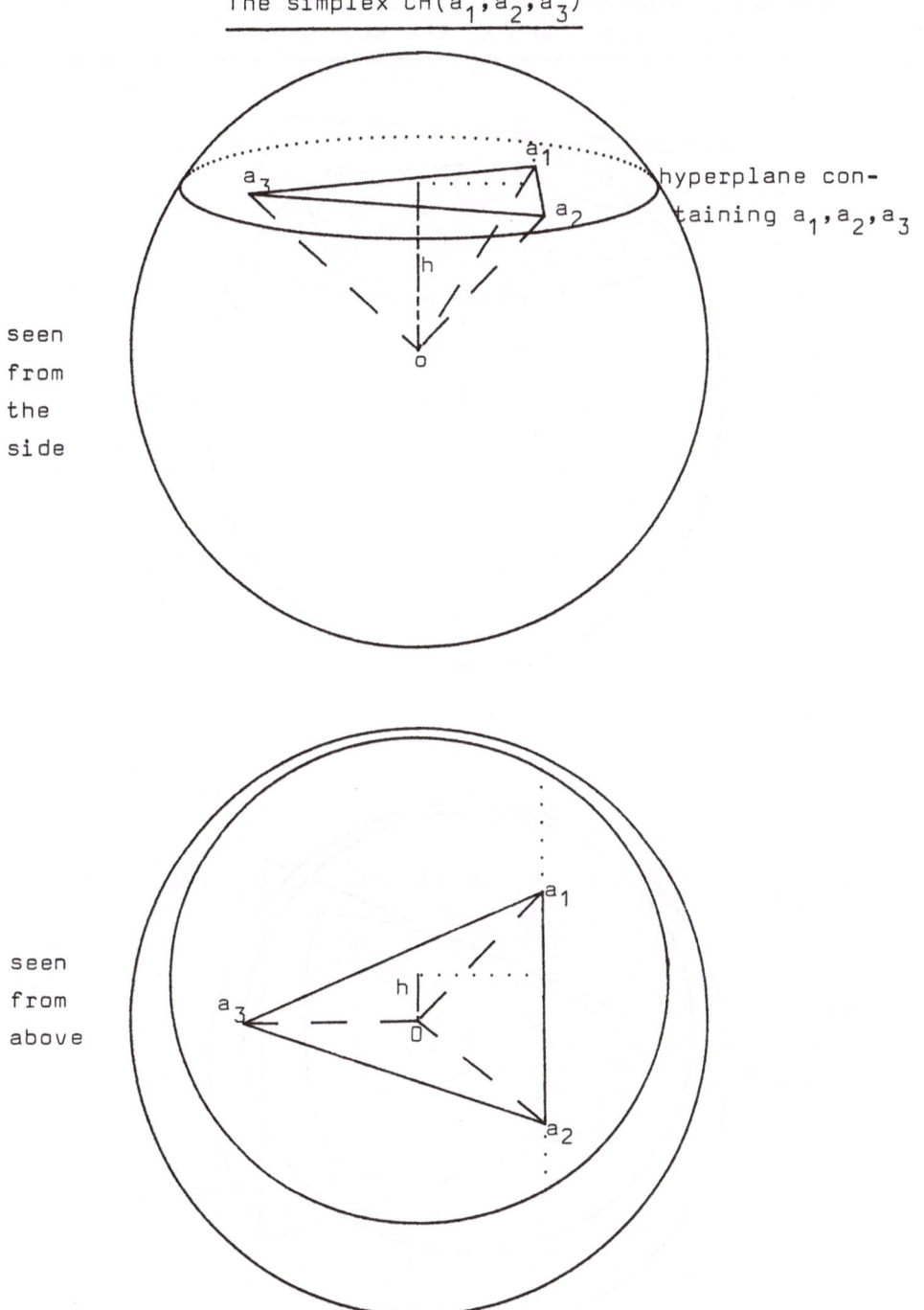

Figure 0.10

A possible estimation for the spherical measure of the
simplex $CH(a_1,a_2,a_3)$ and of the side simplex $CH(a_1,a_2)$

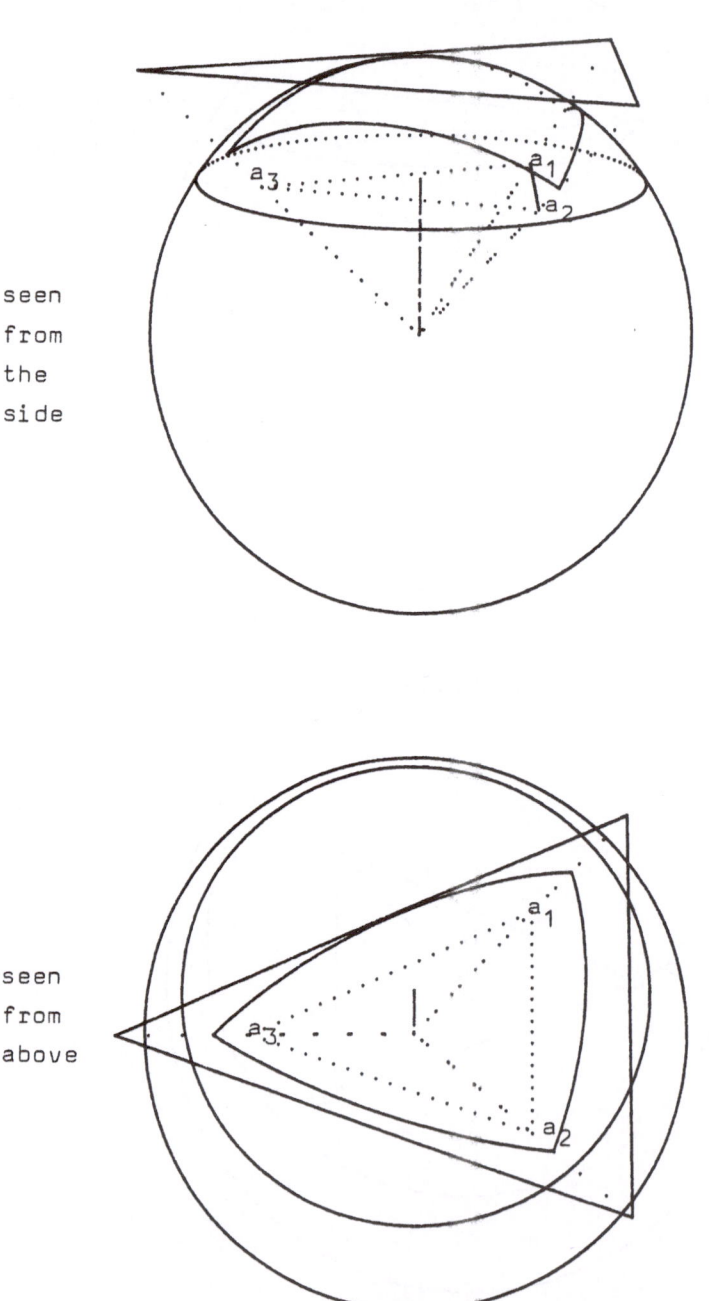

seen
from
the
side

seen
from
above

Theorem.

1) *For every $\delta > 0$ there is a distribution with bounded support according to the rotation-invariance model such that*

(0.11.4) $E_{m,n}(S) = 0(m^\delta)$ *for* $m \to \infty$, *fixed* $n \geq 3$.

2) *For every distribution with bounded support according to the rotation-invariance model we have*

(0.11.5) $E_{m,n}(S) \to \infty$ *for* $m \to \infty$, *fixed* $n \geq 3$.

Note that the second part of this theorem is nontrivial. Recall that Schmidt and Lindberg (1968/1981) have shown that there are examples of RDF's such that

$$E_{m,n}(S) \leq E_{m,n}(V) \leq C(n) \text{ for } m \to \infty, \text{ fixed } n,$$

where V is the number of vertices and where $C(n)$ is not known explicitly. Their result is based on the RDF

$$F(r) := \begin{cases} 0 & \text{for } r < 1 \\ 1 - \frac{1}{r} & \text{for } r \geq 1. \end{cases}$$

For a certain set of distributions we are even able to derive "almost" explicit asymptotic upper bounds for $E_{m,n}(S)$ of size $0(1)$.

Theorem.

If there are real values $k > 0$, $l = n^2$ and $\bar{r}$ such that for all $r > \bar{r}$ we have $1 - F(r) = \frac{k}{r^l}$, then we know that for $n \geq 3$ fixed

(0.11.6) $\lim\limits_{m \to \infty} \sup E_{m,n}(S) \leq n^{5/2}\, C$ *with* $C \in \mathbb{R}$.

After analyzing such classes of distributions, we deal with interesting special distributions and prove

Theorem.

There are functions $\varepsilon(m, n)$ and $\gamma(n)$, $\eta(n)$

           with $\varepsilon(m, n) \to 0$ for $m \to \infty$, fixed $n \geq 3$,

           $\gamma(n)$, $\eta(n) \to 1$ for $n \to \infty$ such that

1) *For uniform distribution on ω_n (unit sphere)*

(0.11.7) $E_{m,n}(S) \leq m^{1/(n-1)}\; n^2\; 2\gamma(n)\; (1 + \varepsilon(m, n))$

(0.11.8) $$E_{m,n}(S) \geq m^{1/(n-1)} \; n^2 \; 2\eta(n) \; (1 - \varepsilon(m,n))$$

2) *For uniform distribution on* Ω_n *(unit ball)*

(0.11.9) $$E_{m,n}(S) \leq m^{1/(n+1)} \; n^2 \; 2\gamma(n) \; (1 + \varepsilon(m,n))$$

(0.11.10) $$E_{m,n}(S) \geq m^{1/(n+1)} \; n^2 \; 2\eta(n) \; (1 - \varepsilon(m,n)).$$

As an example for distributions with unbounded support we analyze the influence of Gaussian distribution.

Theorem.

There are functions $\varepsilon(m,n)$, $\gamma(n)$, $\eta(n)$

$\quad$ *with* $\varepsilon(m,n) \to 0$ *for* $m \to \infty$, *fixed* $n \geq 3$,

$\quad$ $\gamma(n)$, $\eta(n) \to 1$ *for* $n \to \infty$ *such that*

$\quad$ *for Gaussian distribution on* $\mathbb{R}^n$

(0.11.11) $$E_{m,n}(S) \leq \sqrt{\ln m} \; n^{3/2} \; 2^{3/2} \; \gamma(n) \; (1 + \varepsilon(m,n))$$

(0.11.12) $$E_{m,n}(S) \geq \sqrt{\ln m} \; n^{3/2} \; 2^{3/2} \; \eta(n) \; (1 - \varepsilon(m,n))$$

These results show that the order of growth in our general upper bound in the case of bounded support cannot be improved.

It seems to be plausible, at least for the asymptotic case, that uniform distribution on ω_n yields the worst average behaviour. And the behaviour depends on the concentration of probability close to the boundary of support — if there is such a boundary. We believe that — in general — bounded support deteriorates the asymptotic average behaviour.

On the other side we guess that in case of unbounded support a steady, but very slow decrease of $1 - F(r)$ for $r \to \infty$ can significantly improve the asymptotic behaviour. At least in Chapter V, we try to extend our theory and results to problems containing nonnegativity constraints as

(0.11.13) $\quad$ Maximize $\quad v^T x$

$\qquad\qquad$ subject to $\quad a_1^T x \leq 1, \ldots, a_m^T x \leq 1, \; x \geq 0.$

The normal vectors of the nonnegativity constraints cannot satisfy rotational symmetry. The proofs become complicated because some case studies are necessary. Here the advantage of rotational symmetry becomes obvious. However, we succeed in deriving our main theorem. For that purpose we apply an algorithm which is very similar to the complete method used before.

Theorem.

The average number of pivot steps required for the solution of problems as (0.11.13) with nonnegativity constraints, where $a_1, \ldots, a_m, v$ are distributed according to the rotation-invariance model is not greater than

$$(0.11.14) \qquad m^{1/(n-1)} \, (n+1)^4 \, \frac{2}{5} \, \pi(1 + \frac{e\pi}{2}).$$

0.12 CONCLUSION AND CONJECTURES

Many questions remain open in my work and in the whole theory.

1) Are the upper polynomial bounds further improvable?

> Whereas the m-term $m^{1/(n-1)}$ is confirmed by upper and lower bounds, there is a certain discrepancy of the order in n for Phase II. We have an upper bound of n^3 and asymptotic upper and lower bounds of n^2. I guess that n^2 is more likely, but I do not have general bounds of that order yet. In addition, I still hope to find a method for Phase I such that the complete solution method has the same n-term as Phase II. This would mean that I expect such a method to have n^2 as the n-term.

2) Is it possible to get sharper polynomial bounds for special distributions?

> Our proof of polynomiality relies on combinations of several worst-cases. Special assumptions could help to improve the results.

3) Is it possible to transfer the results to problems with nonempty feasible regions but without **known** feasible points?

> Here we have to search for a suitable Phase I-variant which satisfies all stochastic assumptions.

4) Is it possible to come to similar results under stochastic models where degeneracy has positive probability?

5) Do the results hold for other variants, too?

6) Which is the best variant?

> – with respect to the worst case?

> – with respect to the average behaviour?

7) Is there any polynomial variant of the Simplex-Algorithm?

8) Is it possible to study the higher probabilistic moments of the distribution of s?

These and related questions may be a very fruitful field for further research.

Chapter 1

THE SHADOW-VERTEX ALGORITHM

This chapter is to explain the algorithm which shall be analyzed. For our probabilistic analysis it is necessary to use a so-called dual interpretation. Most of the readers may not be familiar with that way to describe the procedure. Therefore we start with the commonly used and well-known primal interpretation. In Section 2 we repeat most of the arguments of Section 1 in the new interpretation. And in Section 3 we show how the algorithm (in dual representation) can be realized in a corresponding tableau-form. So it is clear that much of the material is of an expository nature and that a certain amount of repetitions will occur. But the often observed confusion of readers by the different interpretations may be avoided by this way.

In the remaining part of the chapter we explain our method for Phase I, show that it works and give a numerical example for the complete algorithm.

1.1 PRIMAL INTERPRETATION

In this first section we give a primal interpretation of the shadow-vertex algorithm and a proof that it works. Recall that we deal with problems

(1.1.1)
$$
\begin{array}{ll}
\text{Maximize} & v^T x \\
\text{subject to} & a_1^T x \le 1, \ldots, a_m^T x \le 1 \\
\text{where} & v, x, a_1, \ldots, a_m \in \mathbb{R}^n \text{ and } m \ge n.
\end{array}
$$

Such problems are solved by application of the Simplex-Method, which proceeds in two phases.

(1.1.2)   **Phase I:**   Determination and calculation of a start vertex $x_0 \in X$.
                      The algorithm stops if it becomes obvious that a vertex does
                      not exist.

(1.1.3)   **Phase II:** Construction of a sequence $x_0, \ldots, x_s$ of successively adjacent
                      vertices of X, such that $v^T x_{i+1} > v^T x_i$ for $i = 0, \ldots, s-1$.
                      The algorithm stops at x_s

                      if x_s is the optimal vertex

                      if at x_s it becomes obvious that the problem has no solution.

For the beginning we concentrate on Phase II and its number s of pivot steps. Suppose that x_0 already has been given. Theoretically, this can be done by selecting x_0 as the optimal vertex relative to $u^T x$, where u is an arbitrary vector of $\mathbb{R}^n \setminus \{0\}$.

For practical applications — where x_0 is already given — one could take for u any vector out of the convex cone of the active restriction vectors at x_0. In other words:

$$(1.1.4) \qquad u \in \Big\{ y \mid y = \sum_{\substack{i=1 \\ a_i^T x_0 = 1}}^{m} \rho_i a_i \text{ with } \rho_i \geq 0,\ y \neq 0 \Big\}.$$

Then x_0 is the optimal vertex with respect to $u^T x$. This results from the Lemma of Farkas, because one can show that

Lemma 1.1

(1.1.5)   *Let $\bar{x}$ be a vertex. Then $\bar{x}$ is maximal with respect to $w^T x$ ($w \neq 0$) if and only if w is an element of the convex cone spanned by the restriction vectors which are active at $\bar{x}$.*

If we suppose nondegeneracy of our problem, then there are exactly n active constraints

$$a_{\Delta^i} \text{ with } \Delta^1 < \Delta^2 < \ldots < \Delta^n, \ \Delta = \{\Delta^1, \ldots, \Delta^n\} \subset \{1, \ldots, m\}.$$

Our considerations are simplified by the

Assumption of nondegeneracy

(1.1.6)   *Each n-element subset of $\{a_1, \ldots, a_m, u, v\}$ is linearly independent and each subset of $n+1$ elements out of $\{a_1, \ldots, a_m\}$ is in general position.*

Throughout the following, we assume that this condition is valid.

Under this assumption, every vertex of X is the unique solution x_Δ of a system

$$(1.1.7) \qquad a_{\Delta^1}^T x = 1, \ldots, a_{\Delta^n}^T x = 1, \text{ with } \Delta = \{\Delta^1, \ldots, \Delta^n\},$$

and it must be feasible, i. e.

$$(1.1.8) \qquad a_j^T x \leq 1 \text{ for } j = 1, \ldots, m.$$

The vector u determines not only the initial vertex x_0, but also the sequence $x_0, \ldots, x_s$ of the shadow-vertex algorithm (together with the vector v).

In (1.1.4) it is not explained how the improving edge is selected if there are two or more such edges. If the variant is to be the shadow-vertex algorithm, then the selection rule is as follows.

Suppose that the polyhedron X is projected onto $\text{span}(u, v)$, the two-dimensional plane spanned by u and v. Then those vertices of X, whose images $\Gamma(x)$ under the orthogonal projection Γ on $\text{span}(u, v)$ are vertices of $\Gamma(X)$, will be called shadow-vertices.

Under our assumption of nondegeneracy we have two simple equivalent characterizations for shadow vertices.

Lemma 1.2

Let x_Δ be a vertex of X. Then the following three conditions are equivalent (when nondegeneracy holds):

$$(1.1.9) \qquad \begin{array}{ll} i) & \text{x_Δ is a shadow-vertex.} \\ ii) & \text{$\Gamma(x_\Delta) \in \partial\Gamma(X)$ (the boundary of $\Gamma(X)$)} \\ iii) & \text{there is a vector $w \in \text{span}(u, v) \setminus \{0\}$ such that $w^T x_\Delta = \max\limits_{x \in X} w^T x$.} \end{array}$$

Proof. i) $\implies$ ii): Is clear because $\Gamma(x_\Delta)$ is a vertex of $\Gamma(X)$, so it belongs to $\partial\Gamma(X)$.

ii) $\implies$ iii): Let $\Gamma(x_\Delta) \in \partial\Gamma(X)$. $\Gamma(X)$ is convex and hence there is a supporting hyperplane through $\Gamma(x_\Delta)$ in $\text{span}(u, v)$. So we can find a vector $w \in \text{span}(u, v) \setminus \{0\}$ such that $w^T \Gamma(x) \leq w^T \Gamma(x_\Delta)$ for all $x \in X$. Since $w \in \text{span}(u, v)$, this means that $w^T x \leq w^T x_\Delta$ for all $x \in X$.

iii) $\implies$ i): Let x_Δ be maximal with respect to $w^T x$, $w \in \text{span}(u, v) \setminus \{0\}$. Then it is clear that x_Δ cannot be mapped into the interior of $\Gamma(X)$, because $\Gamma(x_\Delta)$ lies on a supporting hyperplane for $\Gamma(X)$ with normal vector w. This means that $\Gamma(x_\Delta)$ lies on an edge of the two-dimensional polyhedron $\Gamma(X)$. We have to show that it is even a vertex of $\Gamma(X)$. Assume that it is not a vertex. Then w is orthogonal to the edge containing $\Gamma(x_\Delta)$ and we find $z \in \text{span}(u, v)$, $z \neq 0$, $z^T w = 0$ such that $\Gamma(x_\Delta) + z$ and $\Gamma(x_\Delta) - z$ belong to the same edge. Now it is clear that $\Gamma(x_\Delta)$ cannot be maximal with respect to any objective function $(w + \varepsilon z)^T x$ with $\varepsilon \neq 0$ in $\Gamma(X)$. Hence x_Δ cannot be optimal for such an objective in X. So we know that w is the only point of the straight line $w + \mathbb{R}z$ which belongs to $\text{CC}(a_{\Delta^1}, \ldots, a_{\Delta^n})$. Now consider the unique representation of the points of $w + \mathbb{R}z$ $\lambda_1(\gamma)a_{\Delta^1} + \ldots + \lambda_n(\gamma)a_{\Delta^n} = w + \gamma z$ $(\gamma \in \mathbb{R})$ by the vectors $a_{\Delta^1}, \ldots, a_{\Delta^n}$. The coefficients $\lambda_i(\gamma)$ are (affine-) linear functions of γ. Since they are continuous, there must be a k such that $\lambda_k(0) = 0$ and $\lambda_k(\gamma) < 0$ for $\gamma > 0$. And there must be an l such that $\lambda_l(0) = 0$ and $\lambda_l(\gamma) < 0$ for $\gamma < 0$. Note that $k \neq l$ as a result of linearity in γ. So we have for w the following representation

$$0 \neq w = \eta u + \rho v = \sum_{\substack{i=1 \\ i \neq k, l}}^n \lambda_i(0) a_{\Delta^i},$$

which shows that the n vectors u, v, a_{Δ^i} $(i \in \{1, \ldots, n\} \setminus \{k, l\})$ are linearly dependent. This is a contradiction to our assumption of nondegeneracy and proves that $\Gamma(x_\Delta)$ must be a vertex of $\Gamma(X)$.

$\square$

Obviously the start vertex x_0 is such a shadow-vertex, if u is chosen as above. The same is true for the vertex x_s if it is optimal. If w traverses the (smaller) angle between u and v, then we obtain a sequence of such solution vertices, which are shadow-vertices, for "Maximize $w^T x$". This is the sequence $x_0, \ldots, x_s$. The vertices $\Gamma(x_i)$ and $\Gamma(x_{i+1})$ are adjacent in $\Gamma(X)$. Such a sequence is constructed in the same way if $v^T x$ has no solution. Then x_s is the solution to the last of the (bounded) objectives $w^T x$, $w \in \text{span}(u, v)$, and hence it is a shadow-vertex. In this case, x_s is adjacent to an unbounded ray which improves $v^T x$. The following lemma transfers properties from $\Gamma(X)$ to X.

Lemma 1.3

(1.1.10)   Let x_i and x_{i+1} be vertices. If $\Gamma(x_i)$ and $\Gamma(x_{i+1})$ are adjacent
         in $\Gamma(X)$, then x_i and x_{i+1} are adjacent in X, too.

Proof. It is clear that $[x_i, x_{i+1}]$ cannot traverse the interior of X. Since $[\Gamma(x_i), \Gamma(x_{i+1})]$ is an edge of $\Gamma(X)$, we have a $w \in \text{span}(u, v)$, such that $w \neq 0$ and

$$w^T \Gamma(x_i) = w^T x_i = 1 = w^T x_{i+1} = w^T \Gamma(x_{i+1}) \text{ is maximal on } X.$$

Following the Lemma of Farkas, w must be an element of the convex cone spanned by the restriction-vectors which are active in all points of $[x_i, x_{i+1}]$.

Figure 1.1a

Set of shadow-vertices under projection on the observation-
plane

S = 8

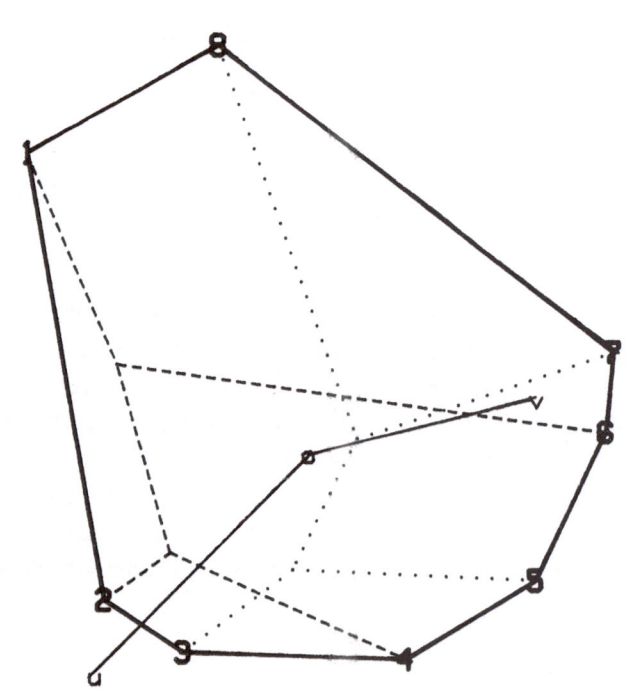

Figure 1.1b

Set of shadow-vertices under a different projection

S=9

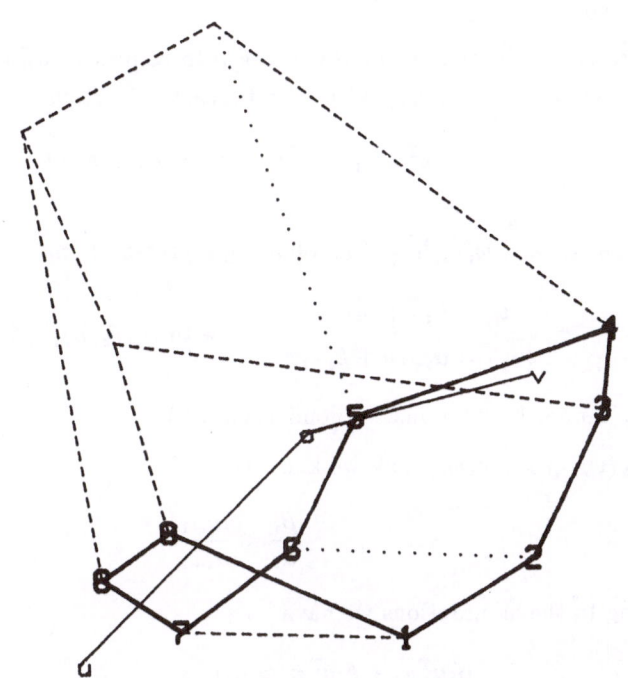

Assume now that this line segment is not an edge of X, i. e. $[x_i, x_{i+1}]$ does not belong to the intersection of $n-1$ hyperplanes. Then w is a (positive) combination of at most $n-2$ vectors $a_{i_1}, \ldots, a_{i_{n-2}}$ and we have

$$w = \lambda u + \rho v = \eta_{i_1} a_{i_1} + \ldots + \eta_{i_{n-2}} a_{i_{n-2}}.$$

But this is a contradiction to (1.1.6). So $[x_i, x_{i+1}]$ is an edge. Consequently, following the boundary of $\Gamma(X)$ means moving on a path of edges in X.

$\square$

Now it remains to show that the objective function $v^T x$ increases when we follow the path $x_0, \ldots, x_s$.

Lemma 1.4

Let $x_0, x_1, \ldots, x_s$ be the maximal vertices with respect to $w_0^T x, w_1^T x, \ldots, w_s^T x$, where $w_0 = u$, $\text{arc}(w_i, v) > \text{arc}(w_{i+1}, v)$ for $i = 0, \ldots, s-1$. Then

$$(1.1.11) \qquad v^T x_{i+1} > v^T x_i \text{ for } i = 0, \ldots, s-1.$$

Proof. Let w_i and w_{i+1} have the following representations

$$w_i = \alpha_i u + \beta_i v$$
$$w_{i+1} = \alpha_{i+1} u + \beta_{i+1} v$$

with $\alpha_i, \beta_i, \alpha_{i+1}, \beta_{i+1} \geq 0$

(α_i, α_{i+1} cannot be 0 because of nondegeneracy).

Since $\text{arc}(w_i, v) > \text{arc}(w_{i+1}, v)$, we know that

$$\frac{\beta_i}{\alpha_i} < \frac{\beta_{i+1}}{\alpha_{i+1}}.$$

According to the assumptions we have

$$\alpha_i u^T x_i + \beta_i v^T x_i > \alpha_i u^T x_{i+1} + \beta_i v^T x_{i+1}$$
$$\alpha_{i+1} u^T x_i + \beta_{i+1} v^T x_i < \alpha_{i+1} u^T x_{i+1} + \beta_{i+1} v^T x_{i+1}.$$

So we have

$$v^T(x_{i+1} - x_i)\frac{\beta_i}{\alpha_i} < u^T(x_i - x_{i+1}) < \frac{\beta_{i+1}}{\alpha_{i+1}} v^T(x_{i+1} - x_i)$$

which leads to

$$v^T(x_{i+1} - x_i) > 0 \quad \text{or} \quad v^T x_{i+1} > v^T x_i.$$

So we have proven that the sequence of shadow-vertices $x_0, \ldots, x_s$ actually is a simplex-path according to (1.1.3).

1.2 DUAL INTERPRETATION

Now we are going to explain the shadow-vertex algorithm in a different way. We use a dual interpretation. This dual view has the advantage that all random events can be explained directly by use of the random input vectors $a_1, \ldots, a_m, u, v$. This is important, because it simplifies the evaluation of expectation values. To begin with, let us recall the nondegeneracy assumption.

(1.2.1)   Every n-element subset of $\{a_1, \ldots, a_m, u, v\}$ is linearly independent and every subset of $n + 1$ elements of $\{a_1, \ldots, a_m\}$ is in general position.

Y will be the notation for the dual polyhedron to X.

(1.2.2) $$Y = \{y \in \mathbb{R}^n \mid x^T y \leq 1 \text{ for all } x \in X\}.$$

CH($\ldots$) stands for the convex hull of given vectors or points.

Lemma 1.5

(1.2.3) $$Y = \{y \mid x^T y \leq 1 \text{ for all } x \in X\} = CH(0, a_1, \ldots, a_m).$$

Proof. First let y be an element of $CH(0, a_1, \ldots, a_m)$, i. e. $y = \lambda_1 a_1 + \ldots + \lambda_m a_m$ with $\lambda_1, \ldots, \lambda_m \geq 0$ and $\lambda_1 + \ldots + \lambda_m \leq 1$. Then we know that for an arbitrary point $x \in X$ $x^T y = \lambda_1 a_1^T x + \ldots + \lambda_m a_m^T x \leq \lambda_1 + \ldots + \lambda_m \leq 1$. For the proof of the opposite direction, let $y \in Y$, i. e. $x^T y \leq 1$ for all $x \in X$. Assume that y does not belong to $CH(0, a_1, \ldots, a_m)$. Then there is a z with $z^T y > 1$ and $z^T \bar{y} \leq 1$ for all $\bar{y} \in CH(0, a_1, \ldots, a_m)$. Hence z belongs to X, because $z^T a_i \leq 1$ for $i = 1, \ldots, m$. So y cannot be an element of Y and this is a contradiction.

$\square$

As in the section before let $\Delta := \{\Delta^1, \ldots, \Delta^n\} \subset \{1, \ldots, m\}$ be an n-element set of indices. We had observed a one-to-one correspondence between the intersection points x_Δ and the index sets Δ, where x_Δ is the solution of the system

(1.2.4) $$a_{\Delta^1}^T x = 1, \ldots, a_{\Delta^n}^T x = 1.$$

On the other side there is a one-to-one correspondence between Δ and the $n - 1$-dimensional simplex $CH(a_{\Delta^1}, \ldots, a_{\Delta^n})$, which will also be denoted by $\Sigma(\Delta)$. So we have the unique correspondence

(1.2.5) $$x_\Delta \longleftrightarrow \Delta \longleftrightarrow CH(a_{\Delta^1}, \ldots, a_{\Delta^n}) = \Sigma(\Delta)$$

$\Sigma(\Delta)$ is $n - 1$-dimensional as a result of nondegeneracy. If x_Δ is a vertex of X, then $\Sigma(\Delta)$ is a facet of Y.

Lemma 1.6

Let x_Δ be a vertex. Then

(1.2.6) $$\Sigma(\Delta) = Y \cap \{y \mid y^T x_\Delta = 1\}.$$

Proof. Inclusion $\subset$ follows from the definition. For the opposite direction we notice that $\{y \mid y^T x_\Delta = 1\}$ defines a hyperplane. Let y be a point of Y and the hyperplane. Then we have

$$y = \lambda_1 a_1 + \ldots + \lambda_m a_m \text{ with } \lambda_1, \ldots, \lambda_m \geq 0, \lambda_1 + \ldots + \lambda_m \leq 1.$$

Exploiting $y^T x = 1$ we do even know that $\lambda_1 + \ldots + \lambda_m = 1$. This representation for y is unique, because for all a_i with $i \notin \Delta$ we have $x_\Delta^T a_i < 1$. So

$$y = \sum_{i=1}^{n} \lambda_{\Delta^i} a_{\Delta^i} \text{ with } \lambda_{\Delta^i} \geq 0, \sum_{i=1}^{n} \lambda_{\Delta^i} = 1,$$

and y is an element of $\mathrm{CH}(a_{\Delta^1}, \ldots, a_{\Delta^n}) = \Sigma(\Delta)$.

$\square$

From this Lemma we immediately derive a dual characterization of the shadow-vertex condition.

If $\Sigma(\Delta)$ is entirely in ∂Y, then we call it a boundary simplex of Y.

Lemma 1.7

(1.2.7) x_Δ is a vertex of X if and only if $\Sigma(\Delta)$ is a boundary simplex of Y.

Proof. Let x_Δ be a vertex. Then $a_{\Delta^1}, \ldots, a_{\Delta^n}$ belong to the hyperplane $\{y \mid y^T x = 1\}$ which contains $\Sigma(\Delta)$. x_Δ is feasible and for that reason we have $a_i^T x_\Delta \leq 1$ for all $i \notin \Delta$. Hence this hyperplane is a supporting hyperplane for Y.

$\square$

Now let $\Sigma(\Delta)$ be a boundary simplex. x_Δ is orthogonal and $a_j^T x_\Delta \leq 1$ for all $j \notin \Delta$, $a_i^T x_\Delta = 1$ for $i \in \Delta$. So x_Δ is feasible and n restrictions are active in x_Δ. This means that x_Δ is a vertex of X.

The following lemma translates the shadow-vertex-condition into dual interpretation.

Figure 1.2

The vertices of X and the active constraints

At every vertex of X we have listed the active constraints.

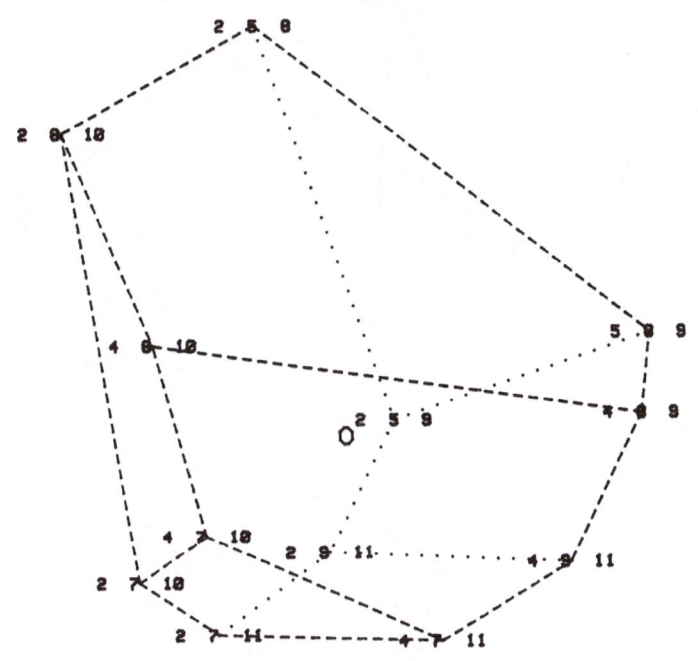

Figure 1.3a

<u>The role of single constraints</u>

In the following figures we have shown the vertices of X where a certain constraint is active (here no. 2).

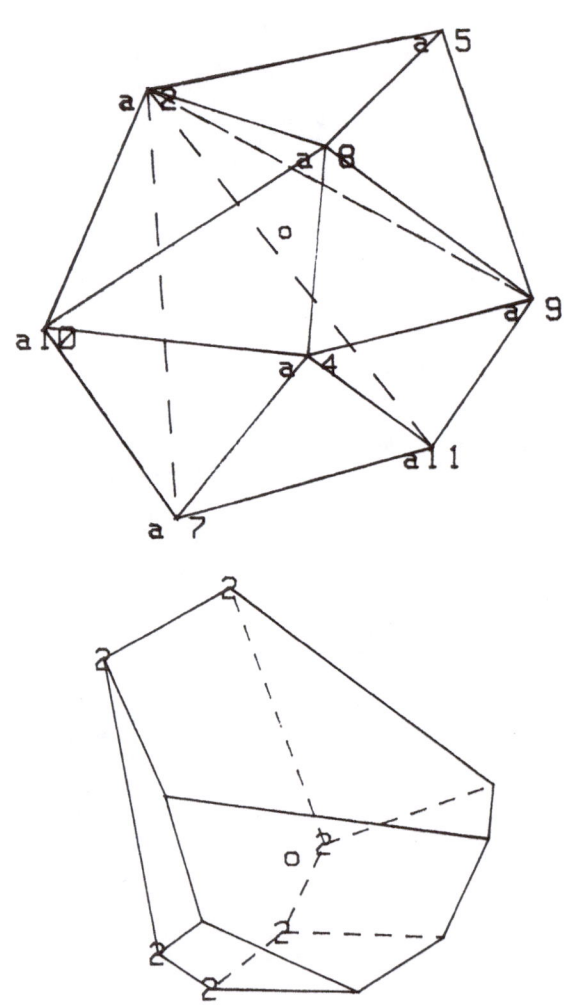

Figure 1.3b

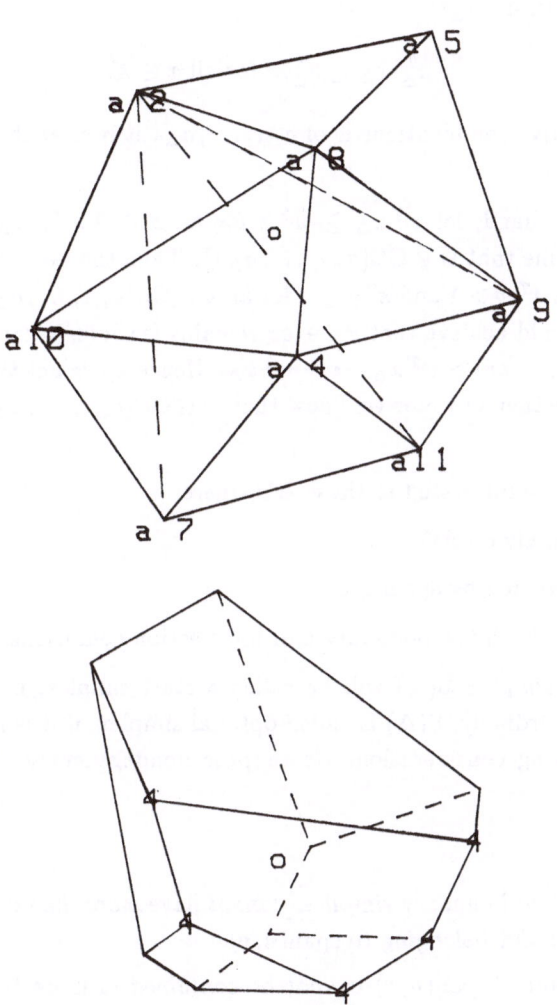

Lemma 1.8

Let x_Δ be a vertex of X and let $w \in \mathbb{R}^n$, $w \neq 0$. Then x_Δ is maximal with respect to $w^T x$ if and only if

$$(1.2.8) \qquad \mathbb{R}^+ w \cap \Sigma(\Delta) \neq \emptyset \quad (\mathbb{R}^+ \text{ stands for } (0, \infty)).$$

Proof. First, let $w \in CC(a_{\Delta^1}, \ldots, a_{\Delta^n})$. (CC denotes the convex cone). Then we know that for arbitrary a_{Δ^i}

$$a_{\Delta^i}^T x_\Delta \geq a_{\Delta^i}^T x \text{ for all } x \in X.$$

Hence for all positive combinations w of $a_{\Delta^1}, \ldots, a_{\Delta^n}$ it is clear that $w^T x_\Delta \geq w^T x$ for all $x \in X$.

On the other hand, let $w^T x_\Delta \geq w^T x$ for all $x \in X$. In x_Δ we have n active restrictions. Assume that $w \notin CC(a_{\Delta^1}, \ldots, a_{\Delta^n})$. Then the Lemma of Farkas yields a vector z such that $z^T w > 1$ and $z^T y \leq 0$ for all $y \in CC(a_{\Delta^1}, \ldots, a_{\Delta^n})$. For sufficiently small $\varepsilon > 0$ we could achieve that $x_\Delta + \varepsilon z$ remains feasible, because $x_\Delta^T a_i < 1$ for all $i \notin \Delta$. And $w^T(x_\Delta + \varepsilon z) > w^T x_\Delta + \varepsilon > w^T x_\Delta$. Hence x_Δ is not the optimal solution. This is a contradiction and now we know that $w \in CC(a_{\Delta^1}, \ldots, a_{\Delta^n})$. $\qquad \square$

From now on we are interested in those Δ's where

- $\Sigma(\Delta)$ lies entirely on ∂Y
- $\Sigma(\Delta)$ is intersected by $\text{span}(u, v)$
- $\Sigma(\Delta)$ satisfies both the boundary and intersection conditions.

A boundary simplex $\Sigma(\Delta)$ will be called a start simplex, if it is intersected by the ray $\mathbb{R}^+ u$. Accordingly, $\Sigma(\Delta)$ is called optimal simplex, if it is intersected by $\mathbb{R}^+ v$. For all the following considerations we suppose nondegeneracy. Now we prove two auxiliary claims.

Proposition.

(1.2.9)   Different boundary simplices cannot have more than one common point belonging to $\text{span}(u, v)$.

(1.2.10)   A point of $\text{span}(u, v)$ cannot be contained in more than two boundary simplices.

Proof. Let $\Sigma(\Delta)$ be a boundary simplex and $w \in \Sigma(\Delta)$. Then the representation

$$(1.2.11) \qquad w = \mu_1 a_1 + \ldots + \mu_m a_m, \ \mu_i \geq 0, \mu_1 + \ldots + \mu_m = 1$$

Figure 1.4

The dual maximality condition

Here the ray $\mathbb{R}^+\omega$ intersects the simplex $CH(a_1, a_2, a_3)$.

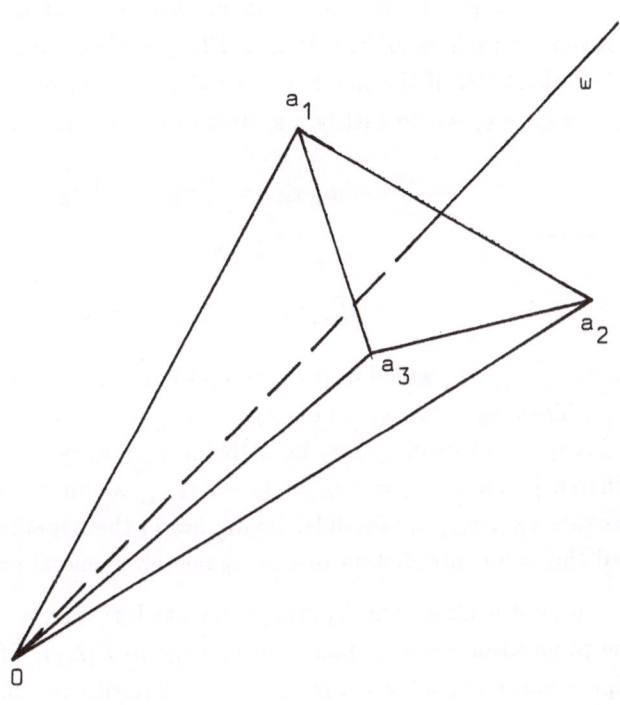

is unique and only the μ_i with $i \in \Delta$ can be positive. Assume now that $w_1, w_2 \in$ span$(u, v) \cap \Sigma(\Delta_1) \cap \Sigma(\Delta_2)$ with $\Delta_1 \neq \Delta_2$. In the representations for w_1 resp. w_2 only the μ_i with $i \in \Delta_1 \cap \Delta_2$ can be positive. Since Δ_1 and Δ_2 are different, at most $n-1$ of the μ_i are positive (let these be $\mu_1, \ldots, \mu_{n-1}$). So we get

$$w_1 = \alpha_1 u + \beta_1 v = \eta_1 a_1 + \ldots + \eta_{n-1} a_{n-1}$$
$$w_2 = \alpha_2 u + \beta_2 v = \rho_1 a_1 + \ldots + \rho_{n-1} a_{n-1}.$$

If $w_1 \neq w_2$ then either α_1 or α_2 is nonzero (for example α_1). Then we obtain

$$\frac{\alpha_2}{\alpha_1} w_1 - w_2 = \frac{\alpha_2}{\alpha_1} \beta_1 v - \beta_2 v = \nu_1 a_1 + \ldots + \nu_{n-1} a_{n-1} \text{ with } \nu_l = \frac{\alpha_2}{\alpha_1} \eta_l - \rho_l.$$

This would be a contradiction to nondegeneracy if $w_1 \neq w_2$. So it is clear that $w_1 = w_2$.

Now we prove the second claim. Assume that $w \in$ span(u, v) belongs to three boundary simplices $\Sigma(\Delta_1), \Sigma(\Delta_2), \Sigma(\Delta_3)$. We show that $\Delta_1 \cap \Delta_2 \cap \Delta_3$ has at most $n-2$ elements. If the number of elements would be greater $n-2$, then the vertices $x_{\Delta_1}, x_{\Delta_2}, x_{\Delta_3}$ would satisfy a system of $n-1$ common equations

$$a_1^T x_{\Delta_1} = a_1^T x_{\Delta_2} = a_1^T x_{\Delta_3} = 1$$
$$\vdots$$
$$a_{n-1}^T x_{\Delta_1} = a_{n-1}^T x_{\Delta_2} = a_{n-1}^T x_{\Delta_3} = 1.$$

So $x_{\Delta_1}, x_{\Delta_2}, x_{\Delta_3}$ would be on a straight line. Let x_{Δ_3} be contained in the interval $x_{\Delta_1}, x_{\Delta_2}$. Then $x_{\Delta_3} = \lambda x_{\Delta_1} + (1-\lambda) x_{\Delta_2}, 0 \leq \lambda \leq 1$. In addition, let the restriction with index i_1 be active in x_{Δ_1}, i_2 be active in x_{Δ_2}, i_3 be active in x_{Δ_3} (where i_1, i_2, i_3 are different). Since $x_{\Delta_3} = \lambda x_{\Delta_1} + (1-\lambda) x_{\Delta_2}$, we know that $a_{i_3}^T x_{\Delta_1} = a_{i_3}^T x_{\Delta_2} = 1$, too, because $x_{\Delta_1}, x_{\Delta_2}$ are feasible. So a_{i_3} lies in the hyperplane through $a_1, \ldots, a_{n-1}, a_{i_1}$, and this is a contradiction to nondegeneracy (general position).

Now it is clear that $\Delta_1 \cap \Delta_2 \cap \Delta_3$ has less than $n-1$ elements. If — contrary to the proposition — $w \in$ span(u, v) belongs to $\Sigma(\Delta_1), \Sigma(\Delta_2), \Sigma(\Delta_3)$, then the unique representation of w has not more than $n-2$ positive coefficients. We obtain an equation of the type

$$w = \alpha u + \beta v = \mu_1 a_1 + \ldots + \mu_{n-2} a_{n-2},$$

which is a contradiction to nondegeneracy (linear independence). This proves (1.2.10).

$\square$

Instead of analyzing a sequence of vertices as in Section 1, we consider here a certain sequence of boundary simplices of Y.

Lemma 1.9

(1.2.12)   The boundary simplices intersecting

$CC(u, v) = \{y \mid y = \lambda u + \rho v, \lambda \geq 0, \rho \geq 0\} \subset \text{span}(u, v)$

can be arranged uniquely in a sequence $\Sigma(\Delta_0), \ldots, \Sigma(\Delta_s)$

such that $\Delta_i \neq \Delta_j$ for $i \neq j$, Δ_i and Δ_{i+1} differ only in one element,

and $\text{arc}(z_i, v) \geq \text{arc}(z_{i+1}, v)$ for every pair (z_i, z_{i+1})

with $z_i \in \Sigma(\Delta_i) \cap \text{span}(u, v)$, $z_{i+1} \in \Sigma(\Delta_{i+1}) \cap \text{span}(u, v)$.

Proof. The intersection set of $CC(u, v)$ and $\Sigma(\Delta)$ (a boundary simplex) is a convex interval and Y is convex. Hence the intersection sets mentioned above can be arranged in a sequence

(1.2.13) $[w_0, w_1], [w_1, w_2], \ldots, [w_s, w_{s+1}]$.

(1.2.9) guarantees that these intervals do not overlap each other.

If we demand that $\text{arc}(w_j, v) > \text{arc}(w_{j+1}, v)$ this sequence is unique, because of (1.2.10), where the uniqueness of the successor interval is shown.

Δ_i and Δ_{i+1} differ only in one element, because the vector w_{i+1} belongs to $\Sigma(\Delta_i)$ and to $\Sigma(\Delta_{i+1})$ at the same time. Else the representation of w_{i+1} by the vectors $a_k (k \in \Delta_i \cap \Delta_{i+1})$ would lead to degeneracy. So assigning $\Sigma(\Delta_i)$ to the interval $[w_i, w_{i+1}]$ leads to a unique sequence.

□

The following lemma shows the connection between dual and primal interpretations.

Lemma 1.10

(1.2.14)   There is a unique correspondence between the sequence of

boundary simplices of Y as constructed above and a sequence

$x_{\Delta_0}, \ldots, x_{\Delta_s}$ of shadow vertices of X. The sequence of vertices

satisfies all the conditions of (1.1.3). So it is a Simplex-Path.

Proof. To every boundary simplex $\Sigma(\Delta)$ we assign the corresponding shadow vertex x_Δ. After that we have a sequence $x_{\Delta_0}, \ldots, x_{\Delta_s}$. Since Δ_i and Δ_{i+1} differ only in one element, $n-1$ restrictions are active in both x_{Δ_i} and $x_{\Delta_{i+1}}$. So these succeeding vertices are adjacent. The vectors $w_0, w_1, \ldots$ are arranged such that $\text{arc}(w_i, v) > \text{arc}(w_{i+1}, v)$. From Lemma 1.3 we conclude $v^T x_{\Delta_i} < v^T x_{\Delta_{i+1}}$. If we set $x_i := x_{\Delta_i}$, our vertex-sequence satisfies all the conditions of a Simplex-Path, which maximizes the value of $v^T x$ and visits only shadow vertices.

□

Figure 1.5a

The boundary simplices of Y which are intersected by the
two-dimensional plane span(u,v)

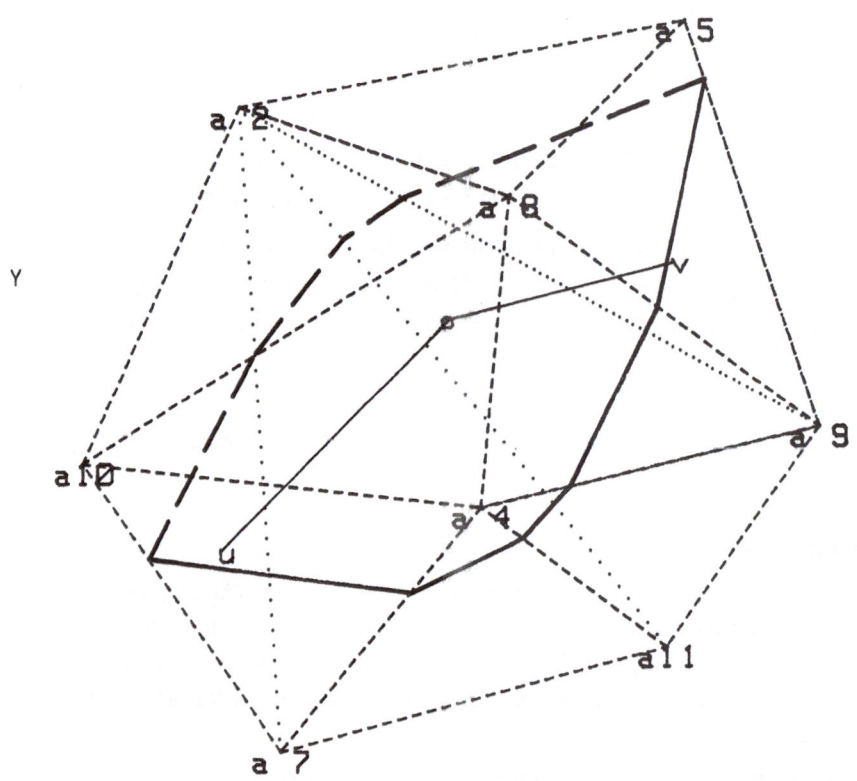

Only 9 of the 12 boundary simplices of Y are intersected.

Figure 1.5b

Foreground of figure 1.5a

Here span(u,v) intersects the boundary simplices
$CH(a_4, a_7, a_{10})$; $CH(a_4, a_7, a_{11})$; $CH(a_4, a_9, a_{11})$;
$CH(a_4, a_8, a_9)$; $CH(a_5, a_8, a_9)$.

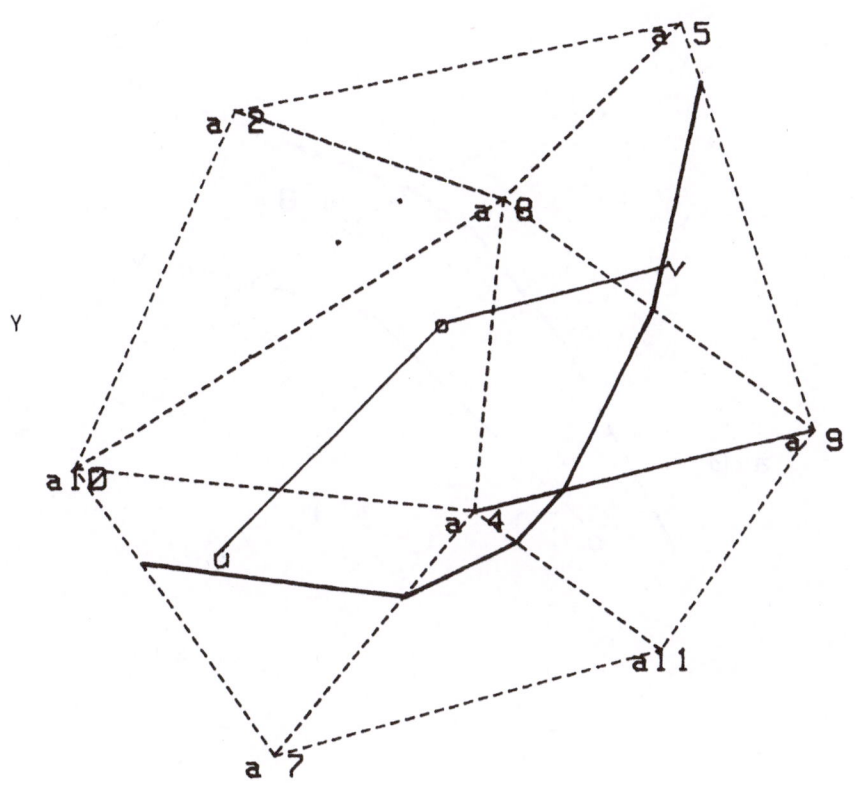

Figure 1.5c

Background of figure 1.5a

span(u,v) intersects the boundary simplices

$CH(a_2,a_5,a_9)$; $CH(a_2,a_9,a_{11})$; $CH(a_2,a_7,a_{11})$; $CH(a_2,a_7,a_{10})$.

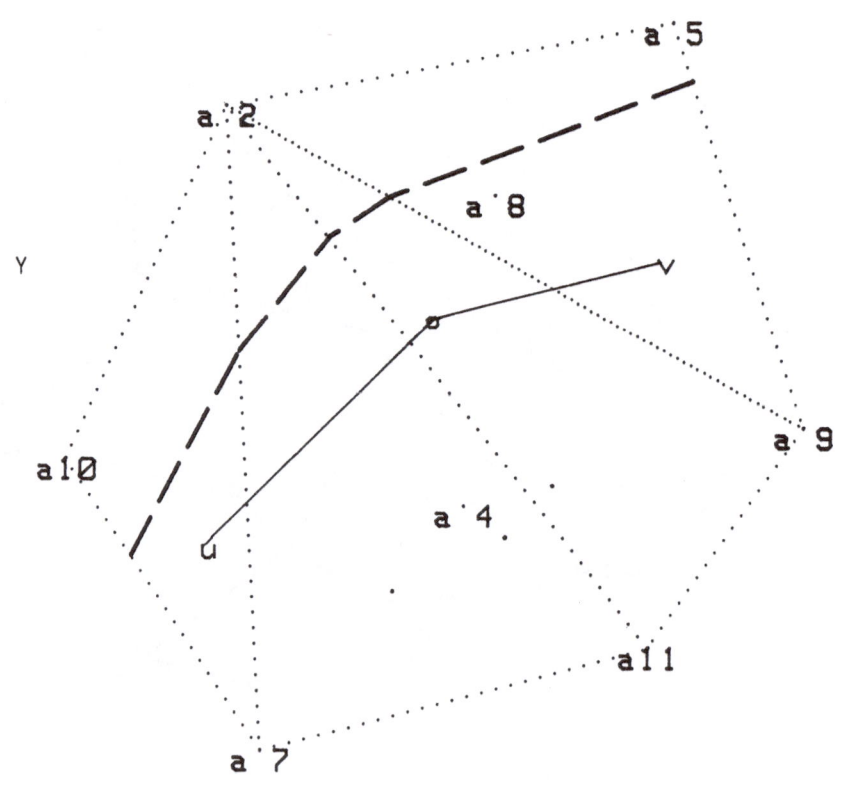

Figure 1.5d

Shadow-vertex-path under projection on span(u,v)

Only 4 pivot steps are necessary. So s=4, S=9.

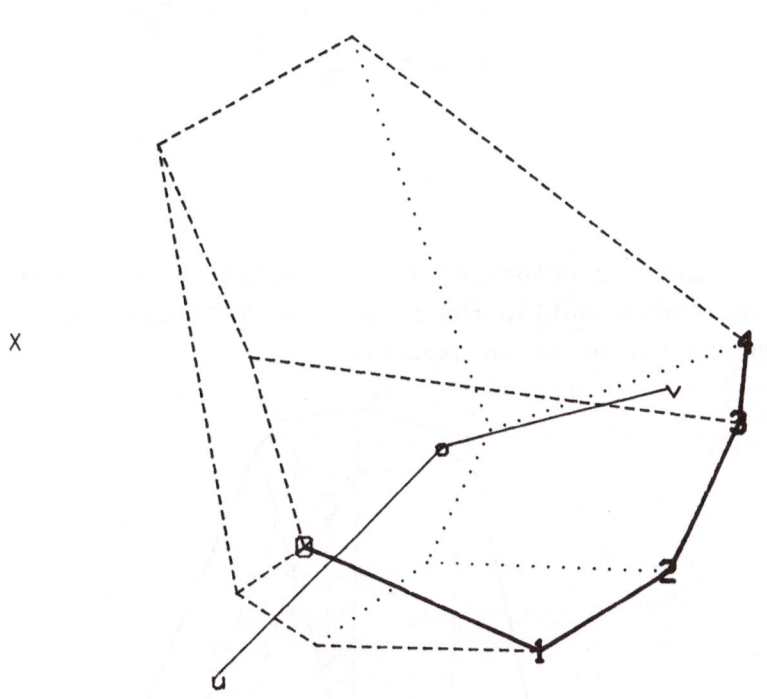

Figure 1.6

<u>Y and X in the case of unbounded region $X \subset \mathbf{R}^3$</u>

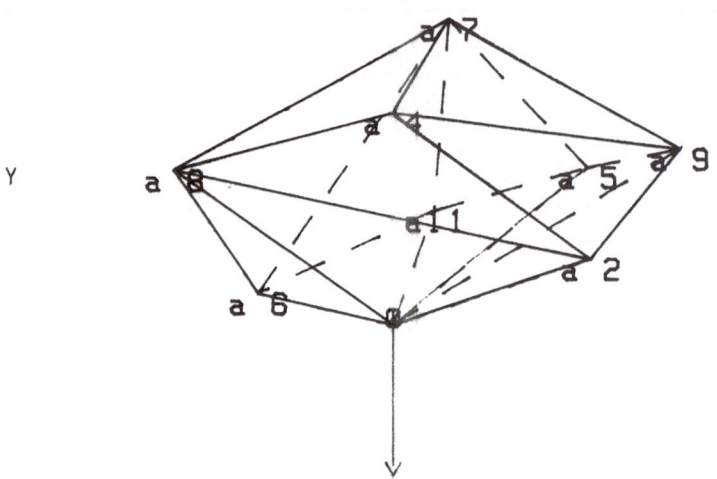

All generating points of Y are located in the upper half-space. Consequently, the polyhedron X is unbounded (or open) in the downward direction v.

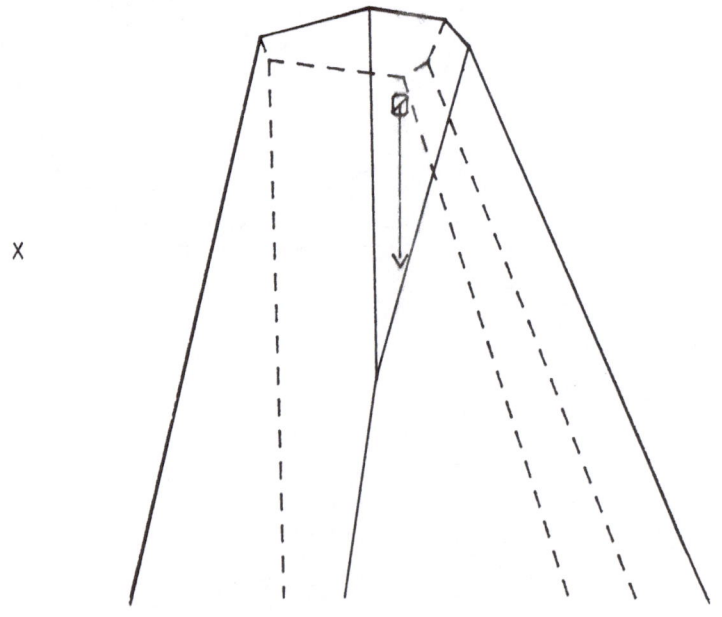

Figure 1.7

Y and unbounded X and the intersection/projection with/on span(u,v)

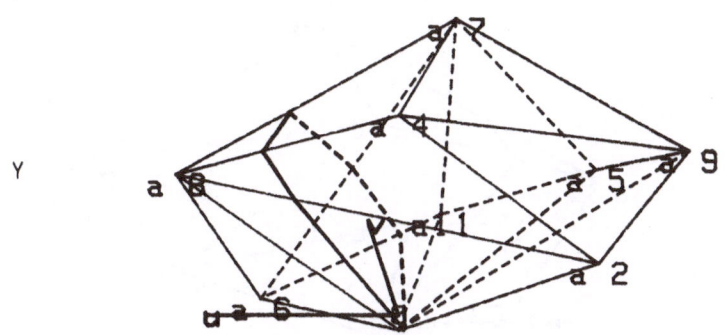

span(u,v) intersects

$CH(0,a_2,a_8)$; $CH(a_2,a_4,a_8)$; $CH(a_4,a_7,a_8)$;
$CH(a_6,a_7,a_8)$; $CH(a_6,a_7,a_{11})$; $CH(0,a_6,a_{11})$.

The first and the last simplex in that sequence do not meet the definition of a boundary simplex. They correspond to

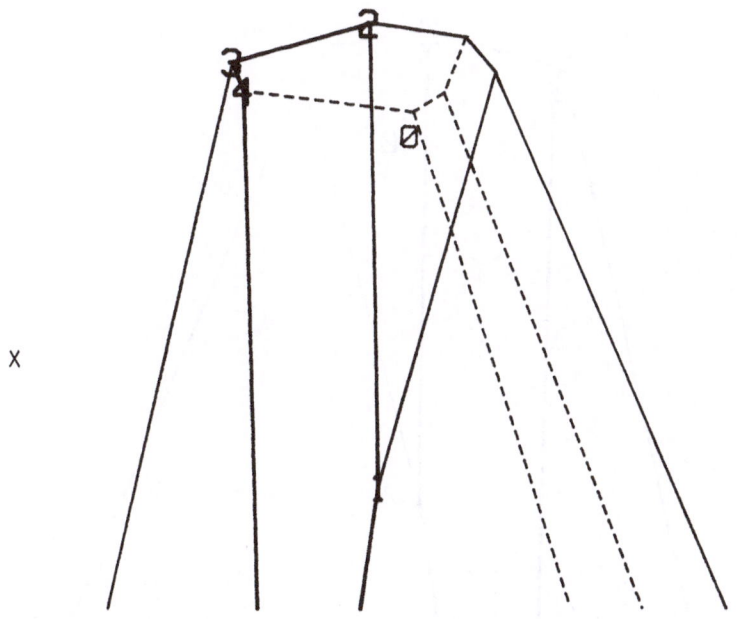

unbounded rays starting in vertex x_1 resp. x_4 of the primal polyhedron X.

Figure 1.7a

<u>Foreground of figure 1.7</u>

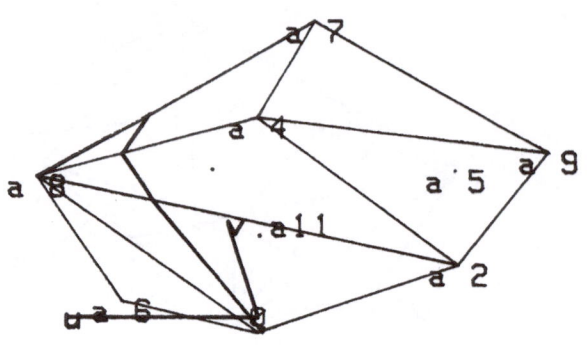

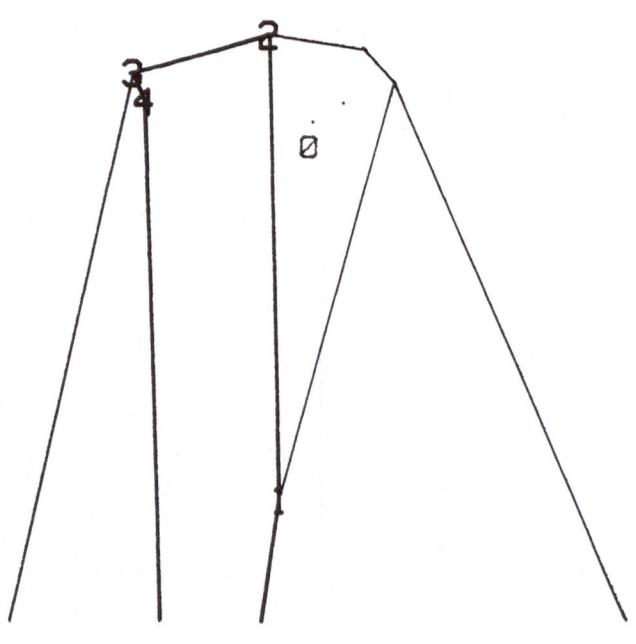

Figure 1.7b

Background of figure 1.7

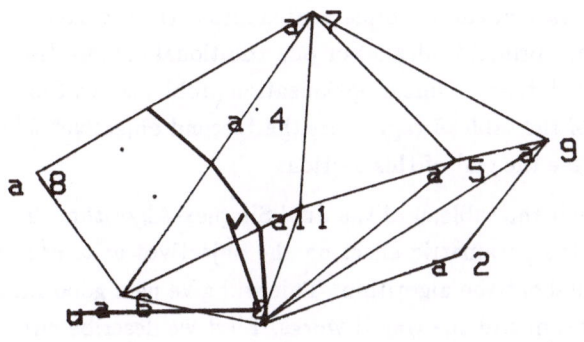

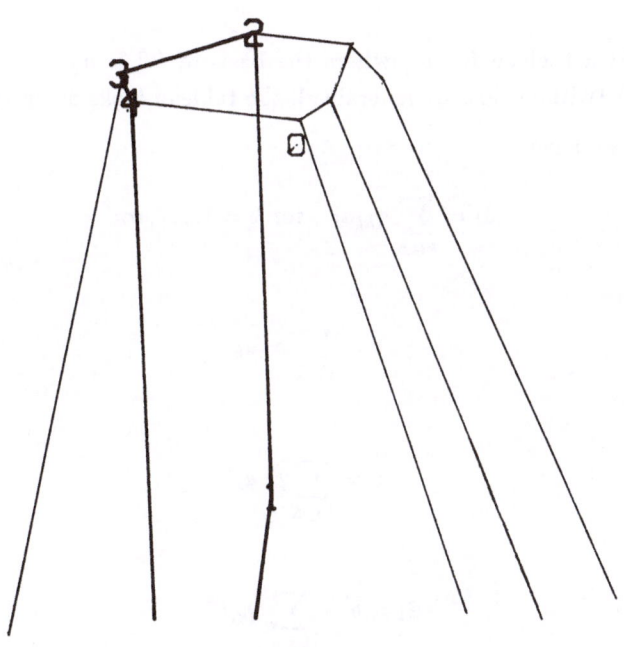

1.3 NUMERICAL REALIZATION
OF THE ALGORITHM

The purpose of this section is to demonstrate how the shadow-vertex algorithm works technically. This variant can be implemented into every tableau-method by adding one additional row (for primal tableaus) or one additional column (for dual tableaus) and can be applied to all types of linear optimization problems as a Phase II-algorithm. The additional part of the tableau represents the "second objective" $u^T x$. For applications in general cases see the end of this section.

Here we prefer the tableau of the dual Simplex-Algorithm. The reason is that we want to control the parametric effect on the objectives $v^T x$, $u^T x$ and on the various restrictions throughout the algorithm. This will give us a good insight into the geometry of the algorithm and the way it works. First we describe our tableau form which can be used for **general variants** of the Simplex-Algorithm.

For the introduction of our tableau suppose that the problem is not degenerate and that a vertex x_q and the corresponding tableau are at hand. Let x_q be the maximal vertex with respect to $w^T x$, where $w = \mu u + v$ ($\mu \geq 0$). Then $w \in CC(u, v)$ and x_q is a shadow-vertex. As in our dual interpretation we are interested in the active restriction vectors $a_{\Delta^1}, \ldots, a_{\Delta^n}$, which form a (cone or dual) basis of $\mathbb{R}^n$, rather than in "basic or nonbasic" variables $x^1, \ldots, x^{n+m}$. These two concepts of "basis" must be distinguished.

We start at a tableau for x_0, where the basis of $\mathbb{R}^n$ is $a_{\Delta_0^1}, \ldots, a_{\Delta_0^n}$. If we set $\Delta_0 = \{1, \ldots, n\}$ (without loss of generality), the tableau looks as on the next page.

In this tableau we have

(1.3.1)
$$a_l = \sum_{k \in \Delta} \gamma_{kl} a_k \quad \text{for } k = 1, \ldots, m$$

(1.3.2)
$$v = \sum_{k \in \Delta} \alpha_k a_k$$

(1.3.3)
$$u = \sum_{k \in \Delta} \beta_k a_k$$

(1.3.4)
$$\Psi_l = b^l - \sum_{k \in \Delta} \gamma_{kl} b^k .$$

Coefficients in the linear combinations of the basic vectors

Basic vectors at x_0	a_1	$\cdots$	a_n	a_{n+1}	$\cdots$	a_m	v	u
a_1	1			$\gamma_{1,n+1}$		$\gamma_{1,m}$	α_1	β_1
a_2	1			$\gamma_{2,n+1}$		$\gamma_{2,m}$	α_2	β_2
a_3	1			$\gamma_{3,n+1}$		$\gamma_{3,m}$	α_3	β_3
$\cdot$							$\cdot$	$\cdot$
$\cdot$							$\cdot$	$\cdot$
$\cdot$							$\cdot$	$\cdot$
a_k		1		$\gamma_{k,1}$			α_k	β_k
$\cdot$							$\cdot$	$\cdot$
$\cdot$							$\cdot$	$\cdot$
$\cdot$								
a_n			1	$\gamma_{n,n+1}$		$\gamma_{n,m}$	α_n	β_n
	0	$\cdots$	0	ψ_{n+1}	$\psi_1 \cdots \psi_m$		Q_v	Q_u

Surplus in the restrictions   Values of the objectives

(The latter is a formula for general right side in the restriction.) In our special case we have $b^l = 1$ for $l = 1, \ldots, m$.

$$(1.3.5) \qquad Q_v = -\sum_{k \in \Delta} \alpha_k b^k \quad (\text{ in our case } -\sum_{k \in \Delta} \alpha_k)$$

$$(1.3.6) \qquad Q_v = -\sum_{k \in \Delta} \beta_k b^k \quad (\text{ in our case } -\sum_{k \in \Delta} \beta_k)$$

Because of nondegeneracy all entries γ_{kl} $(k \in \Delta, l \notin \Delta)$, α_k, β_k $(k \in \Delta)$, Ψ_l $(l \notin \Delta)$ are different from 0. Note that $\gamma_{kk} = 1$ for $k \in \Delta$ and that $\gamma_{kl} = 0$ for $k \in \Delta$, $l \in \Delta$, $k \neq l$, and that $\Psi_l = b^l - b^l = 0$ for $l \in \Delta$.

Q_v and Q_u tell us the negative values of the two objectives at the current vertex. The entries Ψ_l $(l \in \Delta)$ tell the reserves or the surplus in the nonactive restrictions (Ψ_l can be regarded as the value of a slack variable).

To understand these claims, consider the current vertex x_Δ. Let it be the solution of the system

$$a_1^T = b^1, \ldots, a_n^T x = b^n.$$

Then we know that for $l \notin \Delta$

$$\Psi_l = b^l - \sum_{k \in \Delta} \gamma_{kl} b^k = b^l - \sum_{k \in \Delta} \gamma_{kl} a_k^T x_\Delta =$$

$$= b^l - (\sum_{k \in \Delta} \gamma_{kl} a_k)^T x_\Delta = b^l - a_l^T x_\Delta.$$

In the same way, we see that

$$Q_v = -\sum_{k \in \Delta} \alpha_k b^k = \left(-\sum_{k \in \Delta} \alpha_k a_k\right)^T x_\Delta = -v^T x_\Delta$$

and that $Q_u = -u^T x_\Delta$.

The left part of the tableau $(k \in \Delta, l \in \Delta)$ is not necessary. For that reason it is recommended to omit this submatrix in order to save storage capacity and time. But here, for demonstration, we are going to use the extended tableau.

Now the usual method proceeds as follows. We are looking for a tableau such that $\alpha_k \geq 0$ for all $k \in \Delta$. Starting in x_0, we begin with a tableau having $\beta_k \geq 0$ for all $k \in \Delta_0$. Now suppose that we are at a vertex x_q. Then several cases can occur in the current tableau.

(a) If already $\alpha_k \geq 0$ for all $k \in \Delta$, then x_q is optimal and we are ready.

(b) If there are entries $\alpha_k < 0$ ($k \in \Delta$), then x_q is not optimal, because v is not contained in the convex cone of the basic vectors. So we have to look for another cone. This is done by replacing one basic vector by a nonbasic vector. The leaving vector is determined by a fixed rule, which characterizes the variant. In our dual or polar tableau this rule determines the pivot row (in primal tableaus the pivot column). Let i be the index of the pivot row. Now the pivot column has to be calculated. This is done by finding a pivot element γ_{ij} in the pivot row. Such a pivot element must satisfy the following conditions

$$(1.3.7) \qquad \alpha_i < 0, \; \gamma_{ij} < 0 \quad \text{and} \quad \frac{\Psi_j}{\gamma_{ij}} = \max_{\substack{l \notin \Delta \\ \gamma_{il} < 0}} \frac{\Psi_l}{\gamma_{il}}.$$

Geometrically, $\alpha_i < 0$ means that v is not contained in $CC(a_1, \ldots, a_n)$. Now consider the hyperplane through the points $0, a_1, \ldots, a_{i-1}, a_{i+1}, \ldots, a_n$. It divides $\mathbb{R}^n$ into two halfspaces. v and a_i do not lie in the same of these halfspaces, because of $\alpha_i < 0$. So we have to admit a_j to the basis, which is situated in the same halfspace as v. Only then there will be a hope that v belongs to the new basic or polar cone. But this condition yields $\gamma_{ij} < 0$. Finally, the maximality statement in (1.3.7) guarantees that the simplex of the new basic vectors $CH(a_1, \ldots, a_{i-1}, a_j, a_{i+1}, a_{i+2}, \ldots, a_n)$ is a boundary simplex. Again, two cases are possible.

ba) There is no entry $\gamma_{ij} < 0$ in row i. Then all the vectors a_l with $l \notin \Delta$ belong to the same halfspace as a_i. But v lies in the opposite halfspace. Hence v cannot belong to any of such basic cones and it is not contained in $CC(a_1, \ldots, a_m)$. Consequently, there is no solution.

bb) If there are entries $\gamma_{il} < 0$, then the index of the entering vector is determined by

$$(1.3.8) \qquad \frac{\Psi_j}{\gamma_{ij}} = \max_{l \in \Delta} \frac{\Psi_l}{\gamma_{il}}.$$

The old basis $\{a_1, \ldots, a_n\}$ is replaced by the new basis $\{a_1, \ldots, a_n\} \setminus \{a_i\} \cup \{a_j\}$. The according tableau is calculated from the old tableau by performing a pivot step with the pivot element γ_{ij} (compare next page).

Now we want to show that the formulae (1.3.1 – 1.3.6) still hold (with respect to the new basis).

1) From

$$a_j = \gamma_{1j} a_1 + \ldots + \gamma_{ij} a_i + \ldots + \gamma_{nj} a_n$$

we conclude

$$a_i = -\frac{\gamma_{1j}}{\gamma_{ij}} a_1 - \ldots + \frac{1}{\gamma_{ij}} a_i - \ldots - \frac{\gamma_{nj}}{\gamma_{ij}} a_n.$$

	a_1	$\cdots$	a_i	$\cdots$	a_n	a_{n+1}	$\cdots$	a_j	$\cdots$	a_m	v	u
a_1	1	$\cdots$	$-\dfrac{\gamma_{kj}}{\gamma_{ij}}$	$\cdots$	0	$\gamma_{kl}-\dfrac{\gamma_{il}\gamma_{kj}}{\gamma_{ij}}$	$\cdots$	0	$\cdots$	$\gamma_{kl}-\dfrac{\gamma_{il}\gamma_{kj}}{\gamma_{ij}}$	$\alpha_k-\dfrac{\alpha_i\gamma_{kj}}{\gamma_{ij}}$	$\beta_k-\dfrac{\beta_i\gamma_{kj}}{\gamma_{ij}}$
a_j	0	$\cdots$	$\dfrac{1}{\gamma_{ij}}$	$\cdots$	0	$\dfrac{\gamma_{il}}{\gamma_{ij}}$	$\cdots$	1	$\cdots$	$\dfrac{\gamma_{il}}{\gamma_{ij}}$	$\dfrac{\alpha_i}{\gamma_{ij}}$	$\dfrac{\beta_i}{\gamma_{ij}}$
a_n	0	$\cdots$	$-\dfrac{\gamma_{kj}}{\gamma_{ij}}$	$\cdots$	1	$\gamma_{kl}-\dfrac{\gamma_{il}\gamma_{kj}}{\gamma_{ij}}$	$\cdots$	0	$\cdots$	$\gamma_{kl}-\dfrac{\gamma_{il}\gamma_{kj}}{\gamma_{ij}}$	$\alpha_k-\dfrac{\alpha_i\gamma_{kj}}{\gamma_{ij}}$	$\beta_k-\dfrac{\beta_i\gamma_{kj}}{\gamma_{ij}}$
	0	$\cdots$	$-\dfrac{\psi_j}{\gamma_{ij}}$	$\cdots$	0	$\psi_l-\dfrac{\gamma_{il}\psi_j}{\gamma_{ij}}$	$\cdots$	0	$\cdots$	$\psi_l-\dfrac{\gamma_{il}\psi_j}{\gamma_{ij}}$	$Q_v-\dfrac{\psi_j\alpha_i}{\gamma_{ij}}$	$Q_u-\dfrac{\psi_j\beta_i}{\gamma_{ij}}$

(k and l denote the index of the current row resp. column)

This is exactly the formula of the tableau.

2) For a vector a_l (nonbasic in the old and new tableau) we have

$$a_l = \gamma_{1l} a_1 + \ldots + \gamma_{il} a_i + \ldots + \gamma_{nl} a_n.$$

This is equivalent to

$$a_l = \left[\gamma_{1l} - \frac{\gamma_{1j}\gamma_{il}}{\gamma_{ij}} \right] a_1 + \ldots + \frac{\gamma_{il}}{\gamma_{ij}} a_j + \ldots + \left[\gamma_{nl} - \frac{\gamma_{nj}\gamma_{il}}{\gamma_{ij}} \right] a_n,$$

shown in the tableau.

3) The representations for v and u are confirmed analogously.

4) Ψ_i (of the old tableau) is replaced by $-\dfrac{\Psi_j}{\gamma_{ij}}$ in the new tableau. Note that we had

$$\Psi_j = b^j - \gamma_{1j} b^1 - \ldots - \gamma_{ij} b^i - \ldots - \gamma_{nj} b^n.$$

This leads to

$$-\frac{\Psi_j}{\gamma_{ij}} = b^i + \frac{\gamma_{1j}}{\gamma_{ij}} b^1 + \ldots - \frac{1}{\gamma_{ij}} b^j - \ldots + \frac{\gamma_{nj}}{\gamma_{ij}} b^n.$$

Here we have the new formula for Ψ_i.

5) In the new tableau we have for $l \notin \Delta_q$, $l \notin \Delta_{q+1}$ $\Psi_l - \frac{\gamma_{il}\Psi_j}{\gamma_{ij}}$. Inserting the definitions of Ψ_l, Ψ_j leads to

$$b^l - \sum_{\substack{k=1 \\ k \neq j}}^{n} \left\{ \gamma_{kl} - \left[\frac{\gamma_{il}\gamma_{kj}}{\gamma_{ij}} \right] \right\} b^k - \frac{\gamma_{il}}{\gamma_{ij}} b^j$$

the formula shown in the tableau.

6) Similarly, we see that

$$Q_v - \frac{\alpha_i}{\gamma_{ij}} \Psi_j = - \left[\alpha_1 - \frac{\alpha_i \gamma_{1j}}{\gamma_{ij}} \right] b^1 - \ldots - \frac{\alpha_i}{\gamma_{ij}} b^j - \ldots - \left[\alpha_n - \frac{\alpha_i \gamma_{nj}}{\gamma_{ij}} \right] b^n.$$

And the same is true for the u-column.

Remark. Our transformation rule for the tableau guarantees that during the replacement of Δ_q by Δ_{q+1}:

(1.3.9)   All the entries Ψ_l remain nonnegative.

(1.3.10) Q_v decreases.

Proof. If $l \in \Delta$, the first claim is trivial. So let $l \notin \Delta$. In the new tableau we have $\Psi_l - \frac{\gamma_{il}\Psi_j}{\gamma_{ij}}$ instead of Ψ_l. γ_{ij} was selected in a way such that $\gamma_{ij} < 0$. So the case $\gamma_{il} \geq 0$ is trivial. Among the entries with $\gamma_{il} < 0$ ($l \notin \Delta$) we had determined j such that

$$\frac{\Psi_j}{\gamma_{ij}} = \max_{\substack{l \notin \Delta \\ \gamma_{ij} < 0}} \frac{\Psi_l}{\gamma_{il}}.$$

Hence

$$\Psi_l - \frac{\gamma_{il}\Psi_j}{\gamma_{ij}} \geq \Psi_l - \frac{\gamma_{il}\Psi_l}{\gamma_{il}} = 0.$$

So it is guaranteed, that x_{q+1} is feasible and that $\Sigma(\Delta_{q+1})$ is a boundary simplex. Concerning the second claim, we know that $Q_v > Q_v - \frac{\alpha_i \Psi_j}{\gamma_{ij}}$, because Ψ_j is positive and α_i, γ_{ij} are necessarily negative. Hence

$$v^T x_q < v^T x_{q+1}.$$

All the facts mentioned above hold for general variants of the Simplex-Algorithm. But now we have to concentrate on our rule for finding the (dual) pivot row. This is the characterization of the shadow-vertex concept. Now the use of the second objective column is justified.

Remember that we started at a vertex x_0 such that $u^T x$ is maximized in x_0. So for the initial tableau we had $\beta_k \geq 0$ for all $k \in \Delta$.

If an optimal vertex exists, then we shall have $\alpha_k \geq 0$ for all $k \in \Delta$ in the corresponding tableau. We want to guarantee that only shadow vertices are touched on our walk from x_0 to x_s. Now suppose that we have a tableau with $\mu\alpha_k + \beta_k \geq 0$ for all $k \in \Delta$. This is equivalent to the fact that $CC(u, v)$ intersects $CH(a_{\Delta 1}, \ldots, a_{\Delta n})$.

Setting $\mu = 0$ we have the representation for u. Imagine that we move (with growing μ) on the ray $\{w \mid w = \mu v + u, \mu \geq 0\}$. Then this ray intersects a number of basic cones, which uniquely correspond to shadow vertices. For every w_μ we want to know the optimal vertex, maximizing $w^T x$. Consider such a vertex. It will be optimal on an interval $\{w \mid w = \mu v + u, \underline{\mu} \leq \mu \leq \overline{\mu}\}$ or $\{w \mid w = \mu v + u, \underline{\mu} \leq \mu\}$. In the second case our vertex x_0 is the optimal one, in the first case our ray intersects the basic cone as long as $\mu \in [\underline{\mu}, \overline{\mu}]$. It will leave the cone at $w_{\overline{\mu}} = \overline{\mu}v + u$, which is an element of the $n - 1$-dimensional cone

$$CC(a_{\Delta 1}, \ldots, a_{\Delta i-1}, a_{\Delta i+1}, \ldots, a_{\Delta n}).$$

In the corresponding pivot step the vector $a_{\Delta i}$ has to be replaced by a vector of the opposite halfspace. So our ray and particularly the value of $\overline{\mu}$ will help us to determine the index i. To explain this, we distinguish between four sign combinations of the pair (α_k, β_k) in the x_q-tableau.

Figure 1.8

Intersection and replacement of basis vectors

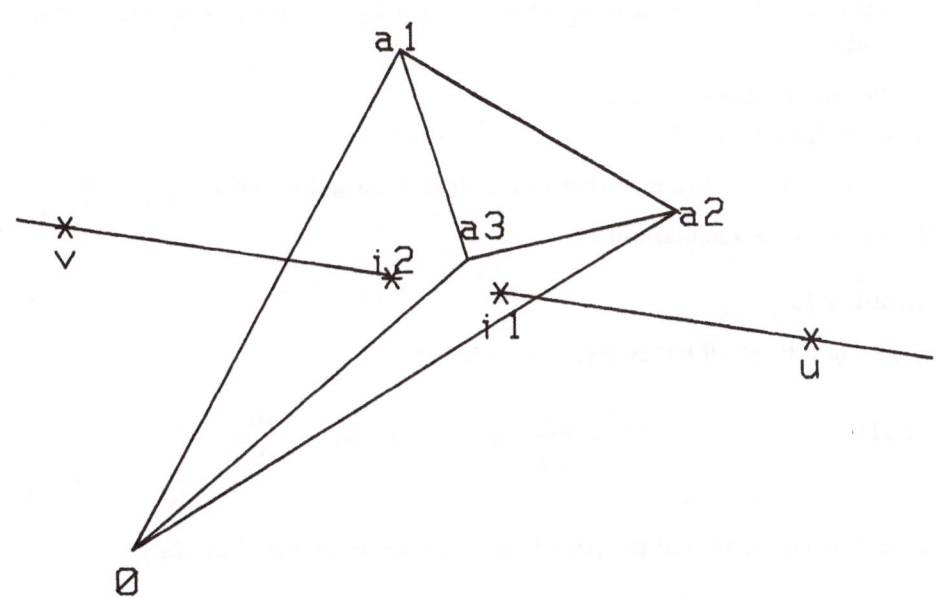

The straight line containing u and v intersects $CC(a_1, a_2, a_3)$.
It enters the cone in i_1 and leaves in i_2.
i_1 and i_2 can be used for calculation of $\underline{u}$ and $\bar{u}$
(and vice versa).
At i_1 the vector a_1 enters the basis.
At i_2 the vector a_2 leaves the basis.

1) $\alpha_k < 0,\ \beta_k > 0$.

This case does not cause any trouble, because $\mu\alpha_k + \beta_k$ remains positive for increasing μ.

2) $\alpha_k < 0,\ \beta_k < 0$.

We had assumed that x_q is a shadow vertex. So this case cannot occur.

3) $\alpha_k > 0,\ \beta_k > 0$.

This case delivers a value $\underline{\mu}_k$, where $\underline{\mu}_k \alpha_k + \beta_k = 0$. For $\mu > \underline{\mu}_k$ the term will be positive.

4) The most interesting case is
$\alpha_k < 0,\ \beta_k > 0$.

Here we have a $\bar{\mu}_k$, such that $\mu\alpha_k + \beta_k$ is nonnegative for $\mu \le \bar{\mu}_k = \frac{-\beta_k}{\alpha_k}$.

These results are summarized in

Lemma 1.11

$\mu_k\alpha_k + \beta_k \ge 0$ for all $k \in \Delta$ is possible only for

$$(1.3.11) \qquad \mu \le \bar{\mu} = \min_{\substack{k \in \Delta \\ \alpha_k < 0 \\ \beta_k > 0}} \bar{\mu}_k \qquad \text{where } \bar{\mu}_k = -\frac{\beta_k}{\alpha_k}.$$

So the leaving vector and the (dual) pivot row are determined by the rule

$$(1.3.12) \qquad \bar{\mu} = -\frac{\beta_j}{\alpha_j} = \min_{\substack{k \in \Delta \\ \alpha_k < 0 \\ \beta_k > 0}} \left(-\frac{\beta_k}{\alpha_k}\right).$$

Remark. Note that $\beta_k > 0$ could be omitted in this minimization rule, since the present tableau does not correspond to the optimal vertex. Let us now recall the essentials of our procedure in algorithmic form.

Theorem 1

In our (dual) tableau form the shadow-vertex algorithm is realized as follows.

1) Find an entry $\alpha_k < 0$ with $k \in \Delta$. If this is impossible, then go to 7).

2) Determine the pivot row and the vector leaving the basis by

$$-\frac{\beta_i}{\alpha_i} = \min_{\substack{k \in \Delta \\ \alpha_k < 0}} -\frac{\beta_k}{\alpha_k}.$$

3) *Find a negative entry in the pivot row (index i). If all γ_{il} with $l \notin \Delta$ are nonneg-ative, then go to 8).*

4) *Determine the pivot column and the vector entering the basis by*

$$\frac{\Psi_j}{\gamma_{ij}} = \min_{\substack{\gamma_{il}<0 \\ l \notin \Delta}} \frac{\Psi_l}{\gamma_{il}}.$$

5) *Perform a pivot step such that $a_{\Delta i}$ leaves and $a_{\Delta j}$ enters the basis.*

6) *Set $q = q+1$ and return to 1).*

7) *x_q is the optimal vertex. Print the active constraints and the value of $v^T x_q = -Q_v$. Go to 9).*

8) *The problem does not have a solution. Print: problem unsolvable.*

9) *STOP.*

$\square$

Remark. The shadow-vertex algorithm can be implemented for any tableau form and for any type of linear programming problem. This can be done by adding a row (into the primal tableau) or a column (dual tableau), which is representing u. This part of the tableau is to be filled in the same way as the region provided for v.

In order to find the pivot column (primal) or the pivot row (dual), we use the following rule

(1.3.13) $$-\frac{\beta_t}{\alpha_t} = \min_{r \in \Theta} -\frac{\beta_r}{\alpha_r}.$$

The index set Θ can take two forms. If nonnegativity of all entries in the v-part is necessary for the solution, then

(1.3.14) $$\Theta = \{r \mid \alpha_r < 0, \beta_r > 0\}.$$

If nonpositivity is desired, then

(1.3.15) $$\Theta = \{r \mid \alpha_r > 0, \beta_r < 0\}.$$

Proof. If we have two analogue tableau-regions for the objectives, then the current vectors appearing in these regions are images of v resp. u under a linear mapping f_T, depending upon the current tableau.

For growing μ the shadow-vertex algorithm calculates the sequence of vertices which are optimal relative to $w^T x$, where $w = \mu v + u$. Let x_q be such a vector and consider the corresponding tableau. If x_q is not optimal for $v^T x$, then $f_T(v)$ cannot satisfy the optimality criterion. But for a certain interval $[\mu, \overline{\mu}]$, where $\mu \geq 0$ and $\overline{\mu} \in \mathbb{R}$, the positive combination $\mu f_T(v) + f_T(u)$ satisfies the criterion. $\overline{\mu}$ can be calculated by applying the formulae given above. And the knowledge about $\overline{\mu}$ gives the desired information on the pivot column resp. pivot row.

1.4 THE ALGORITHM FOR PHASE I

Until now we have started our computation from a **given** vertex x_0 and from the corresponding tableau. Now we want to demonstrate, how the start vertex x_0 and even a shadow-vertex x_0 can be achieved.

In principle, Phase I has the task to deliver an arbitrary vertex of the feasible region. For finding such an arbitrary vertex one may proceed as follows. We define a preliminary problem **with** nonnegativity constraints

$$(1.4.1) \qquad \begin{array}{ll} \text{Maximize} & v^T x \\ \text{subject to} & a_1^T x \leq b^1, \dots, a_m^T x \leq b^m \text{ and } x \geq 0 \\ \text{where} & v, a_1, \dots, a_m, x \in \mathbb{R}^n. \end{array}$$

The feasible region of this problem contains 0 as a vertex. This is a good opportunity for us to choose 0 as start vertex. To unify our notation, we rewrite the restrictions in the following form

$$a_1^T x \leq b^1, \dots, a_m^T x \leq b^m, a_{m+1}^T x \leq b^{m+1}, \dots, a_{m+n}^T x \leq b^{m+n},$$

where $a_{m+i} := -e_i$ (the negative i-th unit vector) and $b^{m+i} := 0$ for $i = 1, \dots, n$. Our start tableau is shown on the next page.

The task of Phase I is to remove all the vectors $-e_k$ ($k = 1, \dots, n$) from the dual basis, while the values Ψ_l ($l = 1, \dots, m$) must remain nonnegative. This will give that all of the original restrictions are satisfied and that in the end n of the original restrictions are active. The sign of the values Ψ_l ($l \geq m$) does not matter, since the generic problem does not contain nonnegativity constraints.

For this purpose the simplest way is as follows. Remove — one after the other — the vectors $-e_k$ from the basis. After r such steps we have a tableau like that below. Now $-e_{r+1}$ is to be removed. While the pivot column is determined, we ignore the additional region of the tableau. So it is guaranteed, that a nonnegativity-restriction vector never reenters the dual basis. Hence n pivot steps are sufficient for finding a vertex of X.

Note that this low effort for Phase I results from the fact that 0 is already a feasible point of the generic problem. If no feasible point is known at the beginning, then Phase I-algorithms usually require more steps.

In the tableau we can choose $i = r + 1$ as the pivot row. Since $\Psi_{m+1}, \dots, \Psi_{m+n}$ need not be negative, we may determine j in two different ways. However, entries with $l \geq m + 1$ do not have a chance for being the pivot element. If we find negative entries in the row $i = r + 1$, then we can determine j by

$$(1.4.2) \qquad \frac{\Psi_j}{\gamma_{ij}} = \max_{\substack{l \leq \Delta \\ l \leq m \\ \gamma_{il} < 0}} \frac{\Psi_l}{\gamma_{il}} \quad \text{as we know from Section 3.}$$

a_1 a_2 $\cdots$ a_m a_{m+1} a_{m+2} $\cdots$ a_{m+n} v u

$-e_1$ $-e_2$ $\cdots$ $-e_n$

$$
\begin{array}{ccccccc|cc}
-a_1^1 & -a_2^1 & \cdots & -a_m^1 & 1 & 0 & \cdots & 0 & -v^1 & -u^1 \\
 & & & & & 1 & & & & \\
 & & & & & & \ddots & & & \\
 & -a_k^l & & & & & & & & \\
 & & \text{main region} & & & & \text{additional region} & & & \\
-a_1^n & -a_2^n & \cdots & -a_m^n & 0 & 0 & \cdots & 1 & -v^n & -u^n \\
\hline
b^1 & b^2 & \cdots & b^m & 0 & & & 0 & 0 & 0 \\
\hline
\psi_1 & \psi_2 & \cdots & \psi_n & \psi_{m+1} & & & \psi_{m+n} & Q_v & Q_u
\end{array}
$$

$a_{m+1} = -e_1$

$a_{m+2} = -e_2$

$\cdots$

$a_{m+n} = -e_n$

Simplex tableau with column headers a_1, a_2, $\ldots$, a_r, a_{r+1}, $\ldots$, a_j, $\ldots$, a_m, $-e_1$, $\ldots$, $-e_r$, $-e_{r+1}$, $\ldots$, $-e_n$, v, u and row labels a_1, $\ldots$, a_r, $-e_{r+1}$, $\ldots$, $-e_n$.

The cells contain entries γ_{kl}, with pivot element $\gamma_{r+1,j}$ (pivot element), and γ_{kl}. The bottom objective rows contain α_1, β_1, $\ldots$, α_r, β_r, α_{r+1}, β_{r+1}, $\ldots$, α_n, β_n, Q_v, Q_u. The right-hand column contains ψ_{r+1}, ψ_m, ψ_{m+1}, ψ_{m+r}.

relevant for finding the pivot column

irrelevant for finding the pivot column

Then it is certain that Ψ_l remains nonnegative for $l \leq m$. If all entries in row i are nonnegative, we use

(1.4.3)
$$\frac{\Psi_j}{\gamma_{ij}} = \min_{\substack{l \in \Delta \\ l \leq m \\ \gamma_{il} > 0}} \frac{\Psi_l}{\gamma_{il}}$$

for finding pivot element and pivot column. Then we obtain nonnegative values Ψ_l for all $l \leq m$, because

- in the a_j-column we get $\Psi_j = 0$
- in all columns belonging to an a_l ($l \leq m$) we have

$$\Psi_l^{\text{new}} = \Psi_l - \frac{\gamma_{il}}{\gamma_{ij}} \Psi_j \geq \Psi_l - \frac{\gamma_{il}}{\gamma_{il}} \Psi_l = 0 \text{ for } l \notin \Delta, \, l \leq m \text{ with } \gamma_{il} > 0.$$

Continuing in this manner we obtain a basis which does not contain any of the vectors $-e_1, \ldots, -e_n$. The according tableau is shown on the next page. Here the values $\Psi_{m+1}, \ldots, \Psi_n$ give us very useful information, because

$$\Psi_{m+r} = b^{m+r} - \sum_{k \in \Delta} \gamma_{kl} b^k = 0 - \sum_{k \in \Delta} \gamma_{kl} b^k.$$

If x_Δ is the solution of $a_1^T x = b^1, \ldots, a_n^T x = b^n$, then

$$x_\Delta^r = +e_r^T x_\Delta = \sum_{k \in \Delta} \gamma_{kr} a_k^T x_\Delta = \sum_{k \in \Delta} \gamma_{kr} b^k = -\Psi_{m+r}.$$

Here we have the reason why we did not drop the additional region. It delivers the current vertex in primal coordinates. For application of the shadow-vertex algorithm it is recommended to add the u-region only after the completion of Phase I. It is necessary to insert nonnegative values (at least one positive) β_k into the entries of the u-region only at the start of Phase II. Then it is clear that

$$u = \sum_{k \in \Delta} \beta_k a_k.$$

After that we continue as in Section 3.

The procedure described above is appropriate if we are completely free in the choice of u. But it cannot be used (unfortunately) for our theoretical stochastic considerations.

$$
\begin{array}{c}
\text{(all nonnegativity restrictions have been removed from the basis)}
\end{array}
$$

In the following chapters we want to analyze the random behaviour of the number of pivot steps under stochastic assumptions. Then it will be necessary to define u as a random variable. So u is determined before Phase I is started. It does not matter, whether u is chosen as a — once and for ever — fixed vector or as a random vector which varies according to a certain distribution. In all cases it is necessary to reach a shadow-vertex with respect to the projection on $\mathrm{span}(u, v)$. Some more complicated conditions on Phase I of an analyzable algorithm will become clear in connection with our stochastic assumptions in Chapter II.

For these reasons we analyze a more complicated, lengthy and somehow crude algorithm for finding a shadow-vertex. It is a mixture of steps used above in the simple algorithm and of steps done in the shadow-vertex algorithm (as in Phase II).

To begin with, we need some notation.

Let $\Pi_k : \mathbb{R}^n \to \mathbb{R}^k$ (for $k = 1, \ldots, n$) be the orthogonal projection from $\mathbb{R}^n$ into $\mathbb{R}^k$ such that

$$(1.4.4) \qquad \Pi_k(x) = \Pi_k \begin{bmatrix} x^1 \\ \cdot \\ x^k \\ \cdot \\ x^n \end{bmatrix} = \begin{bmatrix} x^1 \\ \vdots \\ x^k \end{bmatrix} \qquad \text{for all } x \in \mathbb{R}^n.$$

Also for $k = 1, \ldots, n$ we define the projected programming problems I_k:

$$(1.4.5) \qquad \begin{array}{ll} \text{Maximize} & \Pi_k(x)^T \Pi_k(v) \\ \text{subject to} & \Pi_k(a_1)^T \Pi_k(x) \le 1, \ldots, \Pi_k(a_m)^T \Pi_k(x) \le 1 \\ \text{where} & v, x, a_1, \ldots, a_m \in \mathbb{R}^n \text{ and } m \ge n. \end{array}$$

$\Pi_k(Y)$ is defined as $\mathrm{CH}(\Pi_k(0), \Pi_k(a_1), \ldots, \Pi_k(a_m)) = \Pi_k(\mathrm{CH}(0, a_1, \ldots, a_m))$.

We are now going to give a description of our algorithm for solving the complete problem (Phase I and Phase II). Again, we want to give a primal and a dual interpretation.

Theorem 2

The complete algorithm works as follows.

(1.4.6) Primal interpretation:

1) *Set $k = 2$ and find a vertex of $\Pi_2(X)$ by application of the simple Phase I-algorithm of this section.*

2) *Find the maximal vertex $(x^1, x^2)^T$ of I_2 by means of the shadow-vertex algorithm. If such a solution does not exist, go to 7).*

3) *If $k = n$ then go to 8). Set $k = k + 1$.*

4) *The solution $(x^1, \ldots, x^{k-1})^T$ of I_{k-1} is available. Then vector $(x^1, \ldots, x^{k-1}, 0)^T$ lies on an edge of $\Pi_k(X)$. Find a vertex $(x^1, \ldots, x^k)^T$ on this edge.*

5) *Use the plane $\mathrm{span}(\Pi_k(e_k), \Pi_k(v))$ as projection plane and start the shadow-vertex algorithm in $(x^1, \ldots, x^k)^T$ for finding the optimal vertex of I_k. If there is no solution, go to 7).*

6) *Go to 3).*

7) *A solution of the complete problem does not exist. Print: problem unsolvable. Go to 9).*

8) *$(x^1, \ldots, x^n)^T$ is the solution. Print this vector and $v^T x$.*

9) *STOP.*

(1.4.7) Dual interpretation:

1) *Find a boundary simplex $\Pi_2(\mathrm{CH}(a_{\Delta^1}, a_{\Delta^2}))$ of $\Pi_2(Y)$ using the simplex algorithm.*

2) *Find that boundary simplex of $\Pi_2(Y)$ which is intersected by $\mathbb{R}^+ \Pi_2(v)$. If such a simplex does not exist, go to 7).*

3) *If $k = n$ go to 8). Else set $k = k + 1$.*

4) *The optimal simplex of I_{k-1} is available and has the form*

$$\Pi_{k-1}(\mathrm{CH}(a_{\Delta^1}, \ldots, a_{\Delta^{k-1}})).$$

Look for a vector a_i, $i \notin \{\Delta^1, \ldots, \Delta^{k-1}\}$, such that

$$\Pi_k(\mathrm{CH}(a_{\Delta^1}, \ldots, a_{\Delta^{k-1}}, a_i))$$

is a boundary simplex of $\Pi_k(Y)$.

5) *Use $\mathrm{span}(\Pi_k(e_k), \Pi_k(v))$ as intersection plane and apply the shadow-vertex algorithm starting from $\Pi_k(\mathrm{CH}(a_{\Delta^1}, \ldots, a_{\Delta^{k-1}}, a_i))$. Find the optimal boundary simplex $\Pi_k(\mathrm{CH}(a_{\Delta^1}, \ldots, a_{\Delta^k}))$ which is intersected by $\mathbb{R}^+ \Pi_k(v)$.*

6) *Go to 3).*

7) *There is no solution for the complete problem. Print: problem unsolvable.*

   Go to 9.

8) *The optimal boundary simplex of I_n is available. Print x and $v^T x$.*

9) *STOP.*

Lemma 1.12

*(1.4.8)   The algorithm of Theorem 2 yields the optimal solution of I_n
(if a solution exists) after a finite number of pivot steps. Else
it stops with the information, that the problem is unsolvable.*

Proof. For the proof we concentrate on the dual description. Step 1) works according to the simple algorithm explained at the begin of this section. It requires 2 pivot steps. In Step 2) the normal Simplex-Algorithm is applied to find a solution (for dimension 2 the shadow-vertex algorithm coincides with the usual variants).

If there is no optimal simplex in dimension 2, then all vectors $\Pi_2(a_i)$ with $i = 1, \ldots, m$ and 0 are contained in a halfspace of $\mathbb{R}^2$, which is bounded by a hyperplane through 0 and does not contain $\Pi_2(v)$. Hence v does not belong to $CC(a_1, \ldots, a_n)$ and there cannot be a solution for the original (n-dimensional) problem.

In Step 4) we start with a boundary simplex $\Pi_{k-1}(CH(a_{\Delta^1}, \ldots, a_{\Delta^{k-1}}))$. Then $\Pi_k(CH(a_{\Delta^1}, \ldots, a_{\Delta^{k-1}}))$ is contained in the boundary of $\Pi_k(Y)$. This results from the fact that there is a $z \in \mathbb{R}^{k-1}$, such that $y + \varepsilon z \notin \Pi_{k-1}(Y)$ for all $\varepsilon > 0$ and all y belonging to the $k-1$-dimensional simplex.

So $\Pi_k(CH(a_{\Delta^1}, \ldots, a_{\Delta^{k-1}}))$ is a side simplex of — at least one, at most two — boundary simplices of $\Pi_k(Y)$. Hence there is a vector a_i with $i \notin \{\Delta^1, \ldots, \Delta^{k-1}\}$ such that $\Pi_k(CH(a_{\Delta^1}, \ldots, a_{\Delta^{k-1}}, a_i))$ is a boundary simplex of $\Pi_k(Y)$. The $k-1$-dimensional boundary simplex of $\Pi_{k-1}(Y)$ is intersected by $\mathbb{R}^+\Pi_{k-1}(v)$. Then the $k-1$-dimensional side simplex is intersected by $\text{span}(\Pi_k(v), \Pi_k(e_k))$. Of course, $\Pi_k(CH(a_{\Delta^1}, \ldots, a_{\Delta^{k-1}}, a_i))$ is intersected, too. So the shadow-vertex algorithm can be started. If the intersection point lies in $CC(\Pi_k(v), -\Pi_k(e_k))$, we should use $-\Pi_k(e_k)$ instead of u, not $\Pi_k(e_k)$. If I_k does not have a solution, then $I_{k+1}, \ldots, I_n$ cannot have solutions either (see the argument for I_2). So finally the procedure arrives at the optimal simplex of Y, if an optimal simplex exists.

$\square$

We end this section by demonstrating this procedure in tableau-form. We start with the extended tableau

	a_1	a_m	$-e_1$			$-e_n$	v
$-e_1$	$-a_1^1$	$-a_m^1$	1	0	.....	0	$-v^1$
$-e_2$	$-a_1^2$	$-a_m^2$	0	1	0	0	$-v^2$
$-e_n$	$-a_1^n$	$-a_m^n$	0	0		1	$-v^n$
	ψ_1	ψ_m				ψ_{m+n}	Q_v

First we do the vertex search as mentioned above. For this purpose we take only the first and the second row (according to $-e_1$ and $-e_2$) into regard. The result of this search is the second tableau.

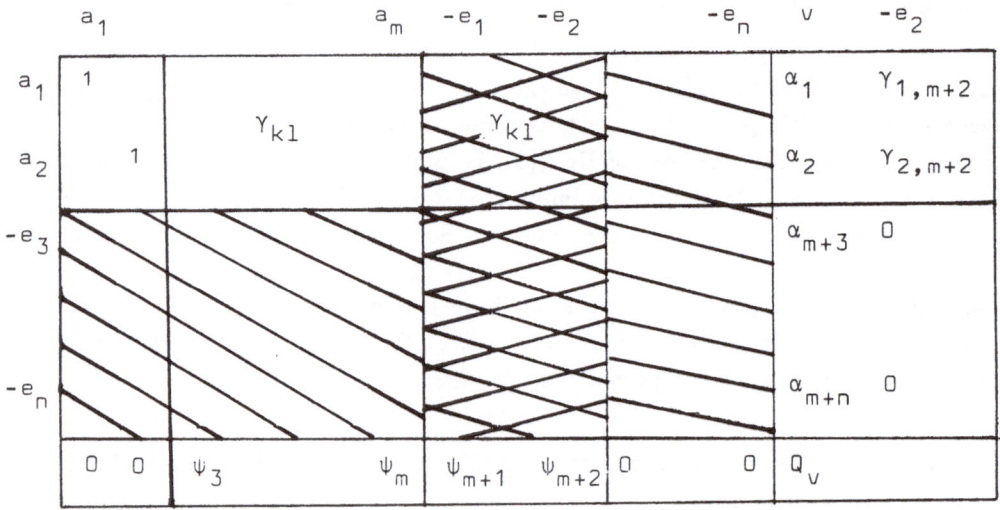

Now we start the shadow-vertex algorithm, ignoring the rows for $-e_3,\ldots,-e_n$ and the columns for $-e_1,-e_2,\ldots,-e_n$. But the complete tableau is to be calculated in every step. The column for u is filled with the entries of the column of $-e_2$ or those of e_2. This does not create a degenerate problem because the $-e_2$ column is ignored in all the following steps. This is also true for the whole double-crossed region of the tableau. This region is used only to obtain the primal entries of the solution vertex. The reason why shall be explained for dimension 3.

We run a number of pivot steps until we have $\alpha_1 > 0$, $\alpha_2 > 0$ or until in one row i all the values Φ_{il} are nonnegative. In the latter case we are ready. In the first case we try to drop the basic vector $-e_3$ and the restriction $x^3 \geq 0$. This can be done by choosing the corresponding row as the pivot row and by finding the pivot element γ_{3j} according to (1.4.2) resp. (1.4.3). After that we have a tableau

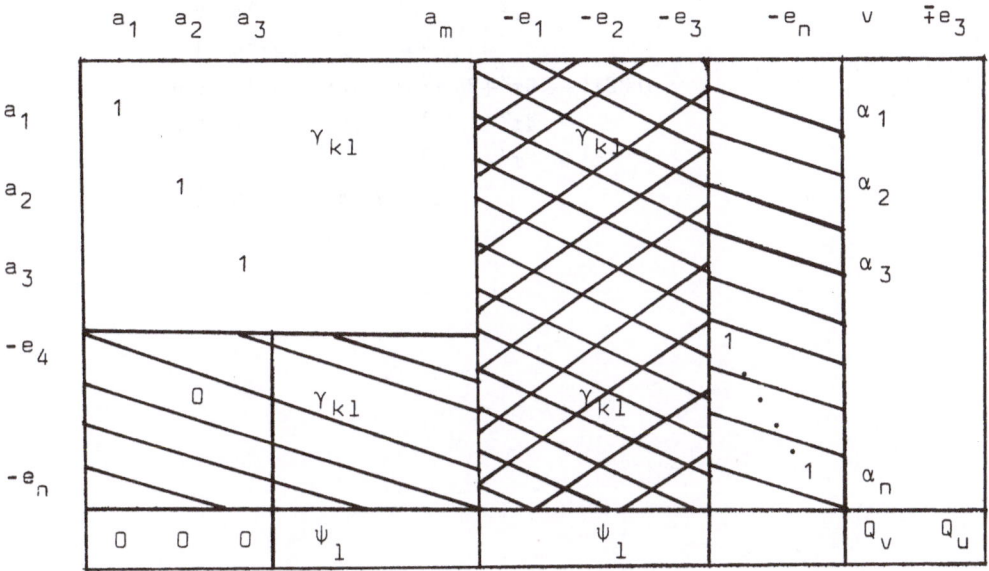

Again, the crossed region can be ignored for the moment. The double-crossed region can be omitted, if one does not want the primal solution. The column for u is filled with the column of $-e_3$ or with the entries of the according vector with opposite sign. This depends on the values α_1, α_2, α_3. It must be guaranteed, that in all components with $\alpha_k < 0$ the entry β_k is positive. This is satisfied either with the representation of $-e_3$ or of $+e_3$, because the given tableau corresponds to a shadow vertex under projection on $\text{span}(e_3, \Pi_3(v))$. Now we start the shadow-vertex algorithm and continue as described before.

Example.   On the following pages the solution of a special problem is demonstrated in tableau form. The problem is

$$
\begin{array}{lrcl}
\text{Maximize} & 2x^1 - \ 5x^2 + 0.3x^3 - 0.75x^4 + 4.8x^5 \\[4pt]
\text{subject to} & x^1 \qquad\quad + \ 5x^3 & & \le 10 \\
& - \ 4x^2 + \ 5x^3 - \ 8x^4 + 0.3x^5 & \le & 3 \\
& -9x^1 + 3.3x^2 - \ 2x^3 + \ 6x^4 + \ 3x^5 & \le & 7 \\
& + \ 2x^2 \qquad\quad - \quad x^4 - \quad x^5 & \le & 2 \\
& - \ 3x^3 + \quad x^4 + 6.9x^5 & \le & 1 \\
& -0.88x^1 + 4.6x^2 + 7.2x^3 + \ 6x^4 - \ 9x^5 & \le & 0.6 \\
& - \ 4x^2 + \ 7x^3 + \ 4x^4 & \le & 0.2 \\
& x^1 + \quad x^2 + \quad x^3 \qquad\qquad - \ 8x^5 & \le & 1 \\
& 5x^3 \ -7.5x^4 & \le & 5\,.
\end{array}
$$

STAGE = 1   STEP = 0   SUM OF STEPS = 0

	1	2	3	4	5	6	7	8	9	10	11	12	13	14		
10	-1.00	0.0	9.00	0.0	0.0	0.88	0.0	-1.00	0.0	1.00	0.0	0.0	0.0	0.0	-2.00	1.00
11	0.0	4.00	-3.30	-2.00	0.0	-4.60	4.00	-1.00	0.0	0.0	1.00	0.0	0.0	0.0	5.00	1.00
12	-5.00	-5.00	2.00	0.0	3.00	-7.20	-7.00	-1.00	-5.00	0.0	0.0	1.00	0.0	0.0	-0.30	1.00
13	0.0	8.00	-6.00	1.00	-1.00	-6.00	-4.00	0.0	7.50	0.0	0.0	0.0	1.00	0.0	0.75	1.00
14	0.0	-0.30	-3.00	1.00	-6.90	9.00	0.0	8.00	0.0	0.0	0.0	0.0	0.0	1.00	-4.80	1.00
	10.00	3.00	7.00	2.00	1.00	0.60	0.20	1.00	5.00	0.0	0.0	0.0	0.0	0.0	0.0	0.0

STAGE = 1   STEP = 1   SUM OF STEPS = 1

	1	2	3	4	5	6	7	8	9	10	11	12	13	14		
8	1.00	0.0	-9.00	0.0	0.0	-0.88	0.0	1.00	0.0	-1.00	0.0	0.0	0.0	0.0	2.00	-1.00
11	1.00	4.00	-12.30	-2.00	0.0	-5.48	4.00	0.0	0.0	-1.00	1.00	0.0	0.0	0.0	7.00	-1.00
12	-4.00	-5.00	-7.00	0.0	3.00	-8.08	-7.00	0.0	-5.00	-1.00	0.0	1.00	0.0	0.0	1.70	-1.00
13	0.0	8.00	-6.00	1.00	-1.00	-6.00	-4.00	0.0	7.50	0.0	0.0	0.0	1.00	0.0	0.75	0.0
14	-8.00	-0.30	69.00	1.00	-6.90	16.04	0.0	0.0	0.0	8.00	0.0	0.0	0.0	1.00	-20.80	8.00
	9.00	3.00	16.00	2.00	1.00	1.48	0.20	0.0	5.00	1.00	0.0	0.0	0.0	0.0	-2.00	1.00

STAGE = 2   STEP = 1   SUM OF STEPS = 2

	1	2	3	4	5	6	7	8	9	10	11	12	13	14		
8	0.84	-0.64	-7.02	0.32	0.0	-0.00	-0.64	1.00	0.0	-0.84	-0.16	0.0	0.0	0.0	0.88	0.16
6	-0.18	-0.73	2.24	0.36	0.0	1.00	-0.73	0.0	0.0	0.18	-0.18	0.0	0.0	0.0	-1.28	0.18
12	-5.47	-10.90	11.14	2.95	3.00	-0.00	-12.90	0.0	-5.00	0.47	-1.47	1.00	0.0	0.0	-8.62	1.47
13	-1.09	3.62	7.47	3.19	-1.00	-0.00	-8.38	0.0	7.50	1.09	-1.09	0.0	1.00	0.0	-6.91	1.09
14	-5.07	11.41	33.00	-4.85	-6.90	0.00	11.71	0.0	0.0	5.07	2.93	0.0	0.0	1.00	-0.31	-2.93
	9.27	4.08	12.68	1.46	1.00	0.00	1.28	0.0	5.00	0.73	0.27	0.0	0.0	0.0	-0.11	-0.27

STAGE = 2   STEP = 2   SUM OF STEPS = 3

	1	2	3	4	5	6	7	8	9	10	11	12	13	14		
8	1.00	-0.00	-9.00	0.00	0.0	-0.88	-0.00	1.00	0.0	-1.00	-0.00	0.0	0.0	0.0	2.00	0.00
7	0.25	1.00	-3.07	-0.50	0.0	-1.37	1.00	0.0	0.0	-0.25	0.25	0.0	0.0	0.0	1.75	-0.25
12	-2.25	2.00	-28.52	-3.50	3.00	-17.67	-0.00	0.0	-5.00	-2.75	1.75	1.00	0.0	0.0	13.95	-1.75
13	1.00	12.00	-18.30	-1.00	-1.00	-11.48	-0.00	0.0	7.50	-1.00	1.00	0.0	1.00	0.0	7.75	-1.00
14	-8.00	-0.30	69.00	1.00	-6.90	16.04	0.00	0.0	0.0	8.00	0.00	0.0	0.0	1.00	-20.80	-0.00
	8.95	2.80	16.61	2.10	1.00	1.75	0.00	0.0	5.00	1.05	-0.05	0.0	0.0	0.0	-2.35	0.05

STAGE = 3   STEP = 1   SUM OF STEPS = 4

	1	2	3	4	5	6	7	8	9	10	11	12	13	14		
8	1.11	-0.10	-7.58	0.17	-0.15	-0.00	-0.00	1.00	0.25	-0.86	-0.09	-0.05	0.0	0.0	1.31	0.05
7	0.42	0.84	-0.86	-0.23	-0.23	-0.00	1.00	0.0	0.39	-0.04	0.11	-0.08	0.0	0.0	0.67	0.08
6	0.13	-0.11	1.61	0.20	-0.17	1.00	0.00	0.0	0.28	0.16	-0.10	-0.06	0.0	0.0	-0.79	0.06
13	2.46	10.70	0.23	1.27	-2.95	-0.00	-0.00	0.0	10.75	0.79	-0.14	-0.65	1.00	0.0	-1.31	0.65
14	-10.04	1.52	43.11	-2.18	-4.18	0.00	0.00	0.0	-4.54	5.50	1.59	0.91	0.0	1.00	-8.14	-0.91
	8.75	3.00	13.78	1.75	1.30	0.00	0.00	0.0	4.50	0.78	0.12	0.10			-0.97	-0.10

STAGE = 3   STEP = 2   SUM OF STEPS = 5

	1	2	3	4	5	6	7	8	9	10	11	12	13	14		
8	1.00	-0.00	-9.00	0.00	-0.00	-0.88	-0.00	1.00	0.00	-1.00	-0.00	-0.00	0.0	0.0	2.00	0.00
7	0.25	1.00	-3.07	-0.50	-0.00	-1.37	1.00	0.0	0.00	-0.25	0.25	0.0	0.0	0.0	1.75	0.0
5	-0.75	0.67	-9.51	-1.17	1.00	-5.89	-0.00	0.0	-1.67	-0.92	0.58	0.33	0.0	0.0	4.65	-0.33
13	0.25	12.67	-27.81	-2.17	-0.00	-17.37	-0.00	0.0	5.83	-1.92	1.58	0.33	1.00	0.0	12.40	-0.33
14	-13.17	4.30	3.39	-7.05	-0.00	-24.60	-0.00	0.0	-11.50	1.68	4.02	2.30	0.0	1.00	11.28	-2.30
	9.70	2.13	26.12	3.27	0.00	7.64	0.00	0.0	6.67	1.97	-0.63	-0.33			-7.00	-0.33

STAGE = 4   STEP = 1   SUM OF STEPS = 6

	1	2	3	4	5	6	7	8	9	10	11	12	13	14		
8	0.99	-0.64	-7.59	0.11	-0.00	-0.00	-0.00	1.00	-0.30	-0.90	-0.08	-0.02	-0.05	0.0	1.37	0.05
7	0.23	0.00	-0.88	-0.33	0.00	-0.00	1.00	0.0	-0.46	-0.10	0.13	-0.03	-0.08	0.0	0.77	0.08
5	-0.83	-3.63	-0.08	-0.43	1.00	-0.00	0.00	0.0	-3.64	-0.27	0.05	0.22	-0.34	0.0	0.45	0.34
6	-0.01	-0.73	1.60	0.12	0.00	1.00	0.00	0.0	-0.34	0.11	-0.09	-0.02	-0.06	0.0	-0.71	0.06
14	-13.53	-13.64	42.78	-3.98	0.00	-0.00	0.00	0.0	-19.76	4.39	1.78	1.83	-1.42	1.00	-6.28	1.42
	9.81	7.71	13.89	2.31	0.00	0.00	0.00	0.0	9.23	1.12	0.06	-0.19	0.44	0.0	-1.54	-0.44

STAGE = 4   STEP = 2   SUM OF STEPS = 7

	1	2	3	4	5	6	7	8	9	10	11	12	13	14		
8	1.00	-0.00	-9.00	0.00	-0.00	-0.88	-0.00	1.00	0.00	-1.00	-0.00	-0.00	-0.00	0.0	2.00	0.00
7	0.23	0.00	-0.88	-0.33	0.00	0.00	1.00	0.0	-0.46	-0.10	0.13	-0.03	-0.08	0.0	0.77	0.08
5	-0.76	-0.00	-8.04	-1.05	1.00	-4.98	-0.00	0.0	-1.97	-0.82	0.50	0.32	-0.05	0.0	4.00	0.05
2	0.04	1.00	-2.20	-0.17	-0.00	-1.37	-0.00	0.0	0.46	-0.15	0.12	0.03	0.08	0.0	0.98	-0.08
14	-13.26	-0.00	12.83	-6.31	-0.00	-18.70	-0.00	0.0	-13.48	2.33	3.49	2.19	-0.34	1.00	7.08	0.34
	9.66	0.00	30.61	3.63	0.00	10.57	0.00	0.0	5.68	2.29	-0.90	-0.39	-0.17	0.0	-9.09	0.17

STAGE = 5   STEP = 1   SUM OF STEPS = 8

	1	2	3	4	5	6	7	8	9	10	11	12	13	14		
8	1.00	-0.00	-9.00	0.00	-0.00	-0.88	-0.00	1.00	0.00	-1.00	-0.00	0.00	-0.00	0.00	2.00	-0.00
7	0.68	0.00	-1.32	-0.11	0.00	0.64	1.00	0.0	-0.00	-0.18	0.01	-0.10	-0.07	-0.03	0.53	0.03
5	1.18	-0.00	-9.92	-0.13	1.00	-2.24	-0.00	0.0	-0.00	-1.16	-0.01	-0.00	-0.00	-0.15	2.96	0.15
2	-0.43	1.00	-1.76	-0.39	-0.00	-2.01	-0.00	0.0	0.00	-0.07	0.24	0.10	0.07	0.03	1.22	-0.03
9	0.98	0.00	-0.95	0.47	0.00	1.39	0.00	0.0	1.00	-0.17	-0.26	-0.16	0.03	-0.07	-0.52	0.07
	4.07	0.00	36.22	0.97	0.00	2.68	0.00	0.0	0.00	3.27	0.57	0.53	-0.31	0.42	-6.10	-0.42

STAGE = 5   STEP = 2   SUM OF STEPS = 9

	1	2	3	4	5	6	7	8	9	10	11	12	13	14		
8	-8.30	-0.00	-0.00	-4.43	-0.00	-14.00	-0.00	1.00	-9.45	0.63	2.45	1.53	-0.24	0.70	6.96	-0.70
7	-0.68	-0.00	-0.00	-0.76	-0.00	-1.28	1.00	0.0	-1.38	0.06	0.36	0.12	-0.10	0.07	1.26	-0.07
5	-9.08	-0.00	-0.00	-5.01	1.00	-16.70	-0.00	0.0	-10.42	0.64	2.69	1.69	-0.27	0.63	8.43	-0.63
2	-2.25	1.00	-0.00	-1.25	-0.00	-4.57	-0.00	0.0	-1.85	0.25	0.72	0.40	0.02	0.17	2.19	-0.17
3	-1.03	-0.00	1.00	-0.49	-0.00	-1.46	-0.00	0.0	-1.05	0.18	0.27	0.17	-0.03	0.08	0.55	-0.08
	41.49	0.00	0.00	18.79	0.00	55.47	0.00	0.0	38.05	-3.29	-9.27	-5.64	0.65	-2.40	-26.07	2.40

OPTIMAL VERTEX = (-3.2936, -9.2723, -5.6393, 0.6465, -2.4007)

OPTIMAL VALUE = 26.0743

Chapter 2

THE AVERAGE NUMBER OF PIVOT STEPS

2.1 THE PROBABILITY SPACE

Let us define what we mean by "average number of pivot steps" precisely. For this purpose consider the matrix of the input data

$$(2.1.1) \qquad \hat{A} := \begin{bmatrix} a_1^T \\ \cdot \\ a_m^T \\ v^T \end{bmatrix} \in \mathbb{R}^{(m+1)n}$$

of a problem

$$(2.1.2) \qquad \begin{array}{ll} \text{Maximize} & v^T x \\ \text{subject to} & a_1^T x \le 1, \ldots, a_m^T x \le 1 \\ \text{where} & v, x, a_1, \ldots, a_m \in \mathbb{R}^n, \ m \ge n. \end{array}$$

We regard this matrix $\hat{A}$ as a random variable in a probability space

$$(2.1.3) \qquad (\mathbb{R}^{(m+1)n}, \mathcal{A}, P),$$

Where $\mathcal{A}$ is the σ-algebra of the Lebesgue-measurable sets of $\mathbb{R}^{(m+1)n}$, and where P is a probability measure defined on $\mathcal{A}$. So we have a probability space containing all the

Figure 2.1

Illustration of different radial distributions in $\mathbf{R}^2$

The numbers at the circles tell the probability that a randomly chosen vector lies inside that circle.

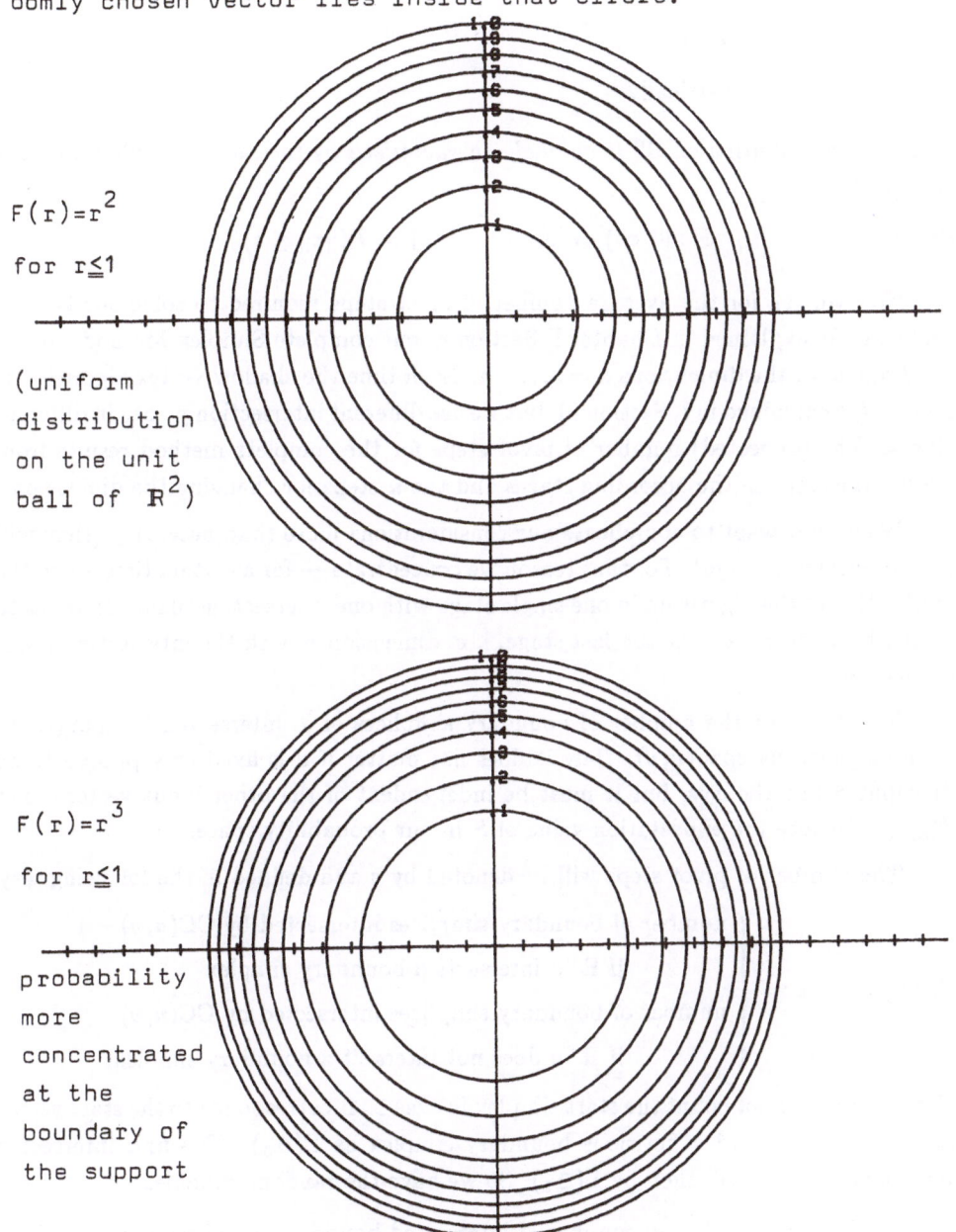

$F(r)=r^2$

for $r \leq 1$

(uniform
distribution
on the unit
ball of $\mathbf{R}^2$)

$F(r)=r^3$

for $r \leq 1$

probability
more
concentrated
at the
boundary of
the support

linear programming problems of our type with m inequalities and n variables. P is a product of $m + 1$ probability measures on $\mathbb{R}^n$, because we postulate that

(2.1.4)   The random vectors $a_1, \ldots, a_m, , v$ are distributed on $\mathbb{R}^n \setminus \{0\}$

         – independently

         – identically

         – symmetrically

The basic distribution on $\mathbb{R}^n$ is uniquely characterized by the radial distribution function (RDF)

$$(2.1.5) \qquad F : [0, \infty) \rightarrow [0, \infty); \quad F(r) := P(\|x\| \leq r).$$

Now we ask for the average number of pivot steps required to solve our type of problem. As explained in Chapter I, Section 4, our complete Simplex-Method runs in $n - 1$ stages of the dimensions $k = 2, \ldots, n$. Each time the shadow-vertex algorithm is applied (as explained in I, Section 3), but we use different intersection planes in different stages. The (expected) number of pivot steps for the complete method results from adding the steps in the according stages and the n steps for changing the dimension.

We do not want to complicate our considerations more than necessary (they will be complicated enough). For that reason we concentrate — for a certain time — on the application of the algorithm in one single stage with one intersection plane. It seems to be the best choice to take the last stage, i. e. dimension n with the intersection plane $\text{span}(e_n, v)$.

Now let S be the number of boundary simplices of Y intersected by $\text{span}(e_n, v)$ or more generally $\text{span}(u, v)$. Here it does not matter if u is fixed or supposed to be distributed like the a_i's, but it must be independent of the other input vectors. Let $E_{m,n}(S)$ denote the expectation value of S in our probability space.

The number of pivot steps will be denoted by s and defined in the following way

$$(2.1.6) \qquad s := \begin{cases} \text{number of boundary simplices intersected by } CC(u, v) - 1 \\ \qquad \text{if } \mathbb{R}^+ u \text{ intersects a boundary simplex} \\ \text{number of boundary simplices intersected by } CC(u, v) \\ \qquad \text{if } \mathbb{R}^+ u \text{ does not intersect a boundary simplex.} \end{cases}$$

Note that we do not count the start simplex in case 1. It corresponds to the start vertex x_0. In case 2 regard a fictitious boundary simplex as $\Sigma(\Delta_0)$. The first intersected boundary simplex will then be $\Sigma(\Delta_1)$. So we have the random variables

$$(2.1.7) \qquad \begin{aligned} S &= S(\hat{A}) = \text{ number of intersected boundary simplices for } \hat{A} \\ s &= s(\hat{A}) = \text{ number of boundary simplices for } \hat{A} \text{ as in (2.1.6).} \end{aligned}$$

Figure 2.2

Critical case: No start simplex, no initial vertex

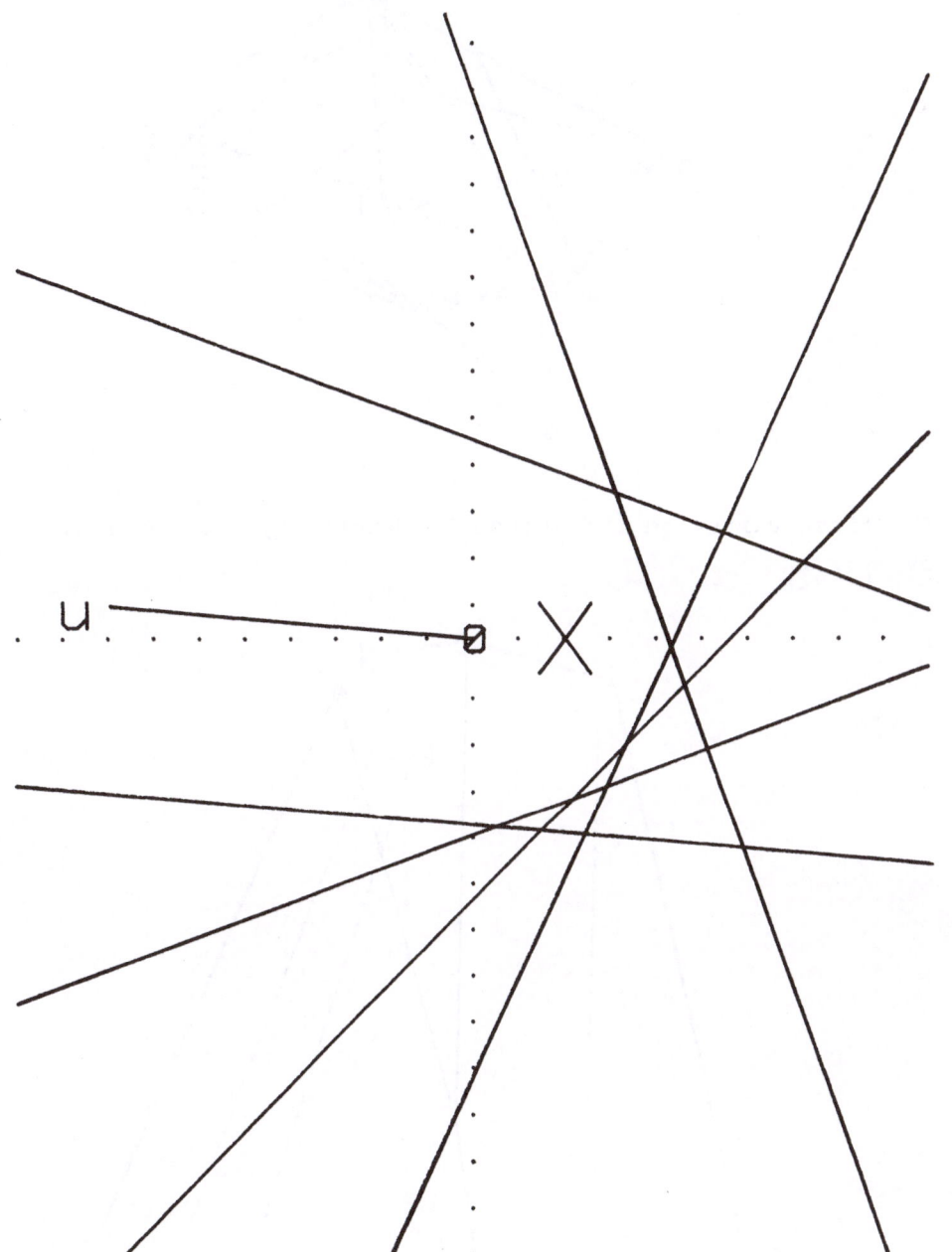

Figure 2.3

Illustration of the definitions of s and S for the case of
unbounded X

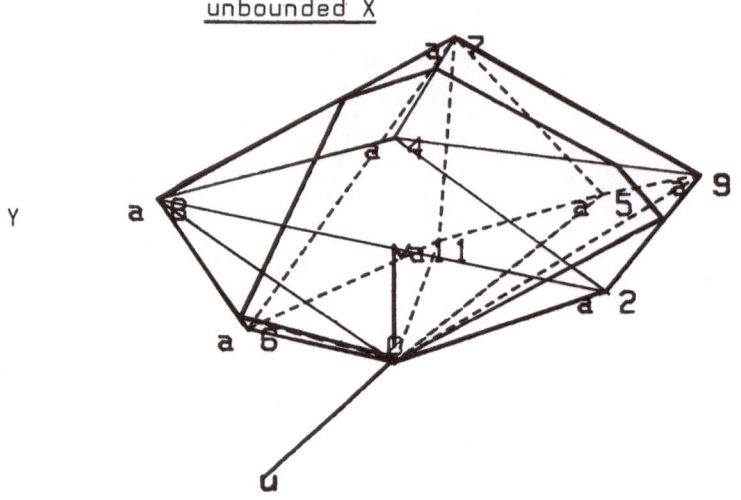

The first vertex on the shadow is denoted by x_1, the last
by $x_s = x_4$.

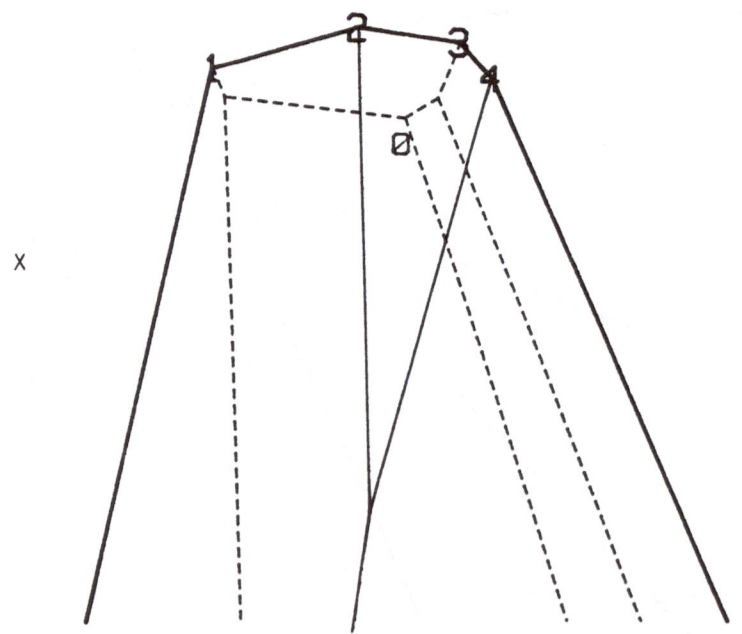

The starting ray is identified with a fictive vertex x_0
for the purpose of calculating s.

Figure 2.3a

Foreground of figure 2.3

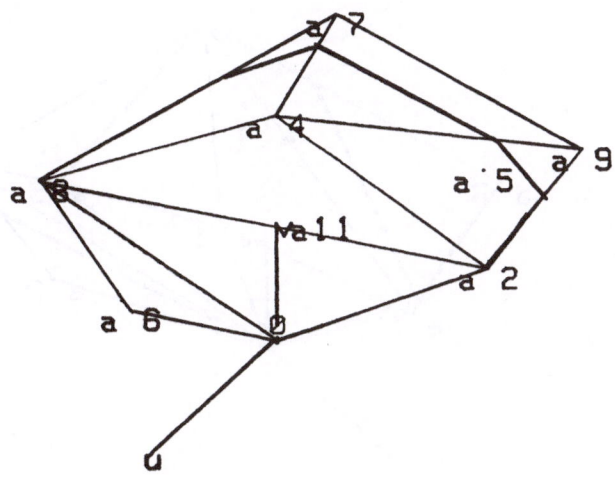

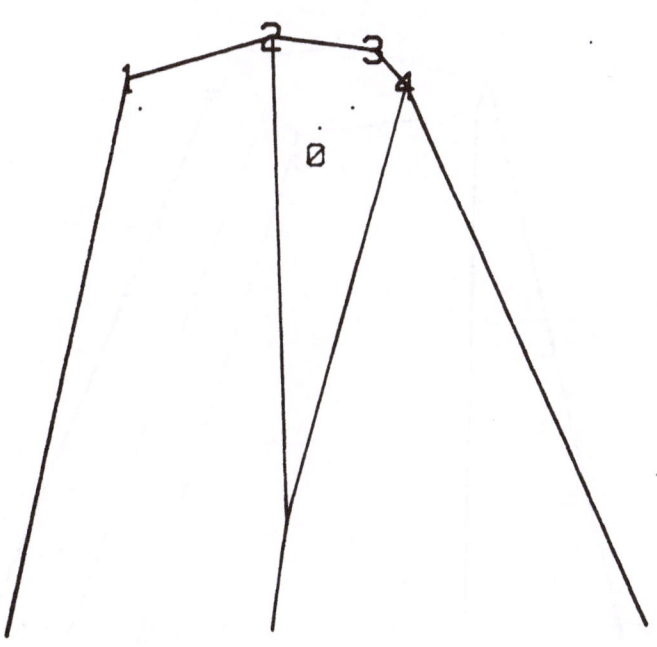

Figure 2.3b

Background of figure 2.3

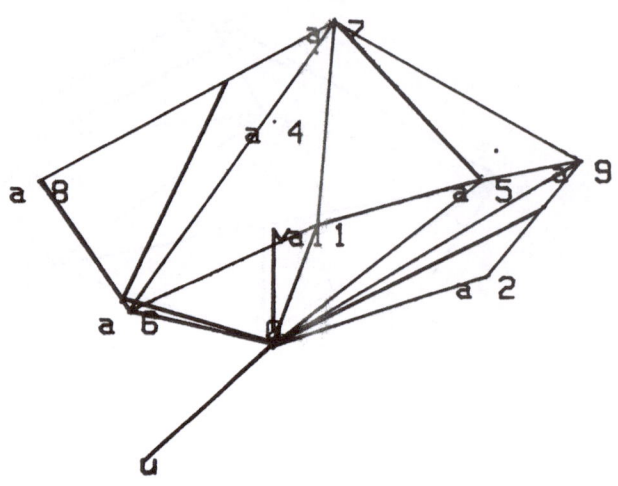

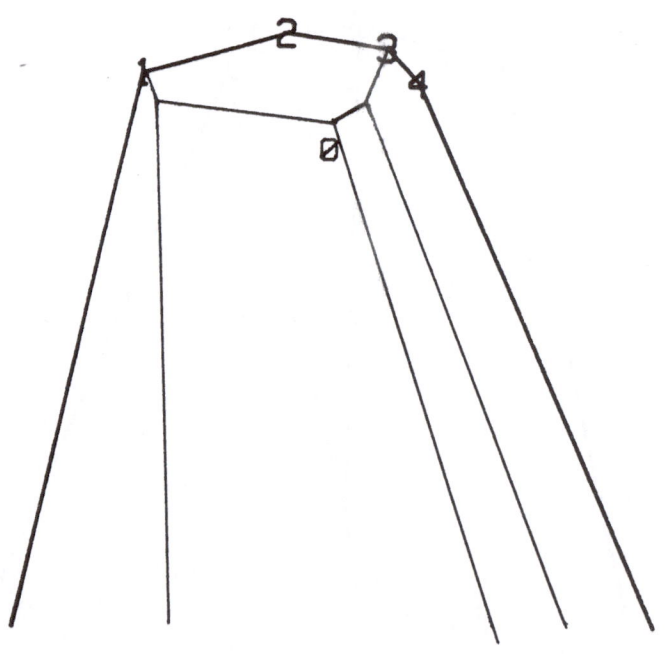

Both are Lebesgue-measurable variables, whose expectation values do exist, since

$$(2.1.8) \qquad\qquad 0 \le s \le S \le \binom{m}{n}.$$

In (2.1.6) S and s are even defined for the case of degeneracy. The expectation values of these variables are not influenced by degeneracy-cases, because the set of degenerate problems has probability 0 in our stochastic model.

Since the origin of $\mathbb{R}^n$ is excluded, we have measure 0 for every hyperplane of $\mathbb{R}^n$. But all degenerate cases imply that at least $n + 1$ of the vectors $0, a_1, \ldots, a_m, v, u$ lie in one hyperplane. So we are allowed to ignore the degenerate cases when we evaluate the expectation values. The following Lemma describes the relation between the two expectation values $E_{m,n}(s)$ and $E_{m,n}(S)$.

Lemma 2.1

$$(2.1.9) \qquad\qquad \frac{1}{4} E_{m,n}(S) = E_{m,n}(s).$$

Proof. Suppose that $a_1, a_2, \ldots, a_m$ and w are randomly chosen vectors (now fixed) and that we have four different problems with different vectors v, u:

1) $v = w, \ u = e_n$

2) $v = e_n, \ u = -w$

3) $v = -w, \ u = -e_n$

4) $v = -e_n, \ u = w$.

Imagine that the shadow-vertex algorithm runs through the resulting cones clockwise. Then the problems require the numbers $s_{u,v}, \ s_{-v,u}, \ s_{-u,-v}, \ s_{v,-u}$ of pivot steps defined as in (2.1.6). We claim that $S = s_{u,v} + s_{-v,u} + s_{-u,-v} + s_{v,-u}$.

Now consider an arbitrary boundary simplex intersected by $\mathrm{span}(u, v)$. Then four cases are possible:

1) $\Sigma(\Delta) \cap \mathrm{span}(u, v)$ lies completely in one of the four cones, e. g. in $CC(u, v)$. Then $\Sigma(\Delta)$ is counted in the value $s_{u,v}$ and not in the others.

2) $\Sigma(\Delta)$ is intersected by $\mathbb{R}^+ u$, but not by $\mathbb{R}^+ v$. Then $\Sigma(\Delta)$ is counted in $s_{-v,u}$ as the optimal simplex; it is not counted in $s_{u,v}$, where it is the start simplex $\Sigma(\Delta_0)$.

3) $\Sigma(\Delta)$ is intersected by $\mathbb{R}^+ v$, not by $\mathbb{R}^+ u$. This case can be treated according to case 2).

Figure 2.4

Illustration of the four cases in the proof
of Lemma 2.1

CASE 1

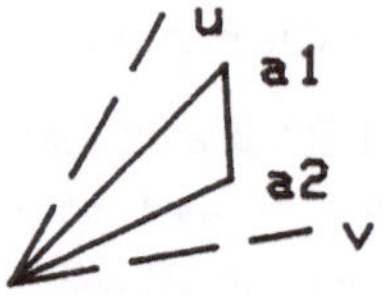

CASE 2

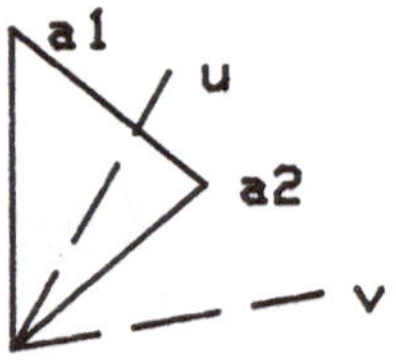

CASE 3

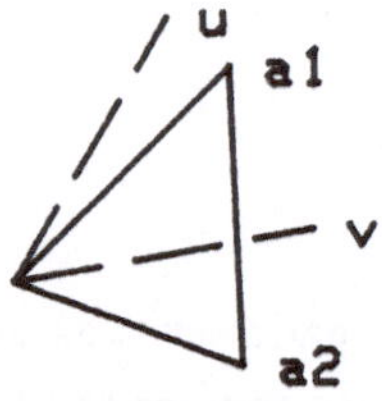

CASE 4

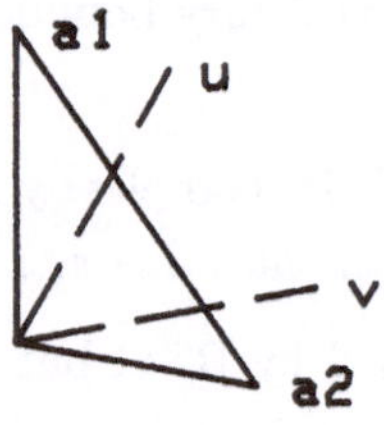

4) $\Sigma(\Delta)$ is intersected by $\mathbb{R}^+ u$ and by $\mathbb{R}^+ v$. Here the boundary simplex is counted in $s_{-v,u}$, but neither in $s_{u,v}$ nor in $s_{v,-u}$.

In S our boundary simplex $\Sigma(\Delta_0)$ is counted once in any case. So the left and the right side in our claim are identical. Since all four cases are equally probable as a result of rotational symmetry, we find

$$E_{m,n}(S) = E_{m,n}(s_{u,v}) + E_{m,n}(s_{v,-u}) + E_{m,n}(s_{-u,-v}) + E_{m,n}(s_{-v,u}).$$

The four expectation values are equal, because v itself is distributed symmetrically under rotations. (u can be held fixed, e. g. $u = e_n$, or it can be distributed analogously). So we have $E_{m,n}(S) = 4E_{m,n}(s)$ and our proposition is proven.

$\square$

This result enables us to deal with the simpler geometrical term S instead of s in the following.

2.2 AN INTEGRAL FORMULA FOR THE EXPECTED NUMBER OF S

In this section an integral formula for $E_{m,n}(S)$ shall be derived. To simplify our considerations, we assume (for the beginning) that

(2.2.1)   The distribution over $\mathbb{R}^n$ posesses a density function $f \in L_1(\mathbb{R}^n)$.

Later we shall show that the results of this section do even apply to distributions without density function. Without loss of generality we set $\Delta = \{1, \ldots, n\}$ and assume that our problem is not degenerate. To begin with, we need some notation.

Let $H := \{x \in \mathbb{R}^n \mid x^n = h\}$, where $h > 0$ is a given real number,

$H^- := \{x \in \mathbb{R}^n \mid x^n \leq h\}$ and

I_{H^-} be the corresponding indicator function.

Analogously, we denote by

$H(a_1 \ldots, a_n)$ the hyperplane through $a_1, \ldots, a_n$;

$H^-(a_1 \ldots, a_n)$ the halfspace which is bounded by $H(a_1, \ldots, a_n)$

and contains the origin;

$I_{H^-(a_1 \ldots, a_n)}$ the corresponding indicator function;

In addition, let $h(a_1, \ldots, a_n)$ be the distance between $H(a_1, \ldots, a_n)$ and the origin.

Now recall what we know about the numbers of intersected boundary simplices and the candidates for that property.

$E_{m,n}(S)$ = expected number of Δ's such that $\Sigma(\Delta)$ is a boundary simplex of Y and intersected by span(u, v)

$\quad$ = (number of candidates Δ) $\cdot$ (probability, that the candidate $\Delta = \{1, \ldots, n\}$ satisfies both conditions)

$\quad$ = $\binom{m}{n} P(\text{CH}(a_1, \ldots, a_n)$ satisfies both conditions)

(2.2.2) $\quad$ = $\binom{m}{n} \int\limits_{\mathbb{R}^n} \ldots \int\limits_{\mathbb{R}^n} P(\text{CH}(a_1, \ldots, a_n)$ satisfies both conditions)$\cdot$

$$\cdot f(a_1) \ldots f(a_m) f(v) \, da_1 \ldots da_m dv.$$

If $a_1, \ldots, a_n$ are fixed, then the random event "boundary simplex" depends only on the position of $a_{n+1}, \ldots, a_m$. The event "intersection" depends only on the position of v and of u. So we have independece.

$$E_{m,n}(S) = \binom{m}{n} \int\limits_{\mathbb{R}^n} \ldots \int\limits_{\mathbb{R}^n} P(\text{CH}(a_1, \ldots, a_n) \text{ is a boundary simplex}) \cdot$$

(2.2.3)

$$\cdot P(\text{CH}(a_1, \ldots, a_n) \cap \text{span}(u, v) \neq \emptyset) \cdot$$

$$\cdot f(a_1) \ldots f(a_m) f(v) \, da_1 \ldots da_m dv.$$

For a more sophisticated study of the intersection event we consider the $n-2$-dimensional side-simplices $\text{CH}(a_1, \ldots, a_{i-1}, a_{i+1}, \ldots, a_n)$ of $\Sigma(\Delta)$.

If $\Sigma(\Delta)$ is intersected, then exactly two of its side-simplices are intersected, too. So we have

$$E_{m,n}(S) = \binom{m}{n} \frac{n}{2} \int\limits_{\mathbb{R}^n} \ldots \int\limits_{\mathbb{R}^n} P(\text{CH}(a_1, \ldots, a_n) \text{ is a boundary simplex}) \cdot$$

(2.2.4)

$$\cdot P(\text{CH}(a_1, \ldots, a_{n-1}) \cap \text{span}(u, v) \neq \emptyset) \cdot$$

$$\cdot f(a_1) \ldots f(a_m) f(v) \, da_1 \ldots da_m dv.$$

For the following considerations we need some more notation. The marginal distribution function and its derivative are

$$G(h) := P(x^n \leq h) = P(x \in H^+)$$

(2.2.5)

$$= \int\limits_{-\infty}^{h} \int\limits_{-\infty}^{\infty} \ldots \int\limits_{-\infty}^{\infty} f(x^1, \ldots, x^n) dx^1 \ldots dx^n.$$

$$g(h) := \frac{\partial G(h)}{\partial h} = \int\limits_{-\infty}^{\infty} \ldots \int\limits_{-\infty}^{\infty} f(x^1, \ldots, x^{n-1}, h) dx^1 \ldots dx^{n-1}.$$

Figure 2.5

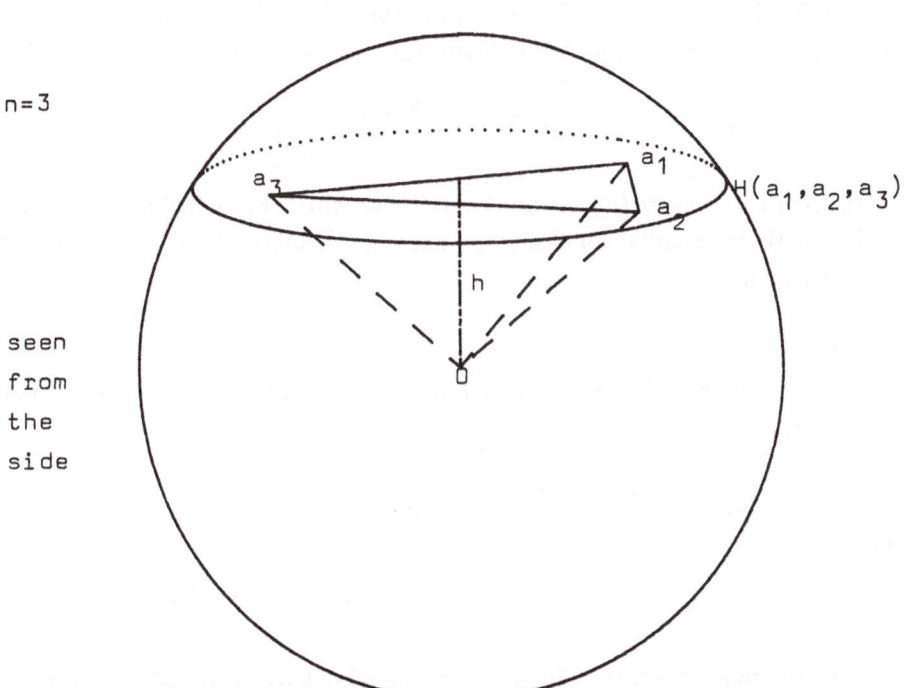

n=3

seen
from
the
side

The simplex $CH(a_1,a_2,a_3)$ and the hyperplane $H(a_1,a_2,a_3)$

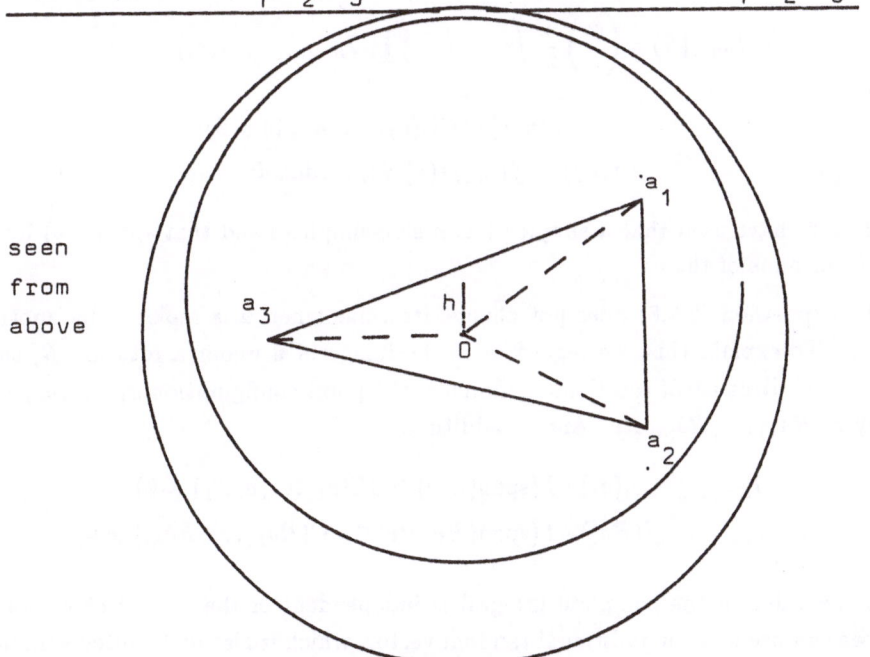

seen
from
above

Let $\Omega_n := \{x \mid \|x\| \le 1\} \subset \mathbb{R}^n$ denote the unit ball of $\mathbb{R}^n$ and let $\omega_n := \{x \mid \|x\| = 1\}$ $\subset \mathbb{R}^n$ stand for the unit sphere of $\mathbb{R}^n$ (surface of the unit ball). (See Appendix, Section 2). Let λ_k be an abbreviation for the Lebesque-measure of dimension k. Now we define

$$(2.2.6) \qquad W(a_1, \ldots, a_{n-1}) = \frac{\lambda_{n-1}(\mathrm{CC}(a_1, \ldots, a_{n-1}) \cap \Omega_n)}{\lambda_{n-1}(\Omega_{n-1})}.$$

This is the sperical measure (normalized) of the cone which is spanned by $a_1, \ldots, a_{n-1}$. The value of W is essential for the intersection probability. This is shown in the following Theorem.

Theorem 3

If u is fixed or distributed independently from $a_1, \ldots, a_m, v$ and symmetrically under rotations, then

$$(2.2.7) \qquad \begin{aligned} E_{m,n}(S) = \binom{m}{n} n \int\limits_{\mathbb{R}^n} \ldots \int\limits_{\mathbb{R}^n} & G(h(a_1, \ldots, a_n))^{m-n} \\ & W(a_1, \ldots, a_{n-1}) f(a_1) \ldots f(a_n)\, da_1 \ldots da_n. \end{aligned}$$

Proof. For the random event "$\mathrm{CH}(a_1, \ldots, a_n)$ is a boundary simplex" it is necessary that for $i = n+1, \ldots, m$ all vectors a_i belong to $H^-(a_1, \ldots, a_n)$. So we have

$$(2.2.8) \qquad \begin{aligned} E_{m,n}(S) = \binom{m}{n} \frac{n}{2} \int\limits_{\mathbb{R}^n} \ldots \int\limits_{\mathbb{R}^n} & \prod_{i=n+1}^{m} I_{H^-(a_1,\ldots,a_n)}(a_i) \cdot \\ & \cdot I(\mathrm{span}(u, v) \cap \mathrm{CC}(a_1, \ldots, a_{n-1}) \ne \emptyset) \cdot \\ & \cdot f(a_1) \ldots f(a_m) f(v)\, da_1 \ldots da_m dv. \end{aligned}$$

Note that we have used that a simplex has n side-simplices and that $\mathrm{span}(u, v)$ intersects two or none of them.

The expression (2.2.8) does not change its value when u is replaced by another vector w. To explain this, we regard w as the image of u under a rotation R, such that $w = Ru$. Because of rotational symmetry the point-configuration $a_1, \ldots, a_m, v$ is as likely as $Ra_1, \ldots, Ra_m, Rv$. And in addition,

$$I_{H^-(a_1,\ldots,a_n)}(a_i) \cdot I(\mathrm{span}(u, v) \cap \mathrm{CC}(a_1, \ldots, a_{n-1}) \ne \emptyset) =$$
$$= I_{H^-(Ra_1,\ldots,Ra_m)}(Ra_i) \cdot I(\mathrm{span}(Ru, Rv) \cap \mathrm{CC}(Ra_1, \ldots, Ra_m) \ne \emptyset).$$

So the value of the complete integral is independent of the special choice of u. Hence we can use u as an additional random vector, which is also distributed symmetrically under rotations and independent of $a_1, \ldots, a_m, v$. Then

Figure 2.6

The side simplex $CH(a_1, a_2)$ and its spherical measure $W(a_1, a_2)$

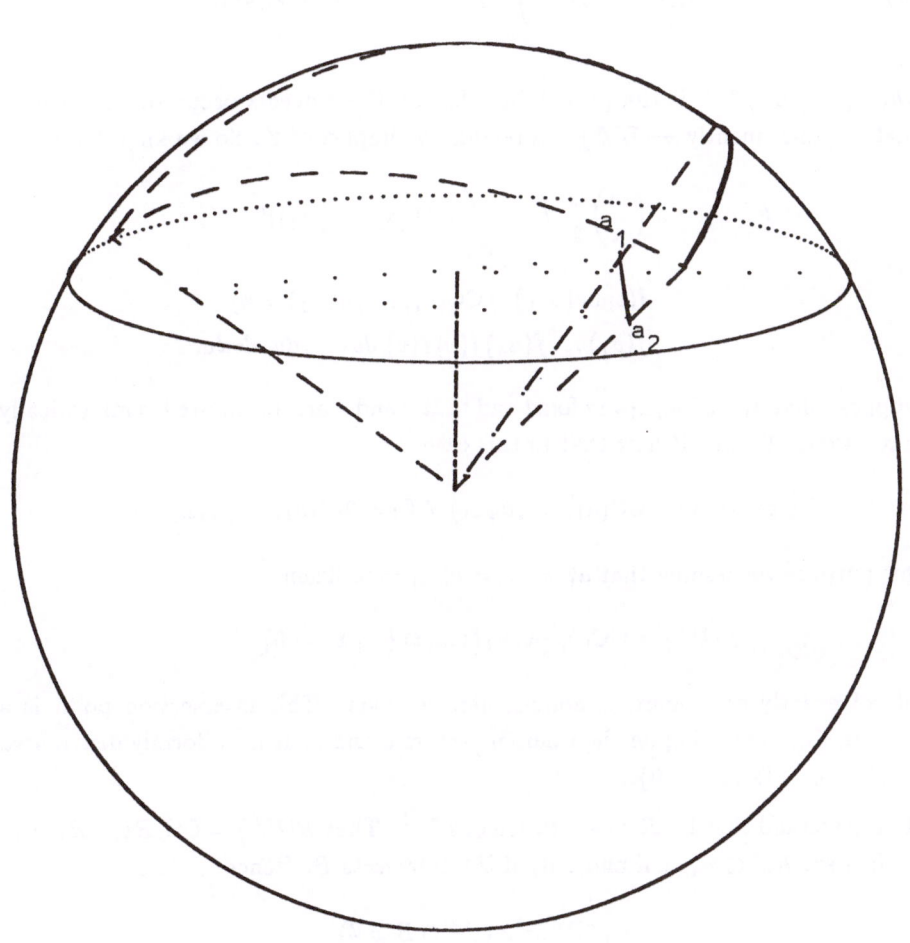

$$E_{m,n}(S) = \binom{m}{n} \frac{n}{2} \int_{\mathbb{R}^n} \cdots \int_{\mathbb{R}^n} \prod_{i=n+1}^{m} I_{H^-(a_1,\ldots,a_n)}(a_i) \cdot$$

(2.2.9)

$$\cdot I(\operatorname{span}(u,v) \cap CC(a_1,\ldots,a_{n-1}) \neq \emptyset) \cdot$$

$$\cdot f(a_1)\ldots f(a_m)f(v)f(u)\, da_1 \ldots da_m dv du.$$

Each a_i $(i = n+1,\ldots,m)$ belongs to $H^-(a_1,\ldots,a_n)$ with probability

(2.2.10) $$G(h(a_1,\ldots,a_n)) = \int_{\mathbb{R}^n} I_{H^-(a_1,\ldots,a_n)}(a_i)f(a_i)da_i.$$

So $G(h(a_1,\ldots,a_n))^{m-n}$ is the probability that all these events occur simultaneously and that — consequently — $\Sigma(\Delta)$ is a boundary simplex of Y. So we know that

$$E_{m,n}(S) = \binom{m}{n} \frac{n}{2} \int_{\mathbb{R}^n} \cdots \int_{\mathbb{R}^n} G(h(a_1,\ldots,a_n))^{m-n} \cdot$$

(2.2.11)

$$\cdot I(\operatorname{span}(u,v) \cap CC(a_1,\ldots,a_{n-1}) \neq \emptyset) \cdot$$

$$\cdot f(a_1)\ldots f(a_n)f(v)f(u)\, da_1 \ldots da_n dv du.$$

Now suppose that $a_1,\ldots,a_{n-1}$ are fixed and that v and u are distributed symmetrically under rotations. We shall show that in this case

$$P(\operatorname{span}(u,v) \cap CC(a_1,\ldots,a_{n-1}) \neq \emptyset) = 2W(a_1,\ldots,a_{n-1}).$$

For this purpose we assume that $a_1^n = \ldots = a_{n-1}^n = 0$. Then

$$U^+ := CC(v,-v,u) \cap \omega_n \cap \{x \mid x^n = 0\}$$

consists of exactly one point, if nondegeneracy holds. This intersection point is a random variable depending on the random vectors u and v. It is uniformly distributed on $\omega_{n-1} := \omega_n \cap \{x \mid x^n = 0\}$.

To understand this, let R be a rotation of $\mathbb{R}^{n-1}$. Then $R(U^+) = CC(Rv,-Rv,Ru)$ intersects a set $RB \subset \omega_{n-1}$ if and only if U^+ intersects B. Hence

$$\mu_+(B) := P(U^+ \cap B \neq \emptyset)$$

is a probability on the Lebesgue-measurable subsets B of ω_{n-1}. This probability is symmetrical under rotations. The same holds for $\mu_-(B) := P(U^- \cap B \neq \emptyset)$, where $U^- := CC(v,-v,-u)$. On the other hand we know that $\mu(B) := \dfrac{\lambda_{n-2}(B)}{\lambda_{n-2}(\omega_{n-1})}$ is the

Figure 2.7a

The boundary simplex condition

Here $CH(a_1, a_2, a_3)$ is a boundary simplex, because all the other points $a_4, \ldots, a_{15}$ lie below $H(a_1, a_2, a_3)$.

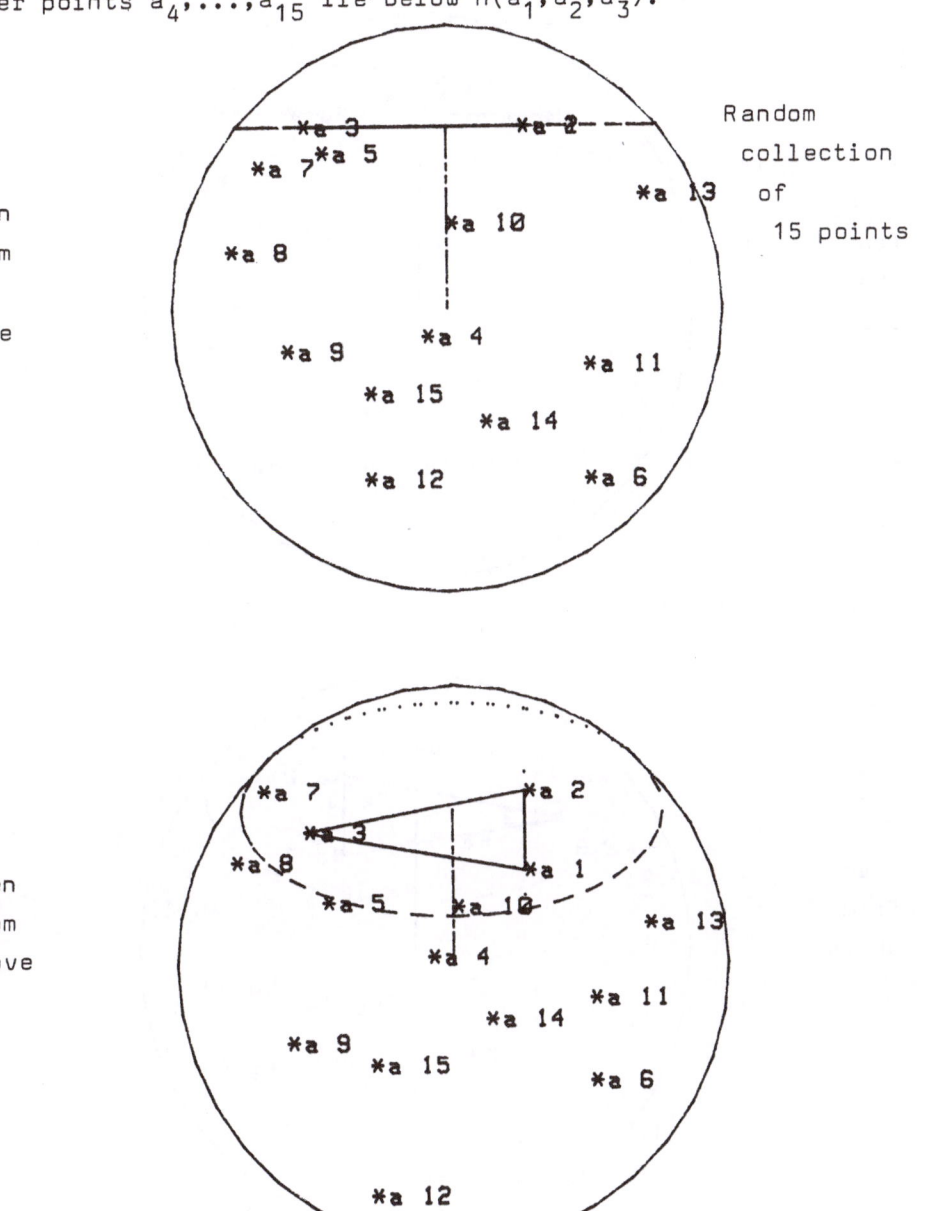

Figure 2.7b

The boundary simplex condition

Here $CH(a_1, a_2, a_3)$ is not a boundary simplex, because a_4, a_5, a_9 lie above $H(a_1, a_2, a_3)$.

seen
from
the
side

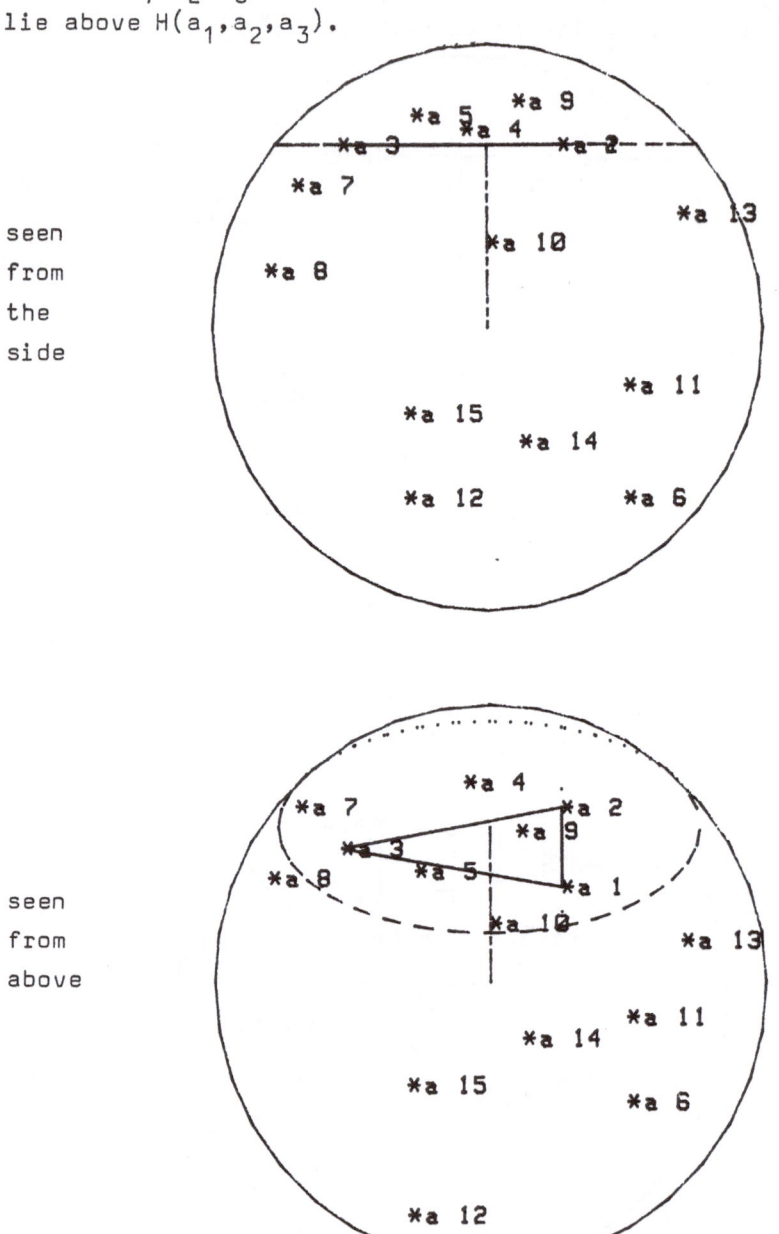

seen
from
above

only probability measure on ω_{n-1} which is symmetrical under rotations (following from the theory of the uniqueness of the Haar-integral on an orthogonal group) (compare NACHBIN (1965)). So we conclude that $\mu(B) = \mu_+(B) = \mu_-(B)$ for all Lebesgue-measurable sets $B \subset \omega_{n-1}$.

Recall that in our case $B = CC(a_1, \ldots, a_{n-1}) \cap \omega_{n-1}$. So $P(\text{span}(u, v) \cap B \neq \emptyset) = P(U^+ \cap B \neq \emptyset) + P(U^- \cap B \neq \emptyset) - P(U^+ \cap B \neq \emptyset \text{ and } U^- \cap B \neq \emptyset)$. Here B is convex. The last term in that sum is 0, since only in degenerate cases there may be an intersection with both sets U^+ and U^-. Consequently,

$$P(\text{span}(u, v) \cap B \neq \emptyset) = 2\frac{\lambda_{n-2}(B)}{\lambda_{n-2}(\omega_{n-1})} = 2\frac{\lambda_{n-1}(CC(B) \cap \omega_{n-1})}{\lambda_{n-1}(\omega_{n-1})} = 2W(a_1, \ldots, a_{n-1}).$$

Insertion of this formula into (2.2.11) leads to the desired formula (2.2.7).

$\square$

Corollary.

If u and v are distributed symmetrically under rotations and if $a_1, \ldots, a_{n-1}$ are fixed, then

$$(2.2.12) \qquad P(\text{span}(u, v) \cap \text{CH}(a_1, \ldots, a_{n-1})) = 2W(a_1, \ldots, a_{n-1}).$$

Note that this formula is not true when u is fixed. Then (2.2.7) can only be obtained by integrating over all possible arrangements of $v, a_1, \ldots, a_{n-1}$.

Figure 2.8a

The intersection condition

Here the plane span(u,v) intersects CH(a_1,a_2,a_3).

We show the intersection of the straight line connecting

u and v with the cone CC(a_1,a_2,a_3).

Two side cones are intersected.

The line enters in CC(a_2,a_3) and leaves in CC(a_1,a_3).

seen
from
above

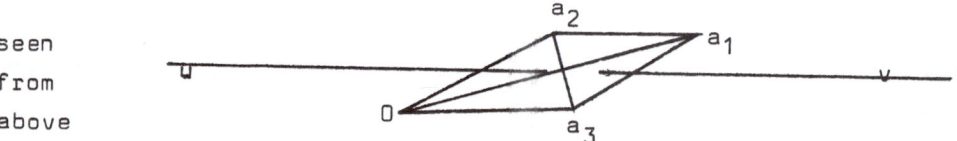

seen
from
the
side

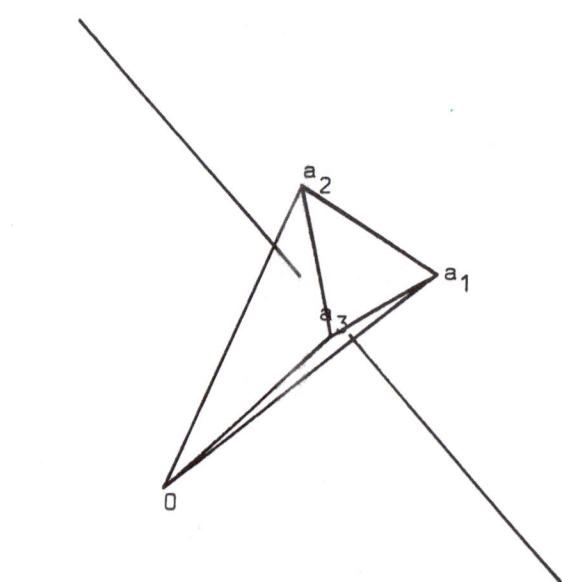

Figure 2.8b

The intersection condition

Here the intersection condition is not satisfied.

The line (and the plane) passes the cone/simplex without

intersection.

seen
from
above

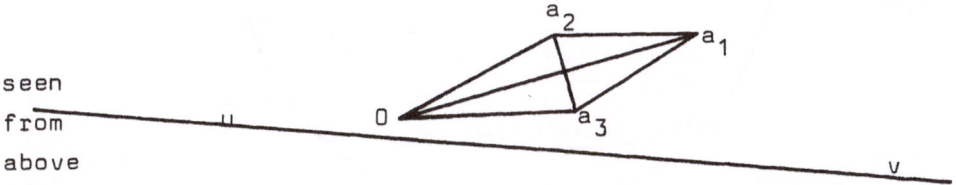

seen
from
the
side

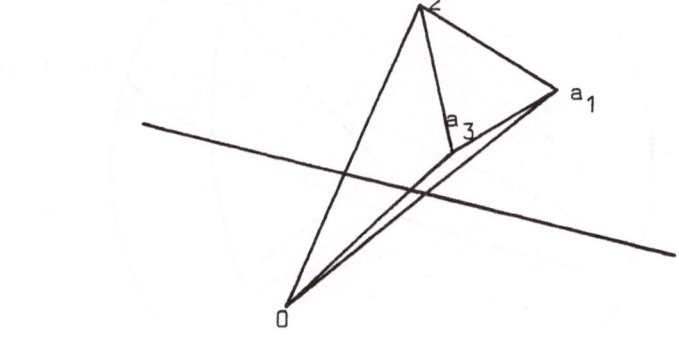

Figure 2.9a

The intersection of span(u,v) with the arc generated by a

side-simplex

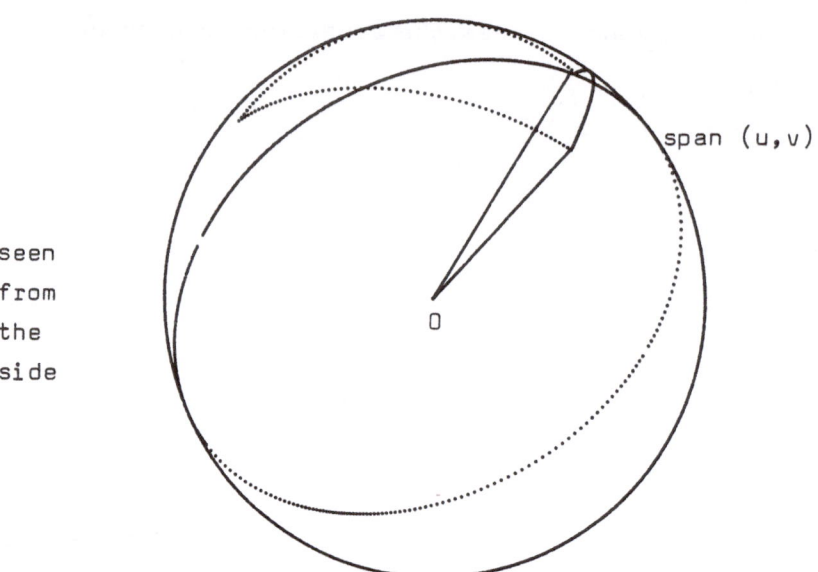

seen
from
the
side

Here span(u,v) intersects the arc generated by the side-
simplex $CH(a_1,a_2)$. If u and v are distributed symmetrically
under rotations, then the probability of the intersection

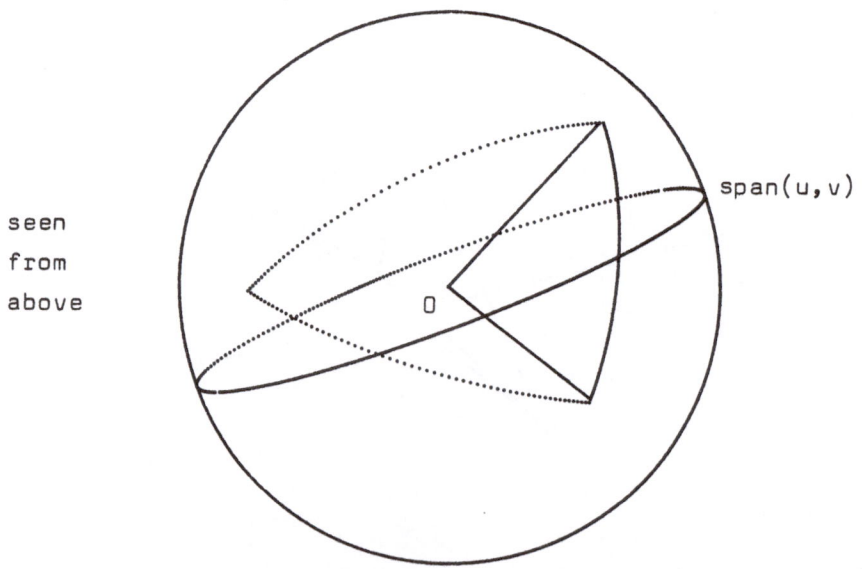

seen
from
above

is proportional to $W(a_1,a_2)$.

Figure 2.9b

The intersection of span(u, v) with the arc generated by a
side-simplex

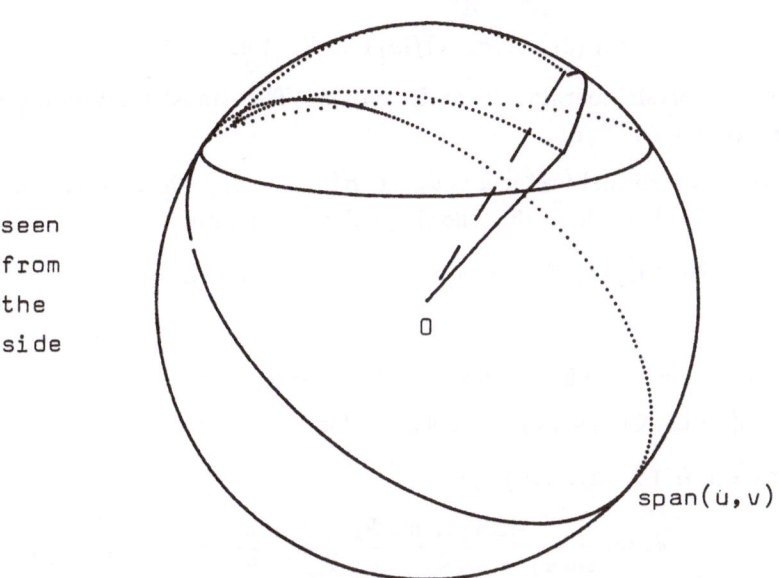

seen
from
the
side

span(u, v)

Here the intersection event does not occur.

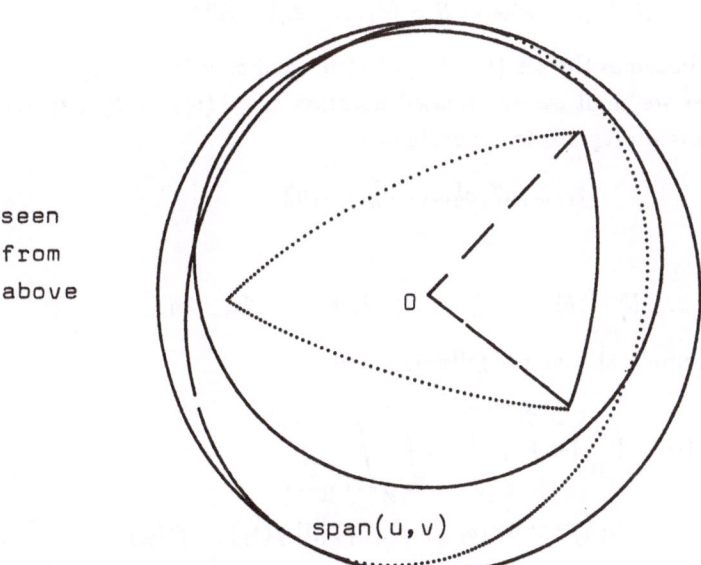

seen
from
above

span(u, v)

2.3 A TRANSFORMATION OF COORDINATES

We can simplify our integral formula

(2.3.1)
$$E_{m,n}(S) = \binom{m}{n} \frac{n}{2} \int\limits_{\mathbb{R}^n} \cdots \int\limits_{\mathbb{R}^n} G(h(a_1,\ldots,a_n))^{m-n} \cdot$$

$$\cdot 2\, W(a_1,\ldots,a_{n-1}) f(a_1) \ldots f(a_n)\, da_1 \ldots da_n$$

by application of a certain coordinate-transformation. This transformation depends on the random vectors $a_1,\ldots,a_n$.

Consider the orthonormal vector $d \in \omega_n$ on $H(a_1,\ldots,a_n)$ which is directed towards this hyperplane. Let d have the following polar coordinates

(2.3.2)
$$\Psi_1 \in [0,2\pi), \Psi_i \in [0,\pi) \quad \text{for } i = 2,\ldots,n-1.$$

Then

(2.3.3)
$$d_n^1 = \sin\Psi_1 \sin\Psi_2 \ldots \sin\Psi_{n-1}, \ d_n^n = \cos\Psi_{n-1} \ \text{and}$$
$$d_n^k = \cos\Psi_{k-1} \sin\Psi_k \ldots \sin\Psi_{n-1}, \ \text{for } 2 \le k \le n-1.$$

Defining the vectors d_k $(k = 1,\ldots,n)$ by

(2.3.4)
$$d_k := \frac{\sin\Psi_1 \ldots \sin\Psi_k}{\sin\Psi_1 \ldots \sin\Psi_k \ldots \sin\Psi_{n-1}} \frac{\partial d_n}{\partial \Psi_k}$$

we obtain an orthonormal vector system $d_1,\ldots,d_n$.

With respect to the new basis $d_1,\ldots,d_n$ the representations for $a_1,\ldots,a_n$ are

(2.3.5)
$$b_i := R^{-1}a_i, \quad \text{where } R = (d_1,\ldots,d_n) \in \mathbb{R}^{n\times n}.$$

Then $H(a_1,\ldots,a_n)$ becomes the set $\{b \mid b^n = h\} \subset \mathbb{R}^n$ and $b_i = (b_i^1,\ldots,b_i^{n-1},h)^T$ for $i = 1,\ldots,n$. Further we shall use the related notation $\bar{b}_i := (b_i^1,\ldots,b_i^{n-1})^T$ for the truncated vector. Now we replace the coordinates

(2.3.6)
$$a_1^1,\ldots,a_1^n, a_2^1,\ldots,a_n^1,\ldots,a_n^n$$

by

(2.3.7)
$$b_1^1,\ldots,b_1^{n-1}, b_2^1,\ldots,b_n^1,\ldots,b_n^{n-1}, \Psi_1,\ldots,\Psi_{n-1}, h.$$

Our main integral formula attains the following form

(2.3.8)
$$E_{m,n}(S) = \binom{m}{n} n \int\limits_0^\infty \int\limits_0^{2\pi} \int\limits_0^\pi \cdots \int\limits_0^\pi \int\limits_{\mathbb{R}^{n-1}} \int\limits_{\mathbb{R}^{n-1}}$$
$$G(h)^{m-n}\, W(b_1,\ldots,b_{n-1})\, |J|\, f(b_1)\ldots f(b_n)$$
$$db_1 \ldots db_n\, d\Psi_{n-1} \ldots d\Psi_2 d\Psi_1\, dh.$$

This is mainly a result of rotational symmetry of f. J is the determinant of the Jacobian for that transformation. The following lemma delivers the value of J. Here we need the matrix

$$(2.3.9) \qquad B := \begin{bmatrix} b_1^1 & \cdots & b_1^{n-1} & 1 \\ \vdots & & \vdots & \vdots \\ b_n^1 & \cdots & b_n^{n-1} & 1 \end{bmatrix}.$$

Lemma 2.2

The absolute value of J, the determinant of the Jacobian belonging to our transformation of coordinates, is

$$|J| = |\det B|(\sin \Psi_2)^1 (\sin \Psi_3)^2 \ldots (\sin \Psi_{n-1})^{n-2}.$$

Proof. The Jacobian matrix Ξ of the coordinate-transformation contains the derivatives of a_j^i with respect to the new variables $b_1^1, \ldots, \Psi_{n-1}, h$ in row number $(j-1)n+i$. So this row looks like

$$\left(\frac{\partial a_j^i}{\partial b_1^1}, \ldots, \frac{\partial a_j^i}{\partial b_1^{n-1}}, \ldots, \frac{\partial a_j^i}{\partial b_n^{n-1}}, \frac{\partial a_j^i}{\partial \Psi_1}, \ldots, \frac{\partial a_j^i}{\partial \Psi_{n-1}}, \frac{\partial a_j^i}{\partial h} \right).$$

Due to our definition for $d_1, \ldots, d_n, \Psi_1, \ldots, \Psi_{n-1}$ we have $\Xi = \Xi_1 \Xi_2$, where

$$\Xi_1 = \begin{bmatrix} R & & & & & & \\ & R & & & & & \\ & & R & & & & \\ & & & R & & & \\ & & & & R & & \\ & & & & & \ddots & \\ & & & & & & R \end{bmatrix} \in \mathbb{R}^{n^2 \times n^2}.$$

and where Ξ_2 is a $(n^2 \times n^2)$ matrix. Its j-th block row (consisting of the rows $(j-1)n+1, \ldots, jn$) has the following entries.

$$
\begin{array}{c}
b_1^1 \;\cdots\cdots\; b_j^1 \;\cdots\cdots\; \cdots \; b_j^{n-1} \;\cdots\; b_n^{n-1} \quad \Psi_1 \;\cdots\; \Psi_{n-1} \quad h \\[4pt]
\begin{array}{c}
a_j^1 \\[30pt] \\[30pt] a_j^n
\end{array}
\left[
\begin{array}{cccccccc}
& 1 & & & & & & 0 \\
& & 1 & & & & & 0 \\
0 & & & 1 & & 0 & R_1 b_j \;\cdots\; R_{n-1} b_j & 0 \\
& & & & 1 & & & 0 \\
& & & & & 1 & & 0 \\
& 0 \;\cdots\cdots\; 0 & & & & & & 1
\end{array}
\right]
\end{array}
$$

Here we have written R_i for $R^{-1}\frac{\partial R}{\partial \Psi_i}$ $(i = 1, \ldots, n-1)$. Then

$$|J| = |\det \Xi| = |\det \Xi_1||\det \Xi_2| = |\det \Xi_2|,$$

because R is an orthogonal matrix and $|\det \Xi_1| = |\det R|^n = 1$.

Now let us deal with $|\det \Xi_2|$. We perform a permutation of rows such that every n-th row of the according block-rows (as mentioned above) is moved to the bottom of the entire matrix Ξ_2. The matrix Ξ_3 generated in this way has the following form.

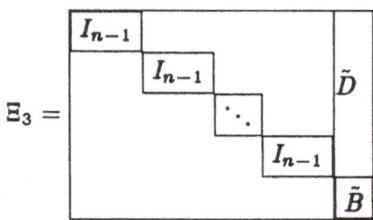

Here we have the unit matrix of rank $n-1$, the entries of $\tilde{D}$ do not matter, and the $(n \times n)$-matrix $\tilde{B} := (\tilde{b}_j^i)$ has the entries

$$\tilde{b}_j^i = e_n^T R_i b_j = e_n^T R^T \frac{\partial R}{\partial \Psi_i} b_j = d_n^T \frac{\partial R}{\partial \Psi_i} b_j$$

for $i = 1, \ldots, n-1$ and $j = 1, \ldots, n$ and $\tilde{b}_j^n = 1$ for $i = n$ and $j = 1, \ldots, n$. Because of $|\det \Xi_2| = |\det \Xi_3| = |\det \tilde{B}|$ we know that $|J| = |\det \tilde{B}|$.

The matrix R consists of the (column)-vectors $d_1, \ldots, d_n$. Its explicit form has been shown before. If we calculate the matrices $\frac{\partial R}{\partial \Psi_i}$ $(i = 1, \ldots, n-1)$ and insert the result in the above formula, we obtain $\tilde{b}_j^i = \sigma_i \tilde{b}_j^i$ for $i = 1, \ldots, n-1$ and $j = 1, \ldots, n$

$\tilde{b}_j^n = 1$ for $j = 1, \ldots, n$; where $\sigma_i = - \sin \Psi_{i+1} \sin \Psi_{i+2} \ldots \sin \Psi_{n-1}$ for $i = 1, \ldots, n-1$. Consequently

$$|J| = |\det B| = |\sigma_1 \ldots \sigma_{n-1}| \, |\det B| = (\sin \Psi_2)(\sin \Psi_3)^2 \ldots (\sin \Psi_{n-1})^{n-2} \, |\det B|.$$

Finally, we use that (see Appendix 2.14)

$$(2.3.10) \quad \lambda_{n-1}(\omega_n) = \int_0^{2\pi} \int_0^{\pi} \ldots \int_0^{\pi} \sin \Psi_2 (\sin \Psi_3)^2 \ldots (\sin \Psi_{n-1})^{n-2} d\Psi_{n-1} \ldots d\Psi_2 d\Psi_1$$

Theorem 4.

$$(2.3.11) \qquad E_{m,n}(S) = \binom{m}{n} n \, \lambda_{n-1}(\omega_n) \int_0^{\infty} G(h)^{m-n} \Lambda_S(h) dh$$

with

$$\Lambda_S(h) := \int_{\mathbb{R}^{n-1}} \ldots \int_{\mathbb{R}^{n-1}} |\det B| \, W(b_1, \ldots, b_{n-1}) f(b_1) \ldots f(b_n) d\bar{b}_1 \ldots d\bar{b}_n.$$

Remark.

Note that

$$(2.3.12) \qquad |\det B| = \lambda_{n-1}(\mathrm{CH}(b_1, \ldots, b_n))(n-1)!.$$

This geometrical formula will be very useful in the following.

2.4 GENERALIZATIONS

Until now we have restricted our considerations to distributions with density functions (absolutely continuous distributions). This has been done to avoid difficulties. But the results concerning $E_{m,n}(S)$ (resp. $E_{m,n}(s)$) do also hold for distributions with arbitrary radial-distribution-functions $F(r) := P(\|x\| \leq r)$.

Remark.

If the given distribution function F has a density function f, then we know that for all $0 < r < \infty$

(2.4.1)

$$F(r) := \lambda_{n-1}(\omega_n) \int_0^r n\, t^{n-1} \hat{f}(t)\,dt \text{ with}$$

$$\hat{f} : [0, \infty) \to \mathbb{R} \text{ and } \hat{f}(\|x\|) = f(x) \text{ for } x \in \mathbb{R}^n.$$

It is possible to approximate a given radial-distribution-function sufficiently well by such absolutely continous RDF's. This makes the following lemma very useful.

Lemma 2.3

Let $(F^{(\nu)})$ be a sequence of RDF's converging pointwise towards the RDF F on $[0, \infty)$. Then we have

(2.4.2)
$$\lim_{\nu \to \infty} E_{m,n}^{(\nu)}(S) = E_{m,n}(S).$$

Here $E_{m,n}^{(\nu)}(S)$ and $E_{m,n}(S)$ represent the expectation values based on the respective RDF's.

Proof. The integral formula (2.2.8) can be transformed as follows.

$$E_{m,n}(S) = \binom{m}{n} n \int_0^\infty \cdots \int_0^\infty T(r_1, \ldots, r_m) dF(r_1) \ldots dF(r_m) \text{ with}$$

(2.4.3)

$$T(r_1, \ldots, r_m) = \int_{\omega_n} \cdots \int_{\omega_n} W(\gamma_1, \ldots, \gamma_{n-1}) \cdot$$

$$\cdot \prod_{i=n+1}^m I_{H^-}(\gamma_1 r_1, \ldots, \gamma_n r_n)(\gamma_i r_i) d_\omega(\gamma_1) \ldots d_\omega(\gamma_n).$$

Here $d_\omega(\cdot)$ denotes the integration element for ω_n, which is normalized by the condition

$$\int_{\omega_n} d_\omega(\gamma) = 1.$$

The indicator function I_{H^-} is defined only when $a_1, \ldots, a_n$ resp. $\gamma_1 r_1, \ldots, \gamma_n r_n$ are linearly independent. In all other cases we set $I_{H^-}(\cdot) := 0$. This definition does not influence the value of $T(r_1, \ldots, r_m)$ as long as $r_1 > 0, \ldots, r_n > 0$, because the degenerate cases form a nullset in $\omega_n \times \omega_n \times \ldots \times \omega_n$ and because $0 \leq I_{H^-} \leq 1$ and $0 \leq 2W(\gamma_1, \ldots, \gamma_{n-1}) \leq 1$. Using Lebesgue's Theorem we can show that T is continuous in every point $(r_1, \ldots, r_m)$ with $r_1 > 0, \ldots, r_m > 0$. So we conclude that T is even uniformly continuous in the m-dimensional interval $[p, q]^m = [p, q] \times \ldots \times [p, q]$, where $0 < p < q < \infty$ and $p, q \in \mathbb{R}$. For any given $\delta > 0$ it is possible to find p_δ, q_δ such that $F(p_\delta) < \delta$ and $F(q_\delta) > 1 - \delta$. Because of the pointwise convergence of $F^{(\nu)})$ towards F there is a number ν_0, such that for all $\nu > \nu_0$ $F^{(\nu)}(p_\delta) < 2\delta$ and $F^{(\nu)}(q_\delta) > 1 - 2\delta$. Hence we know that for $\nu > \nu_0$

$$\int\limits_{p_\delta}^{q_\delta} \ldots \int\limits_{p_\delta}^{q_\delta} dF^{(\nu)}(r_1) \ldots dF^{(\nu)}(r_m) > (1 - 4\delta)^m.$$

We know that $T(\ldots) \le 1$. Consequently the integral over the complementary integration area is bounded from above by $1 - (1 - 4\delta)^m$. So it is sufficient to consider the m-dimensional interval $[p_\delta, q_\delta]^m$ only.

On this area the integrals based on the $F^{(\nu)}$'s converge to the according integral based on F. This is a consequence of the uniform continuity of T and of the fact that $F^{(\nu)} \times \ldots \times F^{(\nu)}$ define probability-measures whose RDF's converge pointwise to $F \times \ldots \times F$. This results from the convergence of $(F^{(\nu)})$ towards F.

Remark.

Every RDF F can be represented as the limit of a pointwise convergent sequence of RDF's $F^{(\nu)}$ which are all absolutely continuous. Hence to every $E_{m,n}^F(S)$ there is a sequence of such expectation values with

(2.4.4) $$\lim E_{m,n}^{F(\nu)}(S) = E_{m,n}^F(S),$$

where all $F^{(\nu)}$ are absolutely continuous.

Consequently, all our previous results and integral formulae do hold even for RDF's without density functions.

Corollary.

The convergence-considerations in the proof of the lemma yield that every RDF can be approximated sufficiently by a sequence of RDF's which are absolutely continuous and which have bounded support (these are RDF's with $F^{(\nu)}(\bar{r}) = 1$ for a point $\bar{r} \in \mathbb{R}$). Hence $E_{m,n}^F(S)$ can even be regarded and represented as the limit of a sequence of expectation values based on absolute continuous RDF's with bounded support.

On the other hand we know that

(2.4.5)
$$S(\hat{A}) = S(\lambda \hat{A}) \text{ for } \lambda > 0, \lambda \in \mathbb{R}$$
$$s(\hat{A}) = s(\lambda \hat{A}).$$

S and s are homogeneous with respect to simultaneous multiplications of all vectors $a_1, \ldots, a_m, v$ with a constant $\lambda > 0$, $\lambda \in \mathbb{R}$. So the expectation values are homogeneous, too. Hence every RDF with bounded support can be "normalized". Let F be a RDF such that $F(r) < 1$ for all $r < \bar{r}$, and that $F(\bar{r}) = 1$. Then we define

$$(2.4.6) \qquad \widehat{F}(\rho) = F(\rho\bar{r}) \text{ for all } \rho \in [0,1] \text{ and } \widehat{F}(1) = 1.$$

Now we know that

$$(2.4.7) \qquad E^F_{m,n}(S) = E^{\widehat{F}}_{m,n}(S).$$

So the claim of the corollary can even be strengthened.

Corollary.

Every expectation value $E^F_{m,n}(S)$ (resp. s) can be regarded as the limit of a sequence of expectation values

$$(2.4.8) \qquad \begin{aligned} &E^F_{m,n}(S) = \lim_{\nu \to \infty} E^{\widehat{F}(\nu)}_{m,n}(S), \text{ where for all } \nu \\ &\widehat{F}^{(\nu)} \text{ is absolutely continuous and} \\ &\widehat{F}^{(\nu)}(1) = 1, \ \widehat{F}^{(\nu)}(\rho) < 1 \text{ for all } r < 1. \end{aligned}$$

As a result we can restrict many of our considerations to RDF's, where Ω_n contains the support and is the smallest ball having this property.

Remark.

But remember that this restriction relies strongly on the fact that m and n are fixed. For different parameters m_i ($i = 1,2$) the convergence of $E^{\widehat{F}(\nu)}_{m,n}$ towards $E^F_{m,n}$ will still be true, but the speed of this convergence may differ significantly. So it is not clear that for $\nu \to \infty$ the asymptotic behaviour of $E^{F(\nu)}_{m,n}(S)$ (i. e. the behaviour for $m \to \infty$, n fixed) approximates the asymptotic behaviour of $E^F_{m,n}(S)$.

For example: we cannot conclude that the maximal possible asymptotic order of growth for $E^F_{m,n}(S)$ with F having bounded support is also an upper bound for all F's (even those with unbounded support). We are better off if we can describe or represent the limit as in (2.4.4) or in (2.4.8) directly without approximation error. In the latter case it is possible to analyze the asymptotic behaviour afterwards, based on the exact representations.

Finally, we want to generalize the formulae for G and g. Here we have for $n \geq 3$

$$(2.4.9) \qquad G(h) = 1 - \frac{\lambda_{n-2}(\omega_{n-1})}{\lambda_{n-1}(\omega_n)} \int\limits_h^\infty \int\limits_{\frac{h}{r}}^1 (1 - \sigma^2)^{(n-3)/2} d\sigma dF(r)$$

$$(2.4.10) \qquad g(h) = \frac{\lambda_{n-2}(\omega_{n-1})}{\lambda_{n-1}(\omega_n)} \int\limits_h^\infty \frac{(r^2 - h^2)^{(n-3)/2}}{r^{n-2}} dF(r).$$

In the first formula we take into regard, that

$$P(x^n > h \mid \|x\| = r) = \frac{\lambda_{n-2}(\omega_{n-1})}{\lambda_{n-1}(\omega_n)} \int\limits_{\frac{h}{r}}^1 (1 - \sigma^2)^{(n-3)/2} d\sigma.$$

Integration over all possible values of r yields the formula given above. The formula for g is for $n \geq 3$ the derivative of G with respect to h.

Chapter 3

THE POLYNOMIALITY OF THE
EXPECTED NUMBER OF STEPS

3.1 COMPARISON OF TWO INTEGRALS

In this chapter we are going to prove that $E_{m,n}(S)$ and also $E_{m,n}(s)$ are polynomial in m and n. The proof is highly technical and must be done very carefully. So I will explain it in detail and I will try to illustrate our considerations as well as possible.

For simplicity, we restrict our considerations to those cases, where the distribution under consideration has a density function and where the distribution's support is contained in Ω_n. This restriction is justified by our results in Chapter 2, Section 4, because we want to have an upper bound for $E_{m,n}(S)$ for fixed m and n.

In addition, all the steps of the proof in Sections 1 – 4 are done under the assumption that $n \geq 3$. For $n = 2$ the resulting estimation is trivial. The most important trick in the proof is the comparison of the expectation value of S with the expectation value of a closely related random variable, denoted by Z.

> While S denotes the number of boundary simplices intersected
> by $\mathrm{span}(u, v)$, Z tells us the number of boundary simplices
> intersected by the ray $\mathbb{R}^+ v$.

A consequence of our stochastic assumptions is that $\mathbb{R}^+ v$ can intersect more than one boundary simplex only on a set with probability 0. On the other hand $Z \leq \binom{m}{n}$.

So it is clear that

(3.1.1) $$E_{m,n}(Z) \leq 1.$$

Remark.

The primal meaning of Z is the number of optimal vertices in the feasible polyhedron X. Exploiting (3.1.1), we could overestimate $E_{m,n}(S)$, if we had an upper bound J for the quotient $\frac{E_{m,n}(S)}{E_{m,n}(Z)}$, because

(3.1.2) $$E_{m,n}(S) \leq \frac{E_{m,n}(S)}{E_{m,n}(Z)} \leq J.$$

In order to investigate this quotient, we need an integral formula for $E_{m,n}(Z)$, which is compatible to

(3.1.3)
$$E_{m,n}(S) = \binom{m}{n} n \int_{\mathbb{R}^n} \cdots \int_{\mathbb{R}^n} G(h(a_1,\ldots,a_n))^{m-n} W(a_1,\ldots,a_{n-1})$$
$$f(a_1)\ldots f(a_n)da_1 \ldots da_n.$$

Such a formula is

(3.1.4)
$$E_{m,n}(Z) = \binom{m}{n} \int_{\mathbb{R}^n} \cdots \int_{\mathbb{R}^n} G(h(a_1,\ldots,a_n))^{m-n} V(a_1,\ldots,a_n)$$
$$f(a_1)\ldots f(a_n)da_1 \ldots da_n.$$

Here we have used the simple fact that for fixed $a_1,\ldots,a_n$

(3.1.5) $$P(\mathrm{CH}(a_1,\ldots,a_n) \cap \mathbb{R}^+ v \neq \emptyset) = V(a_1,\ldots,a_n),$$

where

$$V(a_1,\ldots,a_n) := \frac{\lambda_n\{\mathrm{CC}(a_1,\ldots,a_n) \cap \Omega_n\}}{\lambda_n(\Omega_n)}$$

is the spherical measure of the cone spanned by $a_1,\ldots,a_n$ (which is normalized to 1).

$V(\ldots)$ denotes the spherical measure of an n-dimensional cone, whereas $W(\ldots)$ gives an $n-1$-dimensional (spherical) measure. The two formulae (3.1.4) and (3.1.5) differ only in the terms W resp. V (and the factor n). But in the present form of these integrals we can hardly derive any information on the quotient in question. Again it seems to be useful to apply the (already mentioned) coordinate-transformations. This will help us to separate the influence of the vector (or direction) a_n as well as possible. Recall that a_n belongs to the V-cone but not to the W-cone. So we perform two orthogonal transformations:

Figure 3.1

The simplex $CH(a_1, a_2, a_3)$ and its spherical measure $V(a_1, a_2, a_3)$

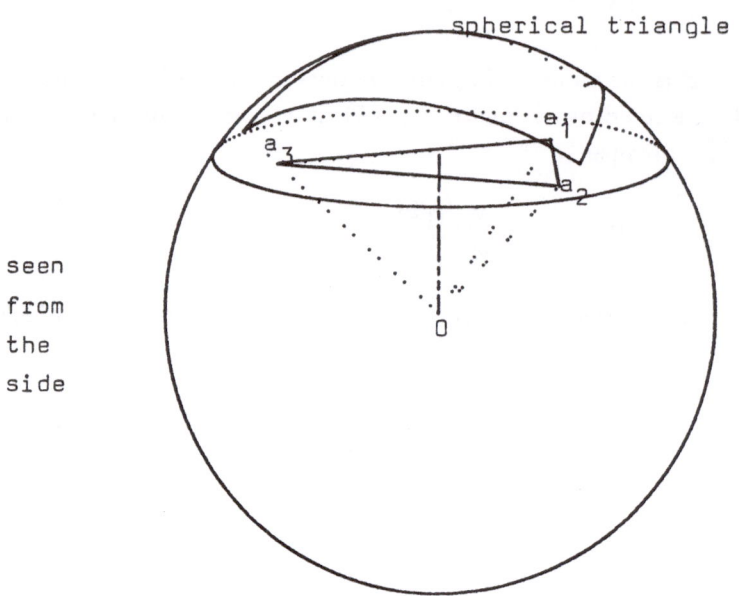

seen
from
the
side

The spherical measure is proportional to the volume of the
spherical triangle.

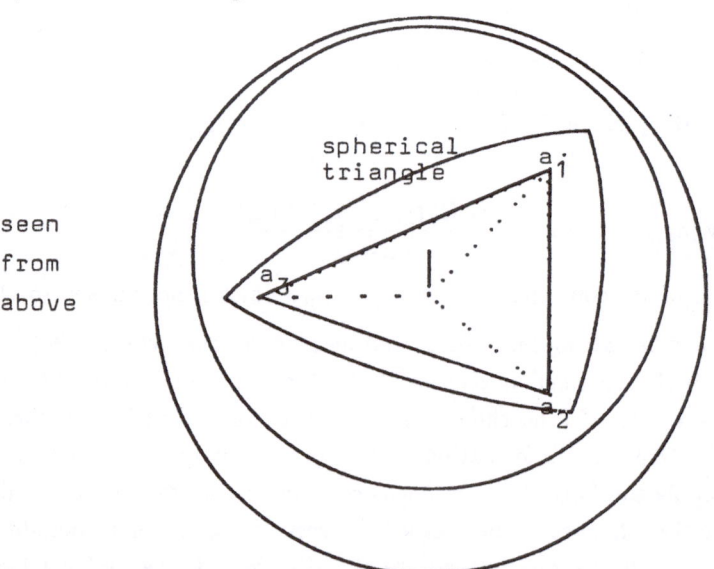

seen
from
above

1) $a_1, \ldots, a_m$ are transformed into $b_1, \ldots, b_m$ such that

(3.1.6)
$$b_1^n = h, \ldots, b_n^n = h.$$

2) $b_1, \ldots, b_m$ are transformed into $c_1, \ldots, c_m$ such that

(3.1.7)
$$c_1^n = h, \ldots, c_n^n = h$$
$$c_1^{n-1} = \theta, \ldots, c_{n-1}^{n-1} = \theta.$$

The second rotation uses the n-th unit vector e_n as a rotational axis. It is a rotation of $\mathbf{R}^{n-1}$. All our corresponding considerations of Chapter 2, Section 3 can be applied. Recall the following notation

$$c = (c^1, \ldots, c^n)^T \in \mathbf{R}^n$$
$$\bar{c} = (c^1, \ldots, c^{n-1})^T \in \mathbf{R}^{n-1}$$
$$\bar{\bar{c}} = (c^1, \ldots, c^{n-2})^T \in \mathbf{R}^{n-2}.$$

Proceeding in that way we obtain

$$E_{m,n}(S) = \binom{m}{n} n \{(n-2)!\}^2 \lambda_{n-1}(\omega_n) \lambda_{n-2}(\omega_{n-1})$$

(3.1.8)
$$\int_0^1 G(h)^{m-n} \int_{\mathbf{R}^{n-1}} \int_0^{\sqrt{1-h^2}} |\theta - c_n^{n-1}| \int_{\mathbf{R}^{n-2}} \cdots \int_{\mathbf{R}^{n-2}}$$
$$|\lambda_{n-2}\{CH(c_1, \ldots, c_{n-1})\}|^2 \, W(c_1, \ldots, c_{n-1})$$
$$f(c_1) \ldots f(c_{n-1}) d\bar{\bar{c}}_1 \ldots d\bar{\bar{c}}_{n-1} d\theta \, f(c_n) d\bar{c}_n dh \ .$$

The corresponding transformations for $E_{m,n}(Z)$ yield

$$E_{m,n}(Z) = \binom{m}{n} \{(n-2)!\}^2 \lambda_{n-1}(\omega_n) \lambda_{n-2}(\omega_{n-1})$$

(3.1.9)
$$\int_0^1 G(h)^{m-n} \int_{\mathbf{R}^{n-1}} \int_0^{\sqrt{1-h^2}} |\theta - c_n^{n-1}| \int_{\mathbf{R}^{n-2}} \cdots \int_{\mathbf{R}^{n-2}}$$
$$|\lambda_{n-2}\{CH(c_1, \ldots, c_{n-1})\}|^2 \, V(c_1, \ldots, c_n)$$
$$f(c_1) \ldots f(c_{n-1}) d\bar{\bar{c}}_1 \ldots d\bar{\bar{c}}_{n-1} d\theta \, f(c_n) d\bar{c}_n dh \ .$$

In both cases we have exploited that

$$|\det B| = \lambda_{n-1}\{CH(b_1, \ldots, b_n)\}(n-1)!$$

Figure 3.2

The two rotations of R^n

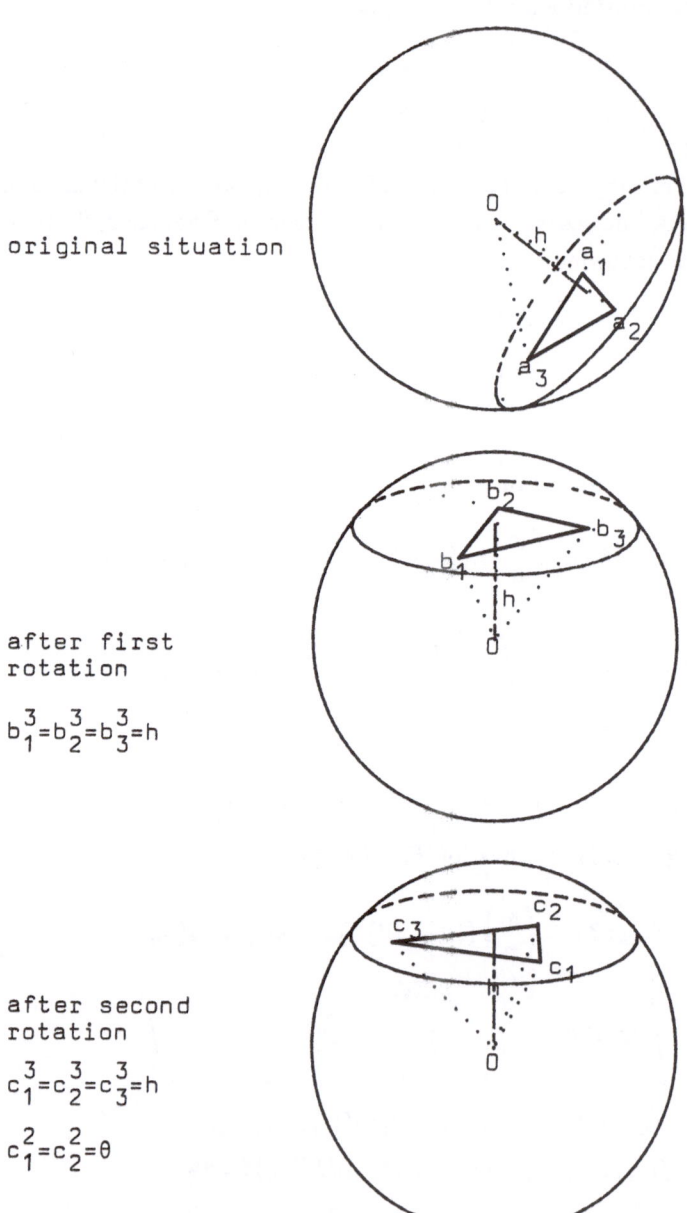

original situation

after first
rotation

$$b_1^3 = b_2^3 = b_3^3 = h$$

after second
rotation

$$c_1^3 = c_2^3 = c_3^3 = h$$

$$c_1^2 = c_2^2 = \theta$$

for dimension n and that the analogue formula for dimension $n - 1$ does hold. For an estimation of the quotient $\frac{E_{m,n}(S)}{E_{m,n}(Z)}$ it is necessary to integrate over all possible configurations of $c_1, \ldots, c_{n-1}, c_n, \theta$ and h.

Here we make use of the fact that the "inner distribution" of $\bar{\bar{c}}$ in the set

$$\{c \mid c^n = h, c^{n-1} = \theta\}$$

only depends on $t = \sqrt{h^2 + \theta^2}$ and is independent of the value of h (if we set $\theta = \sqrt{t^2 - h^2}$). In addition, the weight of this set in our integrals is the same for all pairs (h, θ) with constant value of $t = \sqrt{h^2 + \theta^2}$. And the spherical measure $W(c_1, \ldots, c_{n-1})$ does not depend on $h \in [0, t]$ as long as $\bar{c}_1, \ldots, \bar{c}_{n-1}, \theta = \sqrt{t^2 - h^2}$ are fixed. So it seems to be advantageous to connect and unite all the configurations where $\sqrt{h^2 + \theta^2}$ is constant ($= t$). After that we shall integrate over t in the interval $[0, 1]$.

For the comparison of the resulting numerator and denominator we can apply the following well-known rule.

If $f, g, h > 0$ are functions on $[0, 1]$, then

(3.1.10)
$$\frac{\int\limits_0^1 f(t)g(t)\,dt}{\int\limits_0^1 f(t)h(t)\,dt} \leq \sup_{t \in [0,1]} \frac{g(t)}{h(t)}.$$

Such a pointwise comparison will deliver an estimation for the quotient of integrals.

In both integrals we substitute θ by $\sqrt{t^2 - h^2}$, where t is the new integration variable and $d\theta$ is dropped. From now on we shall represent θ by

$$T = T(t, h) = \sqrt{t^2 - h^2} = \theta.$$

After possible reductions of constants (in n resp. m) we have

$$
\frac{E_{m,n}(S)}{E_{m,n}(Z)} = \frac{n \int_0^1 t \int_0^t G(h)^{m-n} T^{-1} \int_{\mathbb{R}^{n-1}} |T - c_n^{n-1}| \cdot}{\int_0^1 t \int_0^t G(h)^{m-n} T^{-1} \int_{\mathbb{R}^{n-1}} |T - c_n^{n-1}| \cdot}
$$

$$
\frac{\cdot \int_{\mathbb{R}^{n-2}} \cdots \int_{\mathbb{R}^{n-2}} (\lambda_{n-2}\{CH(c_1,\ldots,c_{n-1})\})^2 W(c_1,\ldots,c_{n-1}) \cdot}{\cdot \int_{\mathbb{R}^{n-2}} \cdots \int_{\mathbb{R}^{n-2}} (\lambda_{n-2}\{CH(c_1,\ldots,c_{n-1})\})^2 V(c_1,\ldots,c_n) \cdot}
$$

(3.1.11)

$$
\frac{\cdot f(c_1)\ldots f(c_{n-1}) d\bar{c}_1 \ldots d\bar{c}_{n-1} f(c_n) d\bar{c}_n \, dh \, dt}{\cdot f(c_1)\ldots f(c_{n-1}) d\bar{c}_1 \ldots d\bar{c}_{n-1} f(c_n) d\bar{c}_n \, dh \, dt} \leq
$$

$$
\leq \sup_{t \in [0,1]} \frac{n \int_0^t G(h)^{m-n} T^{-1} \int_{\mathbb{R}^{n-1}} |T - c_n^{n-1}| \cdot}{\int_0^t G(h)^{m-n} T^{-1} \int_{\mathbb{R}^{n-1}} |T - c_n^{n-1}| \cdot}
$$

$$
\frac{\cdot \int_{\mathbb{R}^{n-2}} \cdots \int_{\mathbb{R}^{n-2}} (\lambda_{n-2}\{CH(c_1,\ldots,c_{n-1})\})^2 W(c_1,\ldots,c_{n-1}) \cdot}{\cdot \int_{\mathbb{R}^{n-2}} \cdots \int_{\mathbb{R}^{n-2}} (\lambda_{n-2}\{CH(c_1,\ldots,c_{n-1})\})^2 V(c_1,\ldots,c_n) \cdot}
$$

$$
\frac{\cdot f(c_1)\ldots f(c_{n-1}) d\bar{c}_1 \ldots d\bar{c}_{n-1} f(c_n) d\bar{c}_n \, dh}{\cdot f(c_1)\ldots f(c_{n-1}) d\bar{c}_1 \ldots d\bar{c}_{n-1} f(c_n) d\bar{c}_n \, dh} .
$$

Note that $|T - c_n^{n-1}|$ is the distance of the vector $\bar{c}_n$ (appearing in V, but not in W) from the $n-2$-dimensional hyperplane through $\bar{c}_1,\ldots,\bar{c}_{n-1}$ as observed in the set $\{x \mid x^n = h\} \subset \mathbb{R}^n$.

The remaining work of this chapter is mainly to simplify this frightening quotient and to make it evaluable.

Figure 3.3

Integration over various values of t has highest

priority (most outside integration in (3.1.12))

For fixed t, the inner distributions in the sets with fixed
h,T   are equal.

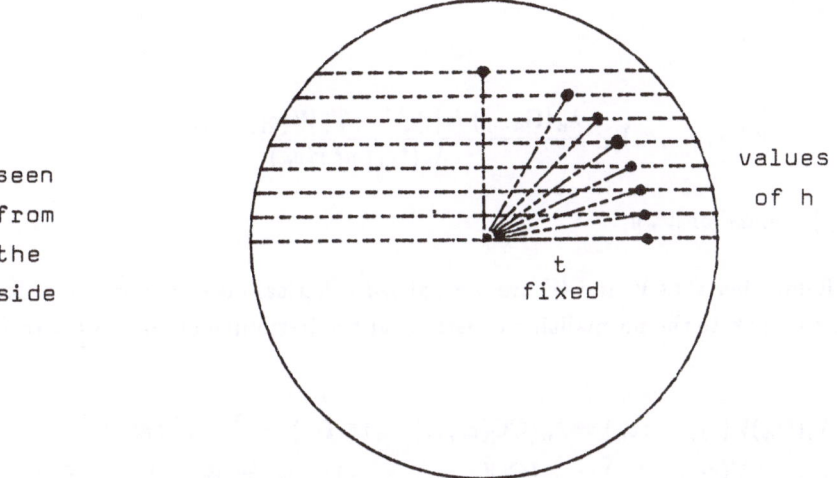

seen
from
the
side

t
fixed

values
of h

So for fixed t and every pair (h,T) any fixed constellation
$(\bar{\bar{c}}_1, \bar{\bar{c}}_2)$ has the same likelihood.

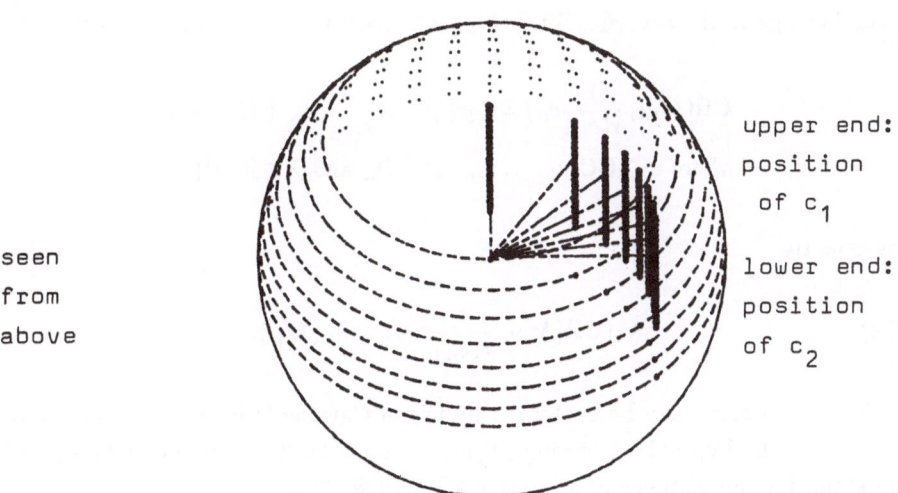

seen
from
above

upper end:
position
of c_1

lower end:
position
of c_2

3.2 AN APPLICATION OF CAVALIERI'S PRINCIPLE

Here we try to estimate the term $V(c_1, \ldots, c_n)$ using the term $W(c_1, \ldots, c_{n-1})$. The result will be the

Lemma 3.1

$$(3.2.1) \qquad V(c_1, \ldots, c_n) \geq \frac{\lambda_n(\Omega_{n-1}) \, h \, |c_n^{n-1} - T| \, W(c_1, \ldots, c_{n-1})}{\lambda_n(\Omega_n) \, nt \, r(c_n)} .$$

Here $r(c_n)$ stands for $\|c_n\|$.

Proof. Remember that V and W are normalized spherical measures of cones. It will be useful to separate the normalizing constants at the beginning of our considerations. Then

$$\lambda_n(\Omega_n)V(c_1, \ldots, c_n) = \lambda_n(CC(c_1, \ldots, c_n) \cap \Omega_n) =: \tilde{V} = \lambda_n(M_V)$$
$$\lambda_n(\Omega_{n-1})W(c_1, \ldots, c_n) = \lambda_n(CC(c_1, \ldots, c_{n-1}) \cap \Omega_n) =: \widetilde{W} = \lambda_{n-1}(M_W) .$$

So we concentrate our considerations on $\tilde{V}$ and $\widetilde{W}$. $\tilde{V}$ is the Lebesgue measure of an n-dimensional set M_V and $\widetilde{W}$ is the Lebesgue measure of the $n-1$-dimensional set M_W. Note that $M_W \subset M_V$. M_W can be regarded as a "facet" of M_V. The point $\frac{1}{r(c_n)}c_n$ belongs to the set M_V. Since M_V is convex, it contains the complete set

$$CH(M_W, \frac{1}{r(c_n)}c_n) = \{y \mid y = \lambda \frac{1}{r(c_n)}c_n + (1-\lambda)z,$$
$$\text{where } z \in CC(c_1, \ldots, c_{n-1}) \cap \Omega_n \text{ and } \lambda \in [0, 1]\}.$$

Consequently

$$(3.2.2) \qquad \lambda_n(CH(M_W, \frac{1}{r(c_n)}c_n)) \leq \lambda_n(M_V).$$

But the first measure can be evaluated by use of Cavalieri's Principle in the following way. Denote the hyperplane through $0, c_1, \ldots, c_{n-1}$ by Ψ. Ψ contains M_W. And let $H(x)$ stand for the distance of a point $x \in \mathbb{R}^n$ to Ψ. Then

$$(3.2.3) \qquad \lambda_n(M_V) = \frac{1}{n}H\left(\frac{1}{r(c_n)}c_n\right)\lambda_{n-1}(M_W).$$

The value of $H(\frac{1}{r(c_n)}c_n)$ can be derived by using elementary geometry (similar triangles). We obtain

(3.2.4)
$$\frac{H(c_n)}{|c_n^{n-1} - T|} = \frac{h}{t} \text{ and } H\left(\frac{1}{r(c_n)}c_n\right) = \frac{1}{r(c_n)}H(c_n).$$

This yields — recalling the normalizing constants —

$$\lambda_n(M_V) = \frac{1}{n}\frac{1}{r(c_n)}\frac{h}{t}\,|c_n^{n-1} - T|\,\lambda_{n-1}(M_W)$$

and

$$V(c_1,\ldots,c_n) \geq \frac{\lambda_n(M_V)}{\lambda_n(\Omega_n)} = \frac{\lambda_{n-1}(\Omega_{n-1})}{\lambda_n(\Omega_n)}\frac{\lambda_{n-1}(M_V)}{\lambda_{n-1}(\Omega_{n-1})}\frac{1}{n}\frac{1}{r(c_n)}\frac{h}{t}|c_n^{n-1} - T|.$$

So the claim of the lemma is true.

$\square$

Our quotient of integrals has the following form

(3.2.5)
$$\frac{E_{m,n}(S)}{E_{m,n}(Z)} \leq \frac{\lambda_n(\Omega_n)n^2 \int_0^t G(h)^{m-n}T^{-1} \int_{\mathbb{R}^{n-1}} |T - c_n^{n-1}|}{\lambda_{n-1}(\Omega_{n-1}) \int_0^t G(h)^{m-n}h(Tt)^{-1} \int_{\mathbb{R}^{n-1}} |T - c_n^{n-1}|^2 \frac{1}{r(c_n)}}$$

$$\left[\frac{\int_{\mathbb{R}^{n-2}} \cdots \int_{\mathbb{R}^{n-2}} W(c_1,\ldots,c_{n-1})(\lambda_{n-1}\{CH(c_1,\ldots,c_{n-1})\})^2}{\int_{\mathbb{R}^{n-2}} \cdots \int_{\mathbb{R}^{n-2}} W(c_1,\ldots,c_{n-1})(\lambda_{n-1}\{CH(c_1,\ldots,c_{n-1})\})^2}\right.$$

$$\left.\frac{f(c_1)\ldots f(c_{n-1})d\bar{c}_1 \ldots d\bar{c}_{n-1}}{f(c_1)\ldots f(c_{n-1})d\bar{c}_1 \ldots d\bar{c}_{n-1}}\right]\frac{f(c_n)d\bar{c}_n\,dh}{f(c_n)d\bar{c}_n\,dh}.$$

The integrals in brackets are identical in the numerator and the denominator and they do not depend in any way on the outer integration variables $\bar{c}_n$ and h. So they can be dropped by reduction of the quotient. What remains, is much more acceptable.

(3.2.6)
$$\frac{E_{m,n}(S)}{E_{m,n}(Z)} \leq \frac{\lambda_n(\Omega_n)n^2 \int_0^t G(h)^{m-n}h(Tt)^{-1}}{\lambda_{n-1}(\Omega_{n-1}) \int_0^t G(h)^{m-n}h(Tt)^{-1}}$$

$$\frac{\int_{\mathbb{R}^{n-1}} |T - c_n^{n-1}|\,f(c_n)d\bar{c}_n\,dh}{\int_{\mathbb{R}^{n-1}} |T - c_n^{n-1}|^2 \frac{1}{r(c_n)}f(c_n)d\bar{c}_n\,dh}.$$

Figure 3.4a

A lower bound for $V(c_1,\ldots,c_n)$ and the application of

Cavalieri's Principle

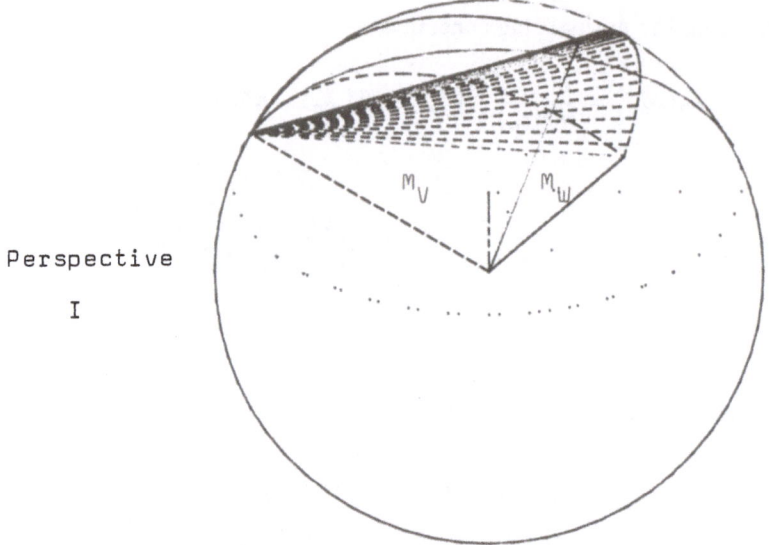

Perspective

I

The spherical triangle generated by $CC(c_1,c_2,c_3)$ is the outside boundary of M_V. It contains the point $c_3 \dfrac{1}{r(c_3)}$.

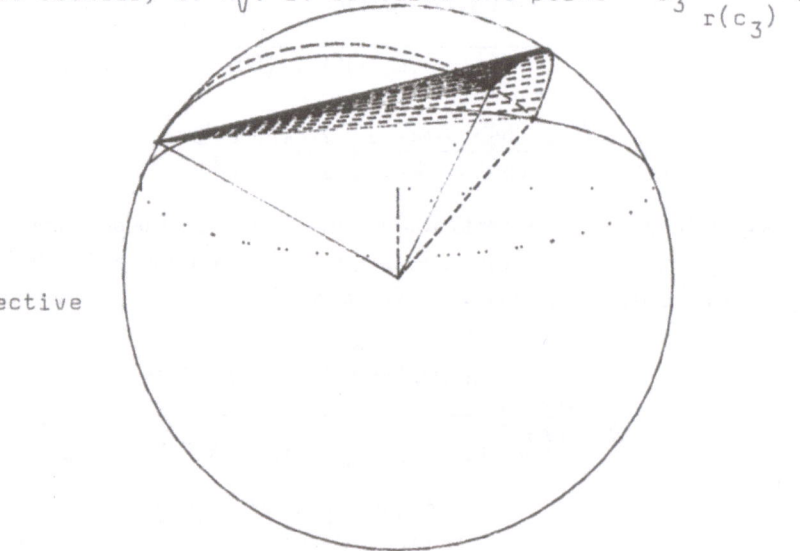

Perspective

II

We connect this point with all points of the arc generated by the side-cone $CC(c_1,c_2)$.

Figure 3.4b

A lower bound for $V(c_1,\ldots,c_n)$ and the application of
Cavalieri's Principle

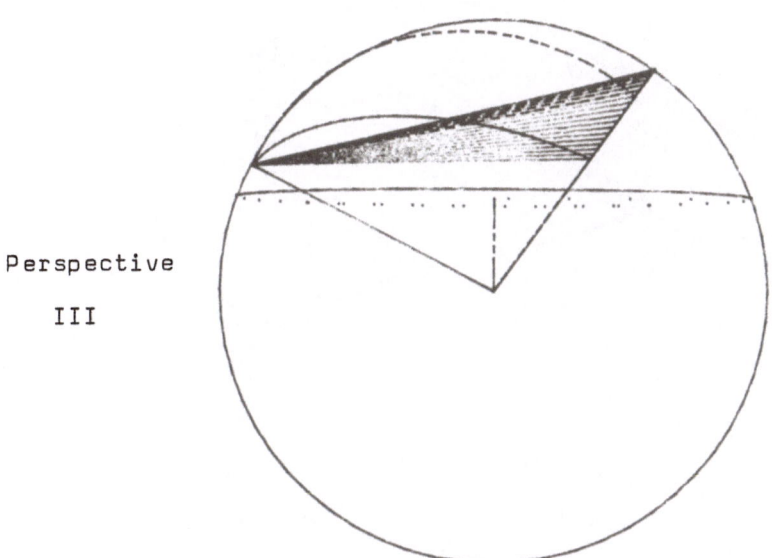

Perspective

III

Since M_V is convex, it contains all these connection lines.

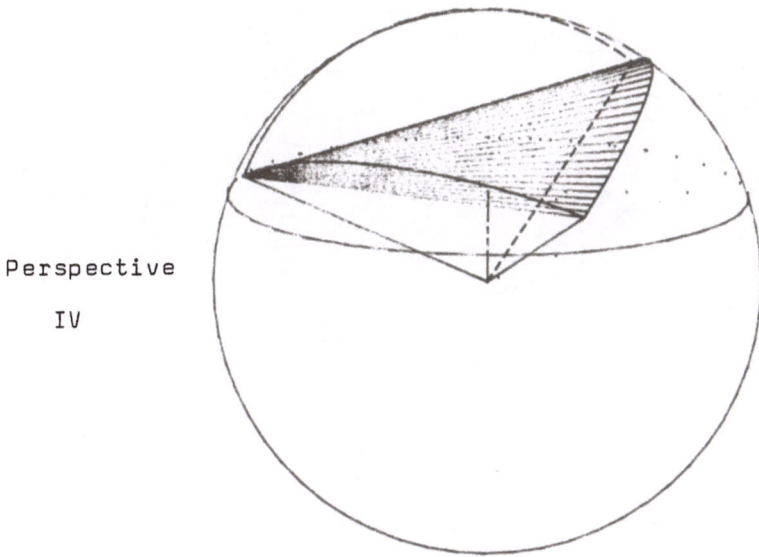

Perspective

IV

The subset of M_V lying below these connection lines can
be used for the derivation of a lower bound.

Figure 3.4c

A lower bound for $V(c_1,\ldots,c_n)$ and the application of

Cavalieri's Principle

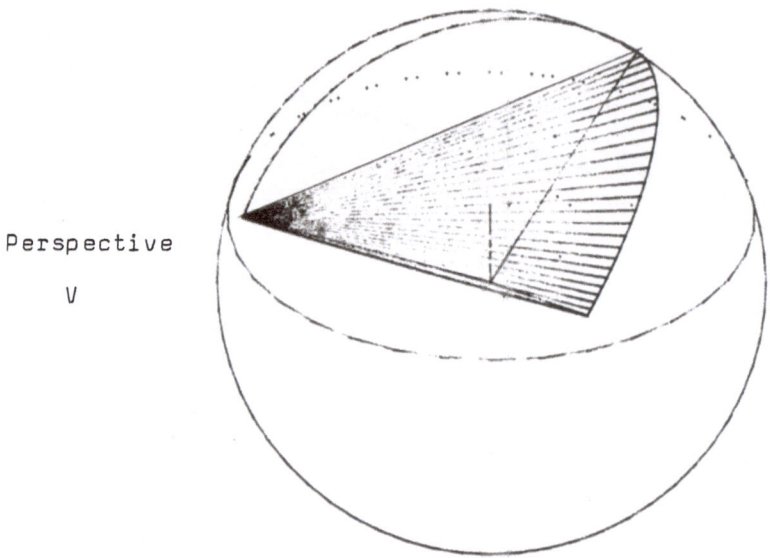

Perspective

V

The subset is the convex hull of a subset of a hyperplane
and one point not belonging to that hyperplane.

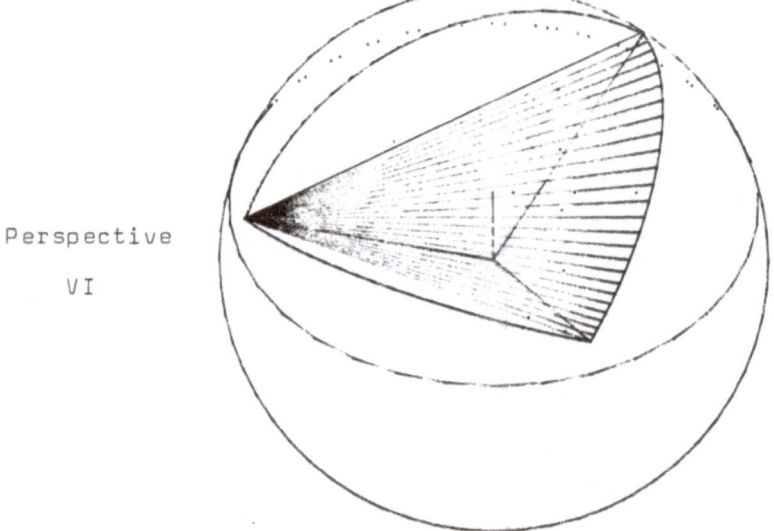

Perspective

VI

So Cavalieri's Principle can be applied.

Figure 3.5a

Illustration and explanation of formula (3.2.4)

The triangles OEF and BAF are similar. So we know that

$$\frac{\overline{BA}}{\overline{BF}} = \frac{\overline{OE}}{\overline{OF}} \quad .$$

Since the triangles OAB and OCD are similar, too, we know

that $\overline{BA} \;\dfrac{\overline{OD}}{\overline{OB}} \; = \overline{DC}$.

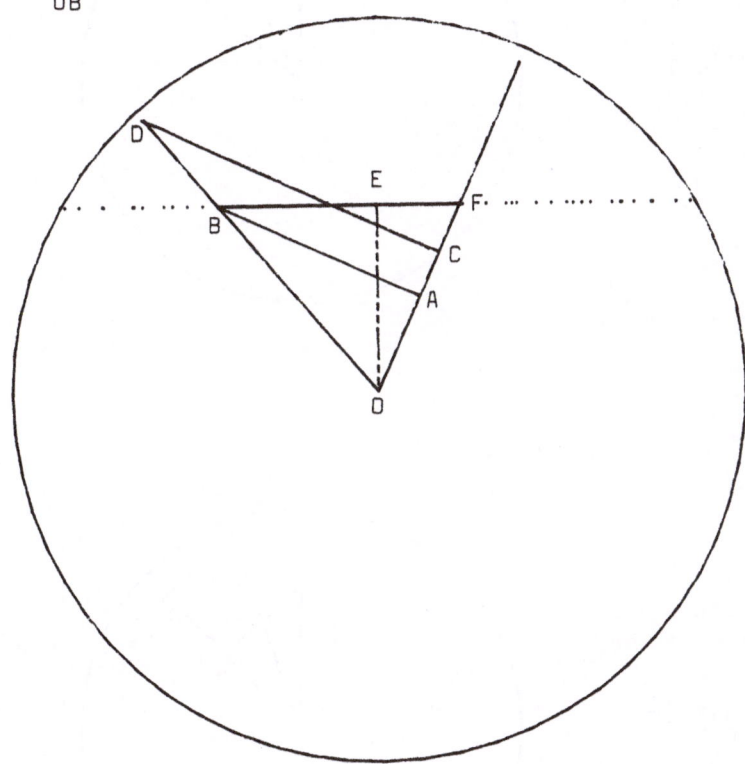

Here $\overline{BE} = -c_n^{n-1}$, $\overline{EF} = T$, $\overline{OE} = h$, $\overline{OF} = t$,

$\overline{BA} = H(c_n)$, $\overline{DC} = H\left(\dfrac{1}{r(c_n)} \; c_n\right)$.

Figure 3.5b

Complementary arrangements

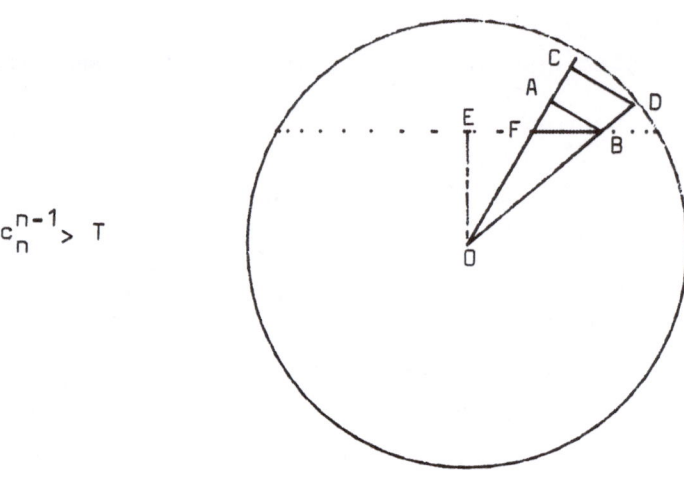

$$c_n^{n-1} > T$$

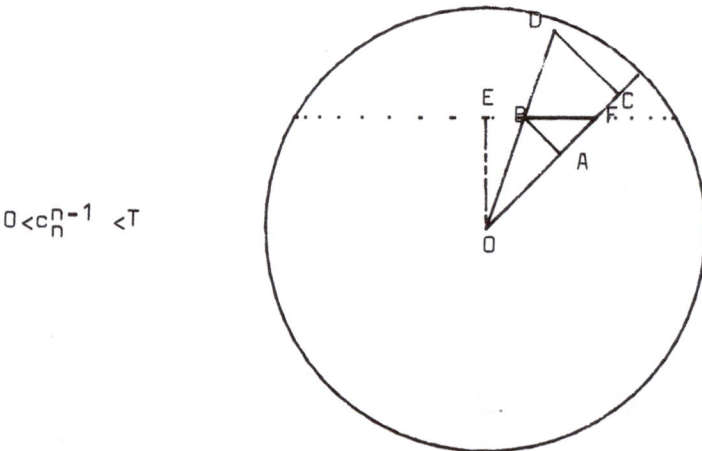

$$0 < c_n^{n-1} < T$$

Note that numerator- and denominator-integral differ only in the term $|T - c_n^{n-1}| \frac{1}{r(c_n)}$, which occurs in the denominator, not in the numerator. So the complete quotient of integrals can be regarded as the expectation value of the variable $\frac{r(c_n)}{|T - c_n^{n-1}|}$ under a certain distribution, determined by the denominator integral.

This is the moment where we should change to polar coordinates, because such an "expectation value" mainly is a result of rotational symmetry of the distribution and f. And polar coordinates are better suited for describing the effects of rotational symmetry.

From now on we use the coordinates

$$r := r(c_n), \; h \text{ and } \gamma(c_n) \in \omega_{n-1}, \text{ such that}$$

(3.2.7) $$c_n = \begin{bmatrix} \sqrt{r^2 - h^2} \; \gamma(c_n) \\ h \end{bmatrix} \quad \text{and} \quad \bar{c}_n = \sqrt{r^2 - h^2} \; \gamma(c_n).$$

For abbreviation we take $R := R(r, h) := \sqrt{r^2 - h^2}$.

Now we exploit four formulae for integration over the set $\Omega_n \cap \{x \mid x^n = h\} \subset \mathbb{R}^n$. These formulae are proven in the Appendix (6.2.31–34).

Lemma 3.2

Let γ denote the surface-integration element for ω_{n-1}. Then for $n \geq 2$

(3.2.8) $$\int_{\omega_{n-1}} d\gamma(x) = \lambda_{n-2}(\omega_{n-1})$$

(3.2.9) $$\int_{\omega_{n-1}} x^{n-1} d\gamma(x) = 0$$

(3.2.10) $$\int_{\omega_{n-1}} |x^{n-1}| d\gamma(x) = \frac{2\lambda_{n-3}(\omega_{n-2})\lambda_{n-2}(\omega_{n-1})}{(n-2)\lambda_{n-2}(\omega_{n-1})}$$

(3.2.11) $$\int_{\omega_{n-1}} (x^{n-1})^2 d\gamma(x) = \frac{1}{n-1}\lambda_{n-2}(\omega_{n-1})$$

if we set $\lambda_0(\omega_1) = 2$.

For a geometrical illustration of these integrations see the attached figure. For fixed $r \; (= r(c_n))$, the denominator contains

$$\int_{\omega_{n-1}(R)} |T - c_n^{n-1}|^2 d\gamma_R(\bar{c}_n),$$

where $\omega_{n-1}(R)$ is the $n-1$-dimensional sphere with radius R and γ_R is the corresponding integration element. Rotational symmetry delivers

$$\int_{\omega_{n-1}(R)} T^2 d\gamma_R(\bar{c}_n) + \int_{\omega_{n-1}(R)} (c_n^{n-1})^2 d\gamma_R(\bar{c}_n) - 2 \int_{\omega_{n-1}(R)} T c_n^{n-1} d\gamma_R(\bar{c}_n) =$$

$$= T^2 \int_{\omega_{n-1}(R)} d\gamma_R(\bar{c}_n) + \int_{\omega_{n-1}(R)} (c_n^{n-1})^2 d\gamma_R(\bar{c}_n) - 0 =$$

$$= T^2 R^{n-2} \lambda_{n-2}(\omega_{n-1}) + R^2 R^{n-2} \frac{1}{n-1} \lambda_{n-2}(\omega_{n-1}) =$$

$$= (T^2 + \frac{1}{n-1} R^2) R^{n-2} \lambda_{n-2}(\omega_{n-1}).$$

In the numerator we have

$$\int_{\omega_{n-1}(R)} |T - c_n^{n-1}| d\gamma_R(\bar{c}_n) =$$

$$= \int_{\substack{\omega_{n-1}(R) \\ c_n^{n-1} < -T}} (R+T) d\gamma_R(\bar{c}_n) + \int_{\substack{\omega_{n-1}(R) \\ T < c_n^{n-1}}} (R-T) d\gamma_R(\bar{c}_n) +$$

$$+ \int_{\substack{\omega_{n-1}(R) \\ -T \le c_n^{n-1} \le 0}} (T+R) d\gamma_R(\bar{c}_n) + \int_{\substack{\omega_{n-1}(R) \\ 0 \le c_n^{n-1} \le T}} (T-R) d\gamma_R(\bar{c}_n) \le$$

$$= \int_{\omega_{n-1}(R)} \max(R,T) d\gamma_R(\bar{c}_n) .$$

In terms of conditional expectations this means for the numerator

(3.2.12) $$E(|T - c_n^{n-1}| \mid \|\bar{c}_n\| = R) \le \max(T, R)$$

and for the denominator

(3.2.13) $$E(|T - c_n^{n-1}|^2 \mid \|\bar{c}_n\| = R) = T^2 + \frac{1}{n-1} R^2 =: T^2 + \nu R^2 \quad \text{with} \quad \nu = \frac{1}{n-1}.$$

Now we have a new upper bound for the quotient.

(3.2.14) $$\frac{E_{m,n}(S)}{E_{m,n}(Z)} = \frac{\lambda_n(\Omega_n) n^2}{\lambda_{n-1}(\Omega_{n-1})}$$

$$\frac{\int_0^t G(h)^{m-n} T^{-1} \int_h^1 R^{n-3} r^{-n+2} \max(T,R) dF(r) dh}{\int_0^t G(h)^{m-n} h(Tt)^{-1} \int_h^1 R^{n-3} r^{-n+2-1} (T^2 + \nu R^2) dF(r) dh} .$$

Figure 3.6a

<u>Averaging over c_n when c_1, c_2, h, R are all fixed</u>

First case: R is less than T $(r < t)$

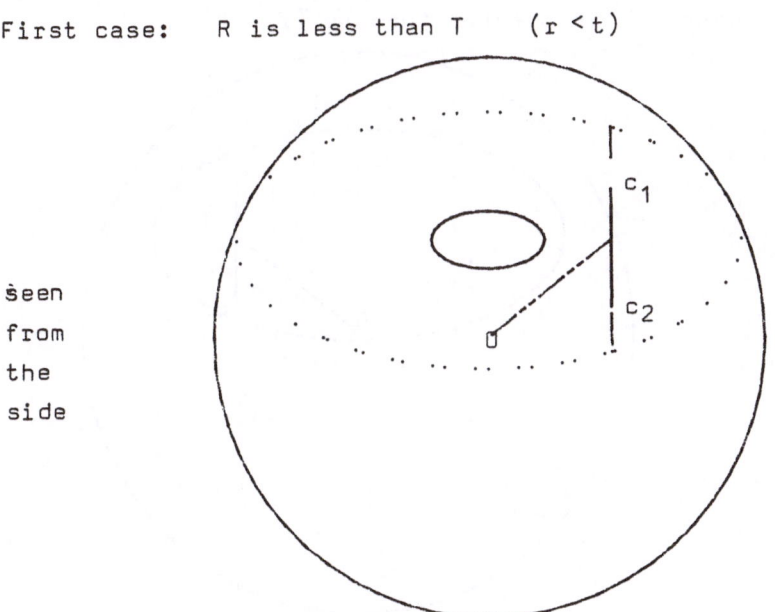

seen
from
the
side

The inner circle shows the possible locations for c_3.

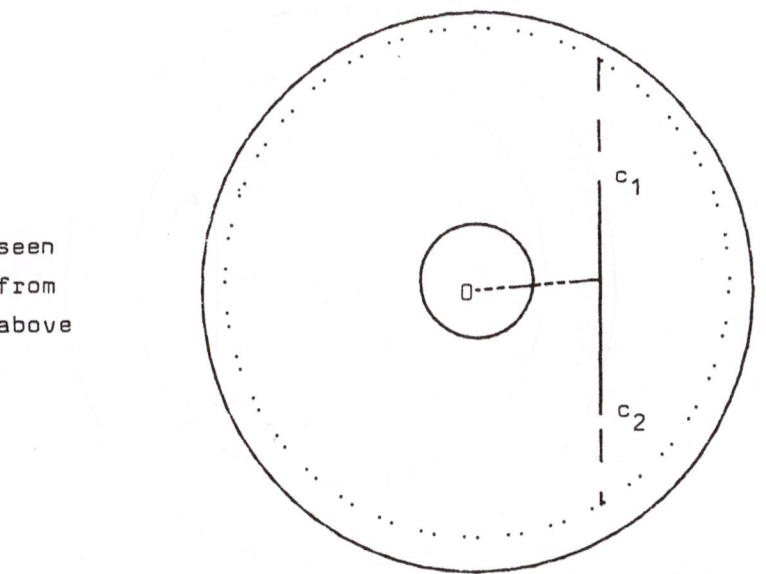

seen
from
above

Figure 3.6b

Averaging over c_n when c_1, c_2, h, R are all fixed

Second case: R is greater than T $(r > t)$

seen
from
the
side

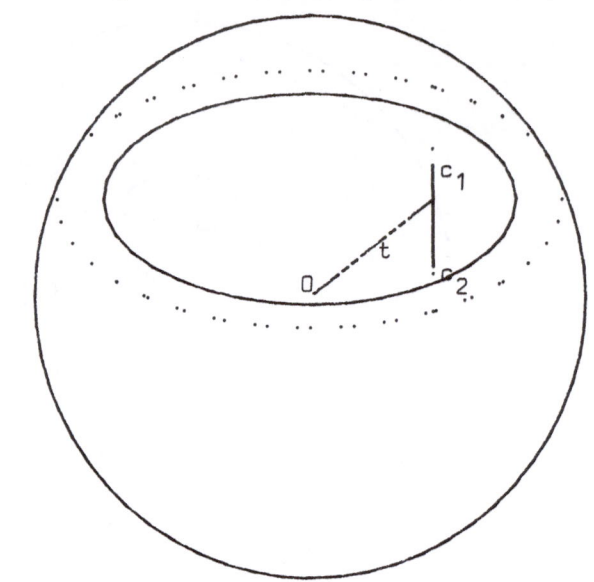

The inner circle shows the possible locations for c_3.

seen
from
above

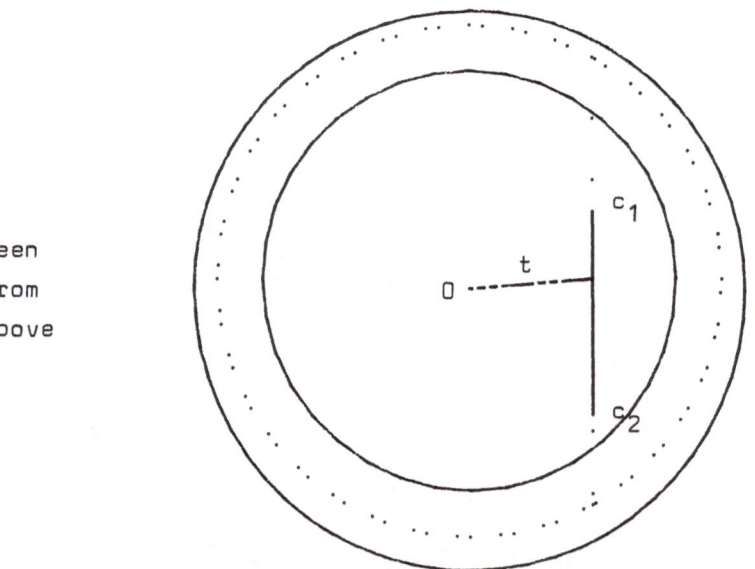

A very pessimistic estimation yields (this and (3.2.13) may be the reason for a factor $n^{\frac{1}{2}}$)

(3.2.15)
$$T^2 + \nu R^2 \geq \nu \max(T^2, R^2).$$

with ν defined as above.

So our bound attains the form

(3.2.16)
$$\frac{E_{m,n}(S)}{E_{m,n}(Z)} \leq \frac{\lambda_n(\Omega_n)n^2(n-1)}{\lambda_{n-1}(\omega_{n-1})} \cdot$$
$$\cdot \frac{\int\limits_0^t G(h)^{m-n}T^{-1} \int\limits_h^1 R^{n-3}r^{-n+2} \max(T, R)dF(r)dh}{\int\limits_0^t G(h)^{m-n}h(Tt)^{-1} \int\limits_h^1 R^{n-3}r^{-n+1} \max(T^2, R^2)dF(r)dh}.$$

Let us denote the quotient outside the integrals by $C(n)$. For technical reasons we define

(3.2.17)
$$\widehat{G}(h) := (G(h) - \frac{1}{2}) \cdot 2 \quad \text{for} \quad h \in [0,1].$$

So we have $\widehat{G}(0) = 0$ and $\widehat{G}(1) = 1$. The use of $\widehat{G}(h)$ will help us to avoid difficult case studies.

Now we know that

$$G(h)^{m-n} = \sum_{k=0}^{m-n} \widehat{G}(h)^k \frac{1}{2^n} \binom{m-n}{k}.$$

Let $k \in \{0, \ldots, m-n\}$ be that value of l which maximizes

(3.2.18)
$$\frac{\int\limits_0^t \widehat{G}(h)^l T^{-1} \int\limits_h^1 R^{n-3}r^{-n+2} \max(T, R)dF(r)dh}{\int\limits_0^t \widehat{G}(h)^l h(Tt)^{-1} \int\limits_h^1 R^{n-3}r^{-n+2-1} \max(T^2, R^2)dF(r)dh}.$$

In explicit form this means

(3.2.19)
$$\frac{E_{m,n}(S)}{E_{m,n}(Z)} \leq C(n)\frac{\begin{array}{c}\int\limits_0^t \int\limits_0^r \widehat{G}(h)^k T^{-1} R^{n-3}r^{-n+2}T \, dh \, dF(r) + \\[6pt] \int\limits_0^t \int\limits_0^r \widehat{G}(h)^k h(Tr)^{-1} R^{n-3}r^{-n+2}T^2 t^{-1} \, dh \, dF(r) + \\[6pt] + \int\limits_t^1 \int\limits_0^t \widehat{G}(h)^k T^{-1} R^{n-3}r^{-n+2}R \, dh \, dF(r)\end{array}}{\begin{array}{c} \\[6pt] \\[6pt] + \int\limits_t^1 \int\limits_0^t \widehat{G}(h)^k h(Tt)^{-1} R^{n-3}r^{-n+2}R^2 r^{-1} \, dh \, dF(r)\end{array}}.$$

Figure 3.8 shows why we prefer to change the sequence of integrations. It would be very convenient to have a RDF with $F(r) = 0$ for all $r \leq t$, because we could drop the first terms in numerator and denominator. For that reason we try to estimate the quotient given above by an corresponding quotient based on a RDF with this nice property.

But before doing that, we should split our quotient into two factors $Q_1(F)$, $Q_2(F)$, where F denotes the RDF under consideration.

$$\frac{E_{m,n}(S)}{E_{m,n}(Z)} = C(n)Q_1(F)Q_2(F) \text{ where}$$

$$Q_1(F) = \frac{\int\limits_0^t \int\limits_0^r \widehat{G}(h)^k T^{-1} R^{n-3} r^{-n+2} T^2 t^{-1} dh\, dF(r) +}{\int\limits_0^t \int\limits_0^r \widehat{G}(h)^k h(Tr)^{-1} R^{n-3} r^{-n+2} T^2 t^{-1} dh\, dF(r) +}$$

$$\frac{+ \int\limits_t^1 \int\limits_0^r \widehat{G}(h)^k T^{-1} R^{n-3} r^{-n+2} R^2 r^{-1} dh\, dF(r)}{+ \int\limits_t^1 \int\limits_0^t \widehat{G}(h)^k h(Tt)^{-1} R^{n-3} r^{-n+2} R^2 r^{-1} dh\, dF(r)} \text{ and}$$

(3.2.20)

$$Q_2(F) = \frac{\int\limits_0^t \int\limits_0^r \widehat{G}(h)^k T^{-1} R^{n-3} r^{-n+2} T\, dh\, dF(r) +}{\int\limits_0^t \int\limits_0^r \widehat{G}(h)^k T^{-1} R^{n-3} r^{-n+2} T^2 t^{-1} dh\, dF(r) +}$$

$$\frac{+ \int\limits_t^1 \int\limits_0^t \widehat{G}(h)^k T^{-1} R^{n-3} r^{-n+2} R\, dh\, dF(r)}{+ \int\limits_t^1 \int\limits_0^t \widehat{G}(h)^k T^{-1} R^{n-3} r^{-n+2} R^2 r^{-1} dh\, dF(r)}.$$

In the next section we shall estimate these factors separately.

Figure 3.7

Illustration of $G(h)$ and $\hat{G}(h)$

$G(h)$ gives the probability of the region below the hyperplane
H relative to the complete space.

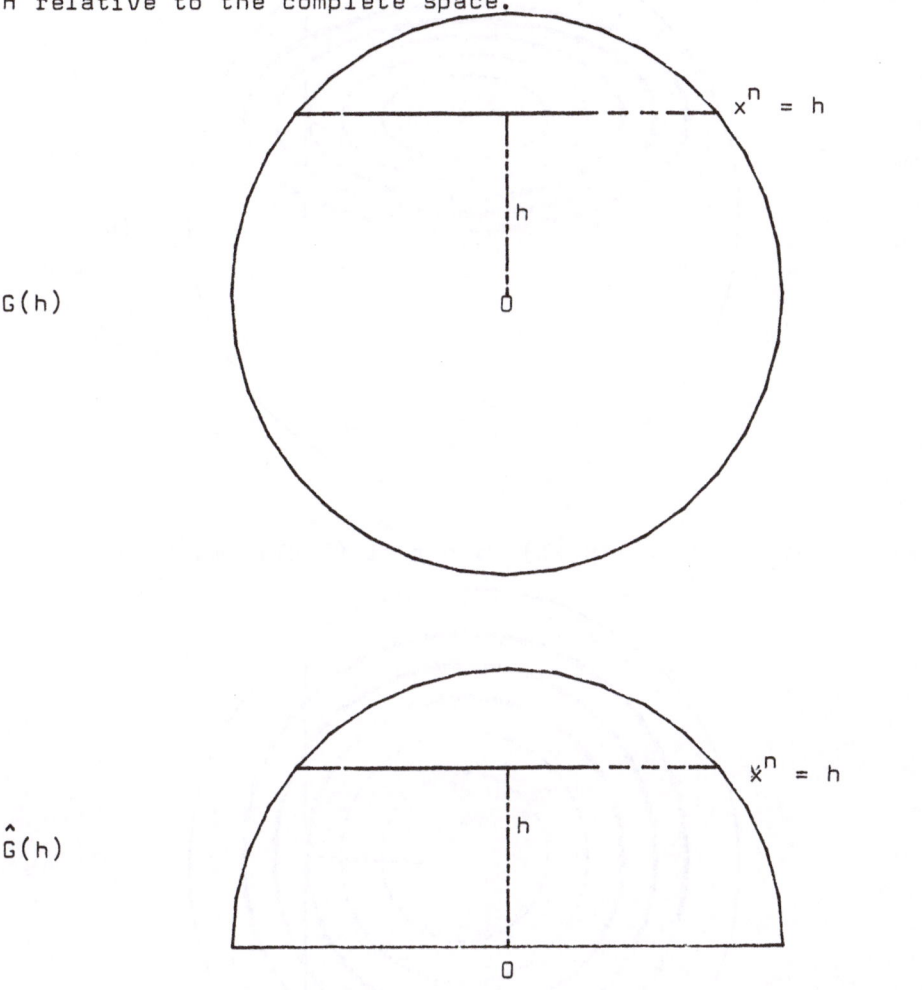

$\hat{G}(h)$ gives the probability of the region below H but with
positive x^n relative to the upper (x^n greater 0) halfspace.

Figure 3.8a

The order of integrations

If we integrate over h with priority (outside integral),

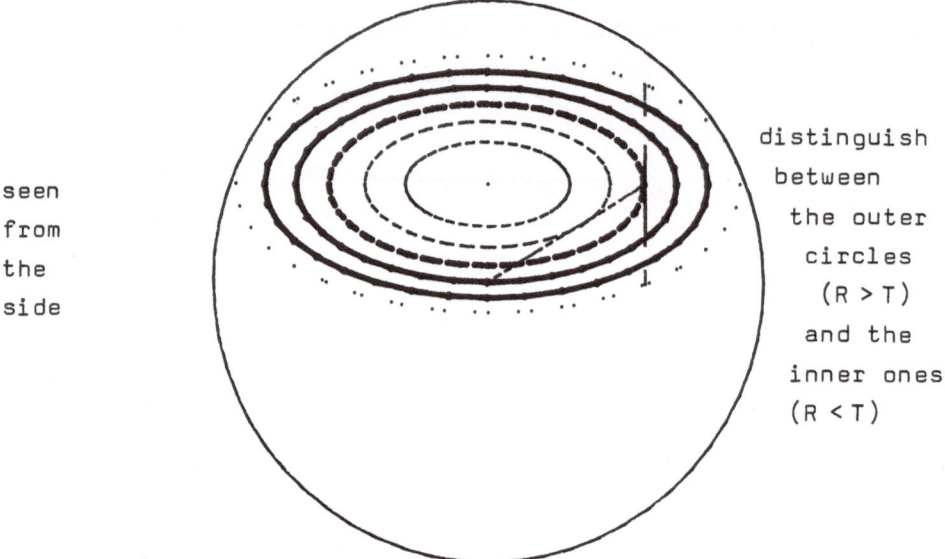

seen
from
the
side

distinguish
between
the outer
circles
(R > T)
and the
inner ones
(R < T)

then the cases r >t (R >T) and r <t (R <T) must be

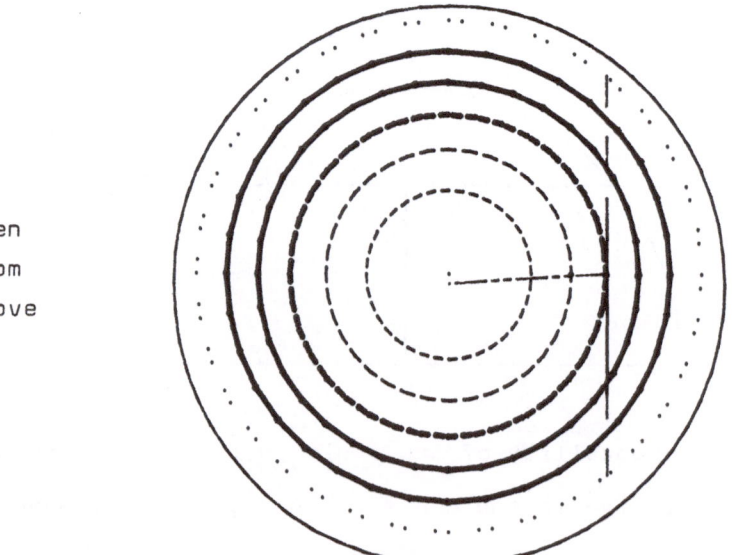

seen
from
above

treated separately.

Figure 3.8b

The order of integrations

If we integrate over r with priority, then the integrand
is simpler.

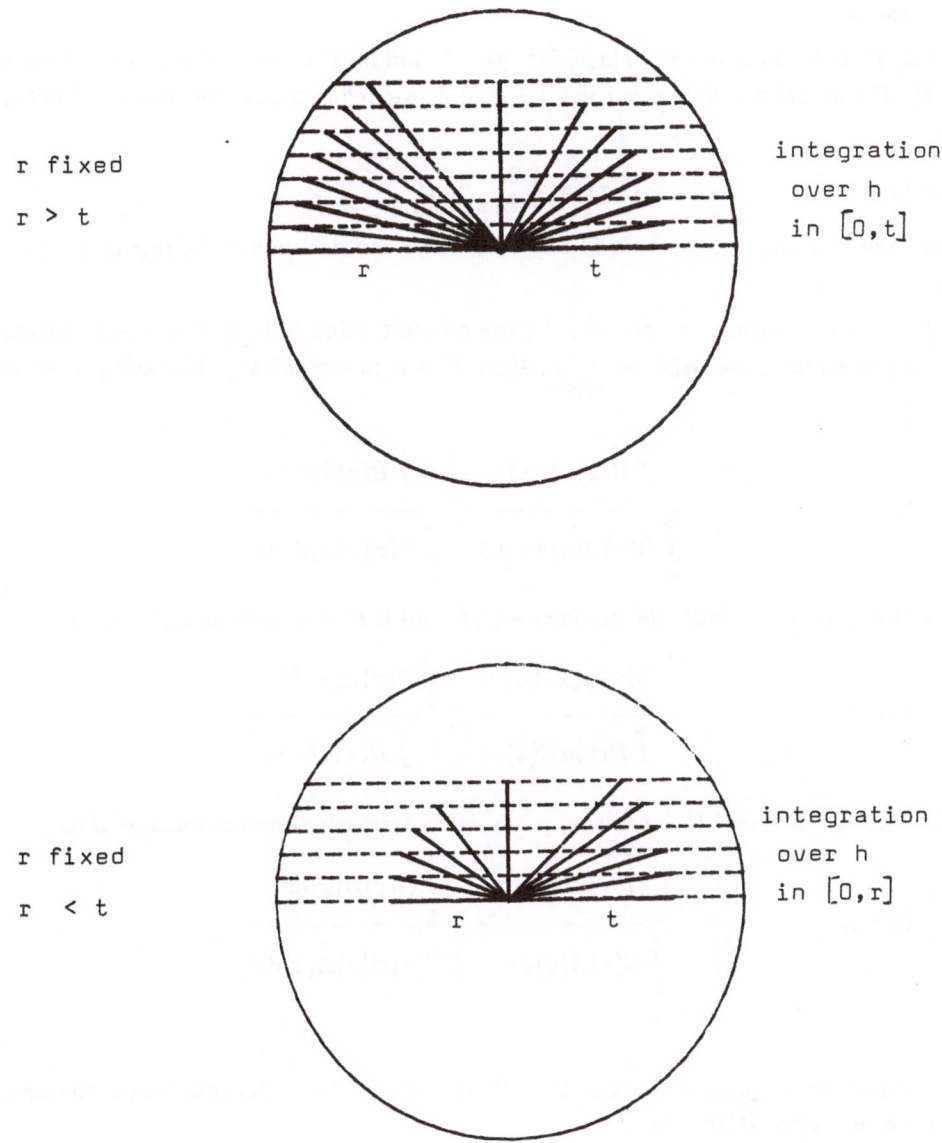

r fixed

r > t

integration
over h
in [0,t]

r fixed

r < t

integration
over h
in [0,r]

It would be convenient if the second case could be avoided.

3.3 THE INFLUENCE OF THE DISTRIBUTION

As announced, we want to continue our estimations in this section. The aim is to do without distributions which require maximum-distinctions. We hope to be able to concentrate on distributions with $F(r) = 0$ for all $r < t$. For this purpose we apply the following lemma, which will be proven in the Appendix (Lemma 6.2).

Lemma 3.3

Let A, B, C be functions from $[a, b]$ into $[0, \infty)$, and let E be a distribution function on $[a, b]$. For an arbitrarily chosen point $\overline{x} \in [a, b]$ we define a new distribution function by

$$(3.3.1) \qquad \overline{E}(x) := \begin{cases} 0 & \text{for } x < \overline{x} \\ E(x) & \text{for } x \leq \overline{x} \end{cases} .$$

Then the following results hold in any case where the denominator-integrals are positive.

1) Let A be continuous and let $\overline{x}$ be the greatest value in $[a, b]$ where $A(x)$ attains its minimum, i. e. $A(\overline{x}) = \min\limits_{x \in [a,b]} A(x)$. If B is monotonically increasing, then we have

$$(3.3.2) \qquad \frac{\int\limits_a^b B(x)dE(x)}{\int\limits_a^b B(x)A(x)dE(x)} \leq \frac{\int\limits_a^b B(x)d\overline{E}(x)}{\int\limits_a^b B(x)A(x)d\overline{E}(x)} .$$

2) If A is monotonically decreasing on $[a, b]$ and if B decreases on $[a, \overline{x}]$, then

$$(3.3.3) \qquad \frac{\int\limits_a^b B(x)A(x)dE(x)}{\int\limits_a^b B(x)dE(x)} \geq \frac{\int\limits_a^b B(x)A(x)d\overline{E}(x)}{\int\limits_a^b B(x)d\overline{E}(x)} .$$

3) If A and B are both increasing (resp. both decreasing) monotonically, then

$$(3.3.4) \qquad \frac{\int\limits_a^b C(x)dx}{\int\limits_a^b C(x)A(x)dx} \geq \frac{\int\limits_a^b C(x)B(x)dx}{\int\limits_a^b C(x)B(x)A(x)dx} .$$

$\square$

Again we are going to exploit the rule (3.1.10). For this purpose we define single-point or extremal RDF's by

$$(3.3.5) \qquad F_{r_1} : [0, 1] \to [0, 1] \quad \text{with} \quad F_{r_1}(r) = \begin{cases} 1 & \text{for } r \geq r_1 \\ 0 & \text{for } r < r_1 \end{cases} .$$

Now we are able to derive an upper bound for $Q_1(F)$.

Proposition 1

$$(3.3.6) \qquad Q_1(F) \leq \frac{\int\limits_0^t T^{n-2} t^{-n+1} dh}{\int\limits_0^t h T^{n-2} t^{-1-n+1} dh}.$$

Proof. From (3.1.10) we know that $Q_1(F)$ (compare 3.2.20) cannot be greater than the maximal quotient which results from using extremal RDF's as defined in (3.3.5).

Applying (3.3.4) we can prove that for fixed $r \leq t$

$$(3.3.7) \qquad \frac{\int\limits_0^r \widehat{G}(h)^k T^{-1} R^{n-3} r^{-n+2} T^2 t^{-1} dh}{\int\limits_0^r \widehat{G}(h)^k h(Tr)^{-1} R^{n-3} r^{-n+2} T^2 t^{-1} dh} \leq \frac{\int\limits_0^r R^{n-2} r^{-n+1} dh}{\int\limits_0^r h r^{-1} R^{n-2} r^{-n+1} dh} =$$

$$= \frac{\int\limits_0^t T^{n-2} t^{-n+1} dh}{\int\limits_0^t h T^{n-2} t^{-n} dh}.$$

This is possible if we set

$$A(h) := h r^{-1}, \quad C(h) := R^{n-2} r^{-n+1} \quad \text{and} \quad B(h) := \widehat{G}(h)^k T t^{-1} R^{-1} r,$$

because A and B are increasing with h.

Analogously, we prove for $r \geq t$

$$\frac{\int\limits_0^t \widehat{G}(h)^k T^{-1} R^{n-3} r^{-n+2} R^2 r^{-1} dh}{\int\limits_0^t \widehat{G}(h)^k h(Tt)^{-1} R^{n-3} r^{-n+2} R^2 r^{-1} dh} \leq \frac{\int\limits_0^t T^{n-2} t^{-n+1} dh}{\int\limits_0^t h t^{-1} T^{n-2} t^{-n+1} dh}.$$

Here we set $A := h t^{-1}$, $C := T^{n-2} t^{-n+1}$ and $B(h) := \widehat{G}(h)^k R^{n-1} T^{-n+1} r^{-n+1} t^{n-1}$. Again, A and B are both monotonically increasing with h. Now we have a rather simple estimation for $Q_1(F)$.

$\square$

For the estimation of $Q_2(F)$ we must be more careful.

Proposition 2

For any given RDF F there is a RDF $\overline{F}$ such that $\overline{F}(r) = 0$ for all $r < t$ and

$$(3.3.8) \qquad Q_2(F) \leq \frac{\int\limits_0^t \widehat{G}_{\overline{F}}(h)^k T^{-1} \int\limits_h^1 R^{n-2} r^{-n+2} d\overline{F}(r) dh}{\int\limits_0^t \widehat{G}_{\overline{F}}(h)^k [1 - G_{\overline{F}}(h)]^{1/(n-1)} T^{-1} \int\limits_h^1 R^{n-2} r^{-n+2} d\overline{F}(r) dh}.$$

Proof. In order to save the reader's effort, we repeat the definition of $Q_2(F)$

$$Q_2(F) = \frac{\begin{aligned}&\int\limits_0^t \int\limits_0^r \widehat{G}_F(h)^k T^{-1} R^{n-3} r^{-n+2} T dh dF(r) + \\ &\int\limits_0^t \int\limits_0^r \widehat{G}_F(h)^k T^{-1} R^{n-3} r^{-n+2} T^2 t^{-1} dh dF(r) + \end{aligned}}{\begin{aligned}&+ \int\limits_t^1 \int\limits_0^t \widehat{G}_F(h)^k T^{-1} R^{n-3} r^{-n+2} R dh dF(r) \\ &+ \int\limits_t^1 \int\limits_0^t \widehat{G}_F(h)^k T^{-1} R^{n-3} r^{-n+2} R^2 r^{-1} dh dF(r) \end{aligned}}.$$

In the proof of the proposition we shall perform four steps. We are trying to replace all terms of $Q_2(F)$, which refer to the given RDF F, by the corresponding term in $\overline{F}$. This has to be done very carefully. The replacements will be made successively.

1st Step

We want to apply (3.3.3). For this purpose we represent $Q_2(F)$ in the following way

$$Q_2(F) = \frac{\int\limits_0^1 B(r) dF(r)}{\int\limits_0^1 B(r) A(r) dF(r)}.$$

Here we have

$$B(r) := \begin{cases} \int\limits_0^r \widehat{G}_F(h)^k T^{-1} R^{n-3} r^{-n+2} T dh & \text{for } r < t \\ \int\limits_0^t \widehat{G}_F(h)^k T^{-1} R^{n-3} r^{-n+2} R dh & \text{for } r \geq t \end{cases}$$

and

$$A(r) := \begin{cases} \dfrac{\int\limits_0^r \widehat{G}_F(h)^k T^{-1} R^{n-3} r^{-n+2} T^2 t^{-1} dh}{B(r)} & \text{for } r < t \\[4ex] \dfrac{\int\limits_0^t \widehat{G}_F(h)^k T^{-1} R^{n-3} r^{-n+2} R^2 r^{-1} dh}{B(r)} & \text{for } r \geq t \end{cases}.$$

Now let $\bar{r}$ be the greatest value in $(0, 1]$ where $A(r)$ attains its minimum, i. e.,

$$A(\bar{r}) = \min_{r \in [0,1]} A(r).$$

And we define

$$\overline{F}(r) := \begin{cases} 0 & \text{for } r < \bar{r} \\ F(r) & \text{for } r \geq \bar{r} \end{cases}.$$

Now all the conditions of (3.3.2) are satisfied, because $B(r)$ is increasing monotonically. This can be shown immediately. In $[0, t]$ we have

$$\int_0^r \widehat{G}_F(h)^k R^{n-3} r^{-n+2} dh = \int_0^1 \widehat{G}_F(qr)^k \sqrt{1-q^2}^{n-3} dq.$$

This expression increases with r since $\widehat{G}_F$ is increasing monotonically.

In $[t, 1]$ it is sufficient that $R^{n-2} r^{-n+2}$ is increasing with r, because $\widehat{G}_F^k$ and T do not depend on r. So we know

$$Q_2(F) \leq \frac{\int_0^1 B(r) d\overline{F}(r)}{\int_0^1 B(r) A(r) d\overline{F}(r)} \qquad \text{according to } (3.3.2),$$

and the first replacement is done.

2nd Step

Now we want to show that the point $\bar{r}$ (where the quotient $A(r)$ attains its minimum) satisfies $\bar{r} \geq t$. In the end this will enable us to drop the integral part with $r < t$. For fixed $r \leq t$ we have

$$\frac{\int_0^r \widehat{G}_F(h)^k T^{-1} R^{n-3} r^{-n+2} T^2 t^{-1} dh}{\int_0^r \widehat{G}_F(h)^k T^{-1} R^{n-3} r^{-n+2} T dh} \geq \frac{\int_0^t \widehat{G}_F(h)^k T^{-1} T^{n-3} t^{-n+2} T^2 t^{-1} dh}{\int_0^t \widehat{G}_F(h)^k T^{-1} T^{n-3} t^{-n+2} T dh}.$$

For the proof we apply (3.3.4) in the equivalent form

$$\frac{\int_a^b C(x) B(x) A(x) dx}{\int_a^b C(x) B(x) dx} \geq \frac{\int_a^b C(x) A(x) dx}{\int_a^b C(x) dx}$$

with

$$A(h) := Tt^{-1}$$

$$B(h) := \begin{cases} R^{n-3}T^{-n+3}r^{-n+2}t^{n-2} & \text{for } h \le r \\ 0 & \text{for } h > r \end{cases}$$

$$C(h) := \widehat{G}_F(h)^k T^{-1} T^{n-3} t^{-n+2} T \ .$$

Then A is decreasing on $[0, t]$ and $B(h)$ behaves as RT^{-1} first.

Because of $r \le t$ we know that $B(h)$ is decreasing on $[0, r]$. On $[r, t]$ we have $B(h) = 0$, so B is monotonically decreasing on $[0, t]$. Now (3.3.4) can be applied.

Consequently we have $\bar{r} \ge t$ and $\overline{F}$ (as defined in Step 1) does not allow values less than t for r. So we conclude

$$Q_2(F) \le \frac{\int\limits_0^1 \int\limits_0^t \widehat{G}_F(h)^k T^{-1} R^{n-3} r^{-n+2} R \, dh \, d\overline{F}(r)}{\int\limits_0^1 \int\limits_0^t \widehat{G}_F(h)^k T^{-1} R^{n-3} r^{-n+2} R^2 r^{-1} \, dh \, d\overline{F}(r)} \ .$$

The fact that the first term in the sums of numerator and denominator have been dropped, makes evaluability more likely. But still we have to pay attention to the factor $\widehat{G}_F^k$. In the following we try to replace it by the corresponding term based on $\overline{F}$.

3rd Step

After having replaced the weighting-function of the outer integration $(dF(r))$, we are allowed to return to the original sequence of integrations. We obtain

$$Q_2(F) \le \frac{\int\limits_0^t \widehat{G}_F(h)^k T^{-1} \int\limits_h^1 R^{n-2} r^{-n+2} \, d\overline{F}(r) \, dh}{\int\limits_0^t \widehat{G}_F(h)^k T^{-1} \int\limits_h^1 R^{n-1} r^{-n+1} \, d\overline{F}(r) \, dh} \ .$$

For the inner denominator-integral there is an interesting estimation

$$\int\limits_h^1 R^{n-1} r^{-n+1} \, d\overline{F}(r) = \int\limits_h^1 \int\limits_{\frac{h}{r}}^1 (1 - \sigma^2)^{(n-3)/2} \sigma \, d\sigma \, d\overline{F}(r)(n-1) \ge$$

$$\ge \int\limits_h^1 \int\limits_{\frac{h}{r}}^1 (1 - \sigma^2)^{(n-3)/2} \, d\sigma \, d\overline{F}(r) \, 2 \frac{\lambda_{n-2}(\omega_{n-1})}{\lambda_{n-1}(\omega_n)} = 1 - \widehat{G}_F(h) \ .$$

The first equation results from pure integration, the second is based on the definition of $G(h)$ in Chapter 2. Now consider the inequality. We know that

$$\frac{\int\limits_{\frac{h}{r}}^{1}(1-\sigma^2)^{(n-3)/2}\sigma d\sigma}{\int\limits_{\frac{h}{r}}^{1}(1-\sigma^2)^{(n-3)/2}d\sigma} \qquad \text{increases with } h.$$

So the minimal quotient will be attained for $h = 0$. Hence the quotient can be underestimated by

$$\frac{\int\limits_{0}^{1}(1-\sigma^2)^{(n-3)/2}\sigma d\sigma}{\int\limits_{0}^{1}(1-\sigma^2)^{(n-3)/2}d\sigma}.$$

Here the numerator is exactly $\frac{1}{n-1}$, the value of the denominator has a geometrical meaning

$$\int\limits_{0}^{1}(1-\sigma^2)^{(n-3)/2}d\sigma = \frac{\lambda_{n-1}(\omega_n)}{2\lambda_{n-2}(\omega_{n-1})} \qquad \text{(see Appendix 2.17).}$$

After integration over r we have

$$\frac{\int\limits_{h}^{1}\int\limits_{\frac{h}{r}}^{1}(1-\sigma^2)^{(n-3)/2}\sigma d\sigma d\overline{F}(r)}{\int\limits_{h}^{1}\int\limits_{\frac{h}{r}}^{1}(1-\sigma^2)^{(n-3)/2}d\sigma d\overline{F}(r)} \geq \frac{2\lambda_{n-2}(\omega_{n-1})}{(n-1)\lambda_{n-1}(\omega_n)}.$$

which justifies the inequality above.

Now consider the quotient

$$\frac{\int\limits_{h}^{1}R^{n-1}r^{-n+1}d\overline{F}(r)}{\int\limits_{h}^{1}R^{n-2}r^{-n+2}d\overline{F}(r)}.$$

The concavity of the function $f(x) = x^{\frac{n-2}{n-1}}$ for $x > 0$ yields

$$\int_h^1 R^{n-2} r^{-n+2} d\overline{F}(r) \leq$$

$$\leq \left[\int_h^1 R^{n-1} r^{-n+1} d\overline{F}(r)\right]^{(n-2)/(n-1)} \left[\int_h^1 d\overline{F}(r)\right]^{1/(n-1)} \leq$$

$$\leq \left[\int_h^1 R^{n-1} r^{-n+1} d\overline{F}(r)\right]^{(n-2)/(n-1)},$$

because $\overline{F}$ is a distribution function. We conclude that

$$\frac{\int_0^1 R^{n-1} r^{-n+1} d\overline{F}(r)}{\int_0^1 R^{n-2} r^{-n+2} d\overline{F}(r)} \geq \left[\int_0^1 R^{n-1} r^{-n+1} d\overline{F}(r)\right]^{1/(n-1)} \geq \left[1 - \widehat{G}_{\overline{F}}(h)\right]^{1/(n-1)}.$$

Our new estimation is

$$Q_2(F) \leq \frac{\int_0^t \widehat{G}_F(h)^k T^{-1} \int_h^1 R^{n-2} r^{-n+2} d\overline{F}(r) dh}{\int_0^t \widehat{G}_F(h)^k \left[1 - \widehat{G}_{\overline{F}}(h)\right]^{1/(n-1)} T^{-1} \int_h^1 R^{n-2} r^{-n+2} d\overline{F}(r) dh}$$

Here we have the chance to replace the remaining index F by application of Lemma 3.3.

4th Step

Now we want to show that the quotient of the last estimation for $Q_2(F)$ does not decrease when we replace $\widehat{G}_F(h)^k$ by $\widehat{G}_{\overline{F}}(h)^k$. For this purpose we remember that $\widehat{G}_F$ resp. $\widehat{G}_{\overline{F}}$ can be represented as convex combinations of extremal RDF's (single-radius-distributions). Let us use the abbreviation $C_2(n)$ for

$$\frac{2\lambda_{n-2}(\omega_{n-1})}{\lambda_{n-1}(\omega_n)}.$$

We know that

$$\widehat{G}_F(h) = C_2(n) \left[\int_0^h \int_0^r (r^2 - \sigma^2)^{(n-3)/2} r^{-n+2} d\sigma dF(r) + \right.$$

$$\left. + \int_h^1 \int_0^h (r^2 - \sigma^2)^{(n-3)/2} r^{-n+2} d\sigma dF(r)\right]$$

and that $G_{\overline{F}}(h)$ has a corresponding representation. The marginal distribution function of an extremal distribution for radius r can be written as

$$\widehat{G}_r(h) := \begin{cases} 1 & \text{for } r \leq h \\ C_2(n) \int\limits_0^h (r^2 - \sigma^2)^{(n-3)/2} r^{-n+2} d\sigma & \text{for } r \geq h. \end{cases}$$

Consider the "normalized" function

$$\widehat{G}_1(h) := \begin{cases} 1 & \text{for } 1 \leq h \\ C_2(n) \int\limits_0^h (1 - \sigma^2)^{(n-3)/2} d\sigma & \text{for } 1 \geq h. \end{cases}$$

Let $r_1 < r_2$, then $\dfrac{\widehat{G}_{r_1}(h)}{\widehat{G}_{r_2}(h)}$ decreases on $[0, r_1]$ monotonically, since $\dfrac{(r_1^2 - \sigma^2)}{(r_2^2 - \sigma^2)}$ decreases with σ on $[0, r_1]$. On $[r_1, r_2]$ the function $\widehat{G}_{r_2}$ is still increasing, while $\widehat{G}_{r_1} = 1$, and on $[r_2, 1]$ the quotient is constant $(= 1)$. So we know that for $h_1 < h_2$

$$\frac{\widehat{G}_{r_1}(h_1)}{\widehat{G}_{r_2}(h_1)} \geq \frac{\widehat{G}_{r_1}(h_2)}{\widehat{G}_{r_2}(h_2)} , \quad \text{which is equivalent to}$$

$$\frac{\widehat{G}_{r_1}(h_1)}{\widehat{G}_{r_1}(h_2)} \geq \frac{\widehat{G}_{r_2}(h_1)}{\widehat{G}_{r_2}(h_2)} , \quad \text{and shows that the quotient}$$

$$\frac{\widehat{G}_r(h_1)}{\widehat{G}_r(h_2)} \quad \text{is monotonically decreasing with } r \text{ on } [0, 1].$$

We apply (3.3.3) with

$$A(r) := \frac{\widehat{G}_r(h_1)}{\widehat{G}_r(h_2)}$$

$$B(r) := \widehat{G}_r(h_2) \text{ which is decreasing for fixed } h_2;$$

$$E(r) := F(r) \text{ and } \overline{E}(r) := \overline{F}(r).$$

So we obtain

$$\frac{\int\limits_0^1 \widehat{G}_r(h_1) dF(r)}{\int\limits_0^1 \widehat{G}_r(h_2) dF(r)} \geq \frac{\int\limits_0^1 \widehat{G}_r(h_1) d\overline{F}(r)}{\int\limits_0^1 \widehat{G}_r(h_2) d\overline{F}(r)} .$$

But this is equivalent to the inequality

$$\frac{\widehat{G}_F(h_1)}{\widehat{G}_{\overline{F}}(h_1)} \geq \frac{\widehat{G}_F(h_2)}{\widehat{G}_{\overline{F}}(h_2)} ,$$

which tells us that $\frac{\widehat{G}_F}{\widehat{G}_{\overline{F}}}$ is a monotonically decreasing function of h as well as the function $[1 - G_{\overline{F}}(h)]^{1/(n-1)}$. Finally, we apply (3.3.4) once more, setting

$$A(h) := \left[1 - \widehat{G}_{\overline{F}}(h)\right]^{1/(n-1)}$$

$$B(h) := \frac{\widehat{G}_F(h)^k}{\widehat{G}_{\overline{F}}(h)}$$

$$C(h) := \widehat{G}_{\overline{F}}(h)^k T^{-1} \int\limits_h^1 R^{n-2} r^{-n+2} d\overline{F}(r) .$$

Then we get

$$Q_2(F) \le \frac{\int\limits_0^t \widehat{G}_F(h)^k T^{-1} \int\limits_h^1 R^{n-2} r^{-n+2} d\overline{F}(r) dh}{\int\limits_0^t \widehat{G}_F(h)^k \left[1 - \widehat{G}_{\overline{F}}(h)\right]^{1/(n-1)} T^{-1} \int\limits_h^1 R^{n-2} r^{-n+2} d\overline{F}(r) dh} \le$$

$$\le \frac{\int\limits_0^t \widehat{G}_{\overline{F}}(h)^k T^{-1} \int\limits_h^1 R^{n-2} r^{-n+2} d\overline{F}(r) dh}{\int\limits_0^t \widehat{G}_{\overline{F}}(h)^k \left[1 - \widehat{G}_{\overline{F}}(h)\right]^{1/(n-1)} T^{-1} \int\limits_h^1 R^{n-2} r^{-n+2} d\overline{F}(r) dh} .$$

At last we have reached our aim and derived the estimation of Proposition 2. Now we can start with the direct evaluation of $Q_1(F)$ and $Q_2(F)$.

3.4 EVALUATION OF THE QUOTIENT

Now we are ready to formulate the main result for Phase II.

Theorem 5

For all distributions satisfying our conditions, i. e.

  – *identical*

  – *independent*

  – *symmetrical under rotations*

we have for $n \ge 2$

(3.4.1) $$E_{m,n}(S) \le m^{1/(n-1)} n^3 \pi \left(1 + \frac{e\pi}{2}\right) .$$

Proof. Let again k be that exponent chosen from $\{0, 1, \ldots, m - n\}$, which leads to the minimum in (3.2.18).

First we estimate $Q_1(F)$ by exploiting Proposition 1 (3.3.6). For the numerator we have

$$\int_0^t T^{n-2} t^{-n+1} dh = \frac{1}{2} \frac{\lambda_{n-1}(\Omega_{n-1})}{\lambda_{n-2}(\Omega_{n-2})} \quad \text{(see Appendix 2.18)}$$

and for the denominator

$$\int_0^t h T^{n-2} t^{-1-n+1} dh = \frac{1}{n}.$$

Hence

(3.4.2) $$Q_1(F) \leq \frac{n}{2} \frac{\lambda_{n-1}(\Omega_{n-1})}{\lambda_{n-2}(\Omega_{n-2})} =: C_3(n).$$

Again, the evaluation of $Q_2(F)$ is more difficult. Here we apply Proposition 2 (3.3.8). Let $z \in [0, 1]$ such that $\widehat{G}_{\overline{F}}(z) = 1 - \frac{1}{k+2}$, or equivalently $1 - \widehat{G}_{\overline{F}}(z) = \frac{1}{k+2}$. If $z \geq t$, then we know that

$$Q_2(\overline{F}) \leq \frac{\int_0^t \widehat{G}_{\overline{F}}(h)^k T^{-1} \int_h^1 R^{n-2} r^{-n+2} d\overline{F}(r) dh}{\int_0^t \widehat{G}_{\overline{F}}(h)^k \left[1 - \widehat{G}_{\overline{F}}(h) \right]^{1/(n-1)} T^{-1} \int_h^1 R^{n-2} r^{-n+2} d\overline{F}(r) dh} \leq (k+2)^{1/(n-1)},$$

because $(1 - \widehat{G}_{\overline{F}}(h))$ is a decreasing function of h. Much more interesting is the case that $z < t$. The term $1 - G_{\overline{F}}(h)$ is not less $\frac{1}{k+2}$ on $[0, z]$ and it is nonnegative everywhere (we need that property on $[z, t]$). In the numerator of the quotient above the interval $[z, t]$ has a value of at most

$$\int_z^t \widehat{G}_{\overline{F}}(h)^k T^{-1} \int_h^1 R^{n-2} r^{-n+2} d\overline{F}(r) dh \leq$$

$$\leq \int_z^t T^{-1} \int_h^1 R^{n-2} r^{-n+2} d\overline{F}(r) dh \leq {}^*$$

$$\leq \int_z^t T^{-1} dh \cdot \int_z^t \int_h^1 R^{n-2} r^{-n+2} d\overline{F}(r) dh \leq {}^{**}$$

$$\leq \frac{\pi}{2} \int_z^t \int_h^1 R^{n-3} r^{-n+2} d\overline{F}(r) dh \leq {}^{***}$$

$$\leq \frac{\pi}{4} \lambda_{n-1}(\omega_n)(\lambda_{n-2}(\omega_{n-1}))^{-1} \left[\widehat{G}_{\overline{F}}(t) - \widehat{G}_{\overline{F}}(z) \right] \leq$$

$$\leq \frac{\pi}{4} \lambda_{n-1}(\omega_n)(\lambda_{n-2}(\omega_{n-1}))^{-1} \frac{1}{k+2} =: \frac{\pi}{4} C_4(n) \frac{1}{k+2}.$$

Here we have used $(*)$ that T^{-1} is an increasing and that $\int_h^1 R^{n-2} r^{-n+2} d\overline{F}(r)$ is a decreasing function of h. In $(**)$ we have exploited that $R \le 1$. And in $(***)$ we have used (2.4.10). The remaining interval $[0, z]$ has a weight of at least

$$
\int_0^z \widehat{G}_{\overline{F}}(h)^k T^{-1} \int_h^1 R^{n-2} r^{-n+2} d\overline{F}(r) dh \ge^*
$$

$$
\ge \int_0^z \widehat{G}_{\overline{F}}(h)^k \int_h^1 R^{n-3} r^{-n+2} d\overline{F}(r) dh
$$

$$
\ge \frac{1}{2} \frac{\lambda_{n-1}(\omega_n)}{\lambda_{n-2}(\omega_{n-1})} \widehat{G}_{\overline{F}}(z)^{k+1} \frac{1}{k+1} \ .
$$

$(*)$ holds because of $\frac{R}{T} \ge 1$ for all values of r which are allowed by $\overline{F}$.

But $\widehat{G}_F(z)^{k+1} = \left[1 - \frac{1}{k+2}\right]^{k+1} \ge e^{-1}$. So we have a weight of at least

$$
\frac{1}{2} \frac{\lambda_{n-1}(\omega_n)}{\lambda_{n-2}(\omega_{n-1})} e^{-1} \frac{1}{k+1} =: \frac{1}{2} C_4(n) e^{-1} \frac{1}{k+1}
$$

for the interval $[0, z]$.

Consequently

$$
Q_2(F) \le \frac{\frac{\pi}{4} C_4(n) \frac{1}{k+2} + \frac{1}{2} C_4(n) e^{-1} \frac{1}{k+1}}{\frac{1}{2} C_4(n) e^{-1} \frac{1}{k+1} \left(\frac{1}{k+2}\right)^{1/(n-1)}} \le
$$

(3.4.3)

$$
\le (k+2)^{1/(n-1)} e \cdot 2(k+1) \left(\frac{\pi}{4} \frac{1}{k+2} + \frac{1}{2e} \frac{1}{k+1}\right) \le
$$

$$
\le (k+2)^{1/(n-1)} \left(\frac{e\pi}{2} + 1\right) \ .
$$

Let us summarize. We know that

$$
\frac{E_{m,n}(S)}{E_{m,n}(Z)} \le C(n) Q_1(F) Q_2(F),
$$

and that

$$
Q_1(F) \le \frac{n}{2} \frac{\lambda_{n-1}(\Omega_{n-1})}{\lambda_{n-2}(\Omega_{n-2})}
$$

$$
Q_2(F) \le \left(1 + \frac{e\pi}{2}\right)(k+2)^{1/(n-1)}
$$

$$
C(n) = \frac{\lambda_n(\Omega_n) n^2 (n-1)}{\lambda_{n-1}(\Omega_{n-1})} \ .
$$

Multiplication yields

(3.4.4)
$$\frac{E_{m,n}(S)}{E_{m,n}(Z)} \leq (k+2)^{1/(n-1)} \frac{\lambda_n(\Omega_n)}{\lambda_{n-2}(\Omega_{n-2})} \frac{n^3(n-1)}{2} \left(1 + \frac{e\pi}{2}\right) =$$
$$(k+2)^{1/(n-1)} \left(1 + \frac{e\pi}{2}\right) \pi n^2(n-1) \quad \text{(see Appendix 2.22)}.$$

Since $k \leq m - n$ and $n \geq 2$ we have $k + 2 \leq m$ and

$$\frac{E_{m,n}(S)}{E_{m,n}(Z)} \leq m^{1/(n-1)} n^3 \pi \left(1 + \frac{e\pi}{2}\right) \quad \text{for } n \geq 3.$$

For $n = 2$ we have $(n - 1) = 1$ and the claim is trivial because of $m^{1/(n-1)} = m$. Now we know that our upper bound for $\frac{E_{m,n}(S)}{E_{m,n}(Z)}$ is also an upper bound for $E_{m,n}(S)$ itself, because of $E_{m,n}(Z) \leq 1$.

$\square$

Corollary.

The conditional expectation value $E_{m,n}(S \mid S > 0)$ — as calculated and analyzed by Haimovich — satisfies

(3.4.5)
$$E_{m,n}(S \mid S > 0) \leq m^{1/(n-1)} n^3 \pi \left(1 + \frac{e\pi}{2}\right).$$

Proof. $E_{m,n}(Z)$ is the probability that $\mathbb{R}^+ v$ intersects a boundary simplex. So we know that $E_{m,n}(Z) \leq P(S > 0)$, because the intersected boundary simplex is counted in S. We conclude that

$$\frac{E_{m,n}(S)}{P(S > 0)} \leq \frac{E_{m,n}(S)}{E_{m,n}(Z)} \leq m^{1/(n-1)} n^3 \pi \left(1 + \frac{e\pi}{2}\right).$$

3.5 THE AVERAGE NUMBER OF STEPS IN OUR COMPLETE SIMPLEX-METHOD

Until now we have dealt only with Phase II. Let us now consider the complete method of Chapter I, Section 4. We begin by restating the algorithm. Let $\Pi_k(x) = \begin{bmatrix} x^1 \\ \vdots \\ x^k \end{bmatrix}$ for $x \in \mathbb{R}^n$ and let I_k be the problem

(3.5.1)
$$\begin{array}{ll} \text{Maximize} & \Pi_k(v)^T \Pi_k(x) \\ \text{subject to} & \Pi_k(a_i)^T \Pi_k(x) \leq 1 \quad \text{for } i = 1, \ldots, m \\ \text{where} & v, x, a_i \in \mathbb{R}^n. \end{array}$$

Let Y_k stand for $\Pi_k(Y) \in \mathbb{R}^k$. Then our algorithm proceeds as follows.

(3.5.2)   1) Calculate and determine a boundary simplex of Y_2 and set $k = 2$.

2) Find a boundary simplex of Y_2, which is intersected by $\mathbb{R}^+\Pi_2(v)$. If this is impossible, go to 7).

3) If $k < n$ set $k = k + 1$; else go to 6).

4) Consider the given boundary simplex of Y_{k-1}

$$\mathrm{CH}(\Pi_{k-1}(a_{\triangle 1}), \ldots, \Pi_{k-1}(a_{\triangle k-1})).$$

Determine a_i such that

$$\mathrm{CH}(\Pi_k(a_{\triangle 1}), \ldots, \Pi_k(a_{\triangle k-1}), \Pi_k(a_i))$$

is a boundary simplex of Y_k.

5) Starting from the given boundary simplex of Y_k apply the shadow-vertex-algorithm using $\mathrm{span}(\Pi_k(e_k), \Pi_k(v))$ as intersection plane and find a boundary simplex which is intersected by $\mathbb{R}^+\Pi_k(v)$. If this is impossible, go to 7); else go to 3).

6) Print the solution, go to 8).

7) Print: problem unsolvable.

8) STOP.

In Chapter I, Section 4, we have shown that and how this algorithm works. A rather naive estimation yields the following result (sufficient from a qualitative point of view).

Theorem 6

(3.5.3)   *The average number of pivot steps for the complete*
                  Simplex-Method is polynomial

Proof. Count the steps which are necessary to perform the method described above.

Step 1) requires two pivot steps.

Step 2) requires (for $n = 2$) at most m pivot steps.

Step 4) requires one step for each change of dimension, that makes $n - 2$ in total.

Step 5) requires at most as much steps as boundary simplices are intersected. But the expectation value of this number is bounded from above (see 3.4.1) by $m^{1/(k-1)}k^3\pi\left(1 + \frac{e\pi}{2}\right)$.

So the average number of steps for the complete method satisfies

$$(3.5.4) \qquad E_{m,n}(s_t) \leq 2 + n - 2 + m + \sum_{k=3}^{n} m^{1/(k-1)} k^3 \pi \left(1 + \frac{e\pi}{2}\right)$$

Now it is clear that $E_{m,n}(s_t)$ is polynomial in m and n.

$\square$

But this result is not satisfying from a quantitative point of view. So we should analyze what happens in the single dimension-stages more carefully.

Remember that in our stochastic model $a_1, \ldots, a_m, v$ are distributed over $\mathbb{R}^n \setminus \{0\}$. Relevant for the problem I_k are the truncated vectors $\Pi_k(a_1), \ldots, \Pi_k(a_m), \Pi_k(v)$, which are distributed according to our conditions on $\mathbb{R}^k \setminus \{0\}$.

At the first glance we observe in stage k a typical (m, k)-problem. But then we notice that only a special set of density functions (or distributions) can appear. Now we shall derive sharper bounds for $E^k_{m,n}(S)$, where $E^k_{m,n}(S)$ denotes the expected number of boundary simplices in Y_k which are intersected by $\mathbb{R}^+ \Pi_k(v)$. Though it would be (theoretically) possible to do this derivation for $k = 2, \ldots, n$ together, we prefer to begin with $k = 3, \ldots, n$ only. After that we shall do the proof for $k = 2$. This will avoid some confusion.

For $k \geq 3$ we have

$$(3.5.5) \qquad \frac{E^k_{m,n}(S)}{E^k_{m,n}(Z)} \leq \frac{\lambda_k(\Omega_k) k^2 \int\limits_0^t G_k(h)^{m-k} T^{-1}}{\lambda_{k-1}(\Omega_{k-1}) \int\limits_0^t G_k(h)^{m-k} h (Tt)^{-1}} \cdot \frac{\int\limits_{\mathbb{R}^{k-1}} |T - c_k^{k-1}| f^k(\Pi_k(c_k)) \, d\overline{\Pi_k(c_k)} \, dh}{\int\limits_{\mathbb{R}^{k-1}} |T - c_k^{k-1}|^2 \sqrt{\frac{1}{(c_k^1)^2 + \ldots + (c_k^k)^2}} f^k(\Pi_k(c_k)) \, d\overline{\Pi_k(c_k)} \, dh}$$

Not every density function f^k over $\mathbb{R}^k$ satisfying our three conditions can occur in such a quotient, because the density function used here must be the result of a projection of $\mathbb{R}^n$ into $\mathbb{R}^k$.

Let $f = f^n$ be the given density function over $\mathbb{R}^n$. Then

$$(3.5.6) \qquad f^k(x^1,\ldots,x^k) = \int\limits_{-\infty}^{\infty} \cdots \int\limits_{-\infty}^{\infty} f^n(x^1,\ldots,x^k,\xi^{k+1},\ldots,\xi^n)d\xi^n \ldots d\xi^{k+1} \ .$$

To understand the remark above, let (for example) $f^n(x)$ be a density function which is symmetrical under rotations of $\mathbb{R}^n$ and which is 0 for $\|x\| \leq \rho < 1$ for a fixed $\rho < 1$ and is 0 for $\|x\| > 1$. Then f^n is the density function of a feasible distribution over $\mathbb{R}^n$. But f^k can never have such a property. This results from the fact that $f^k(x^1,\ldots,x^k) = 0$ is possible only if $f^n(x) = 0$ for all x with $\|x\| \geq \sqrt{(x^1)^2 + \ldots + (x^k)^2}$ (except in null sets).

Consequently, f^n had to be 0 for every $x \in \Omega_n \setminus \{0\}$. This is a contradiction.

So we should exploit the fact that here we have to deal only with a special subset of feasible density functions. This gives us the hope that our method will behave more kindly than in the worst examples of the entire set of distributions.

Figure 3.9a

<u>Projected distributions</u>

The figure shall illustrate the uniform distribution on the unit sphere of R^2

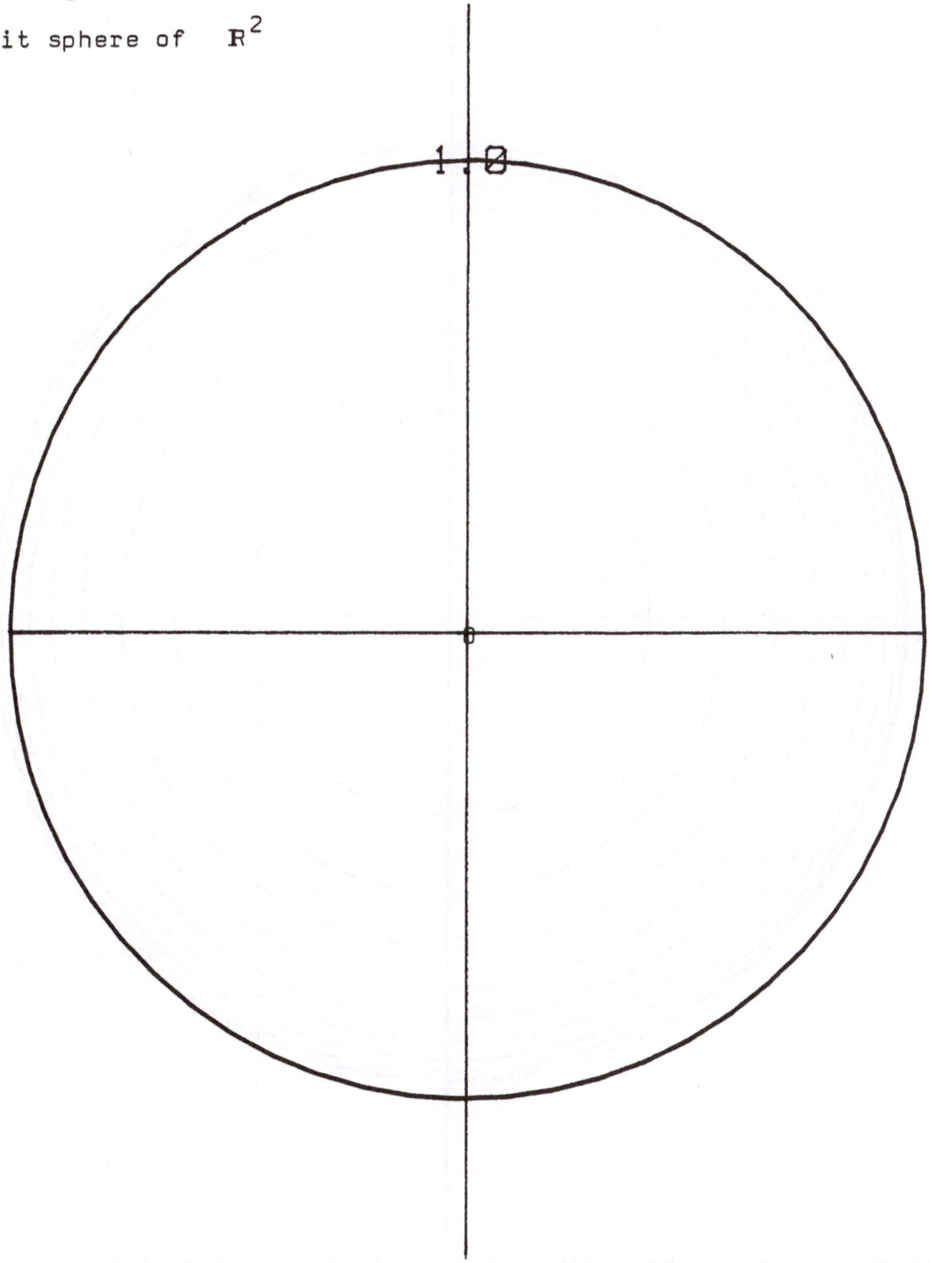

It can also be regarded as an illustration of the corresponding distribution in R^3.

Figure 3.9b

<u>Projected distributions</u>

Here we want to illustrate the result of the projection on

R^2 for the uniform distribution on the unit sphere of R^3.

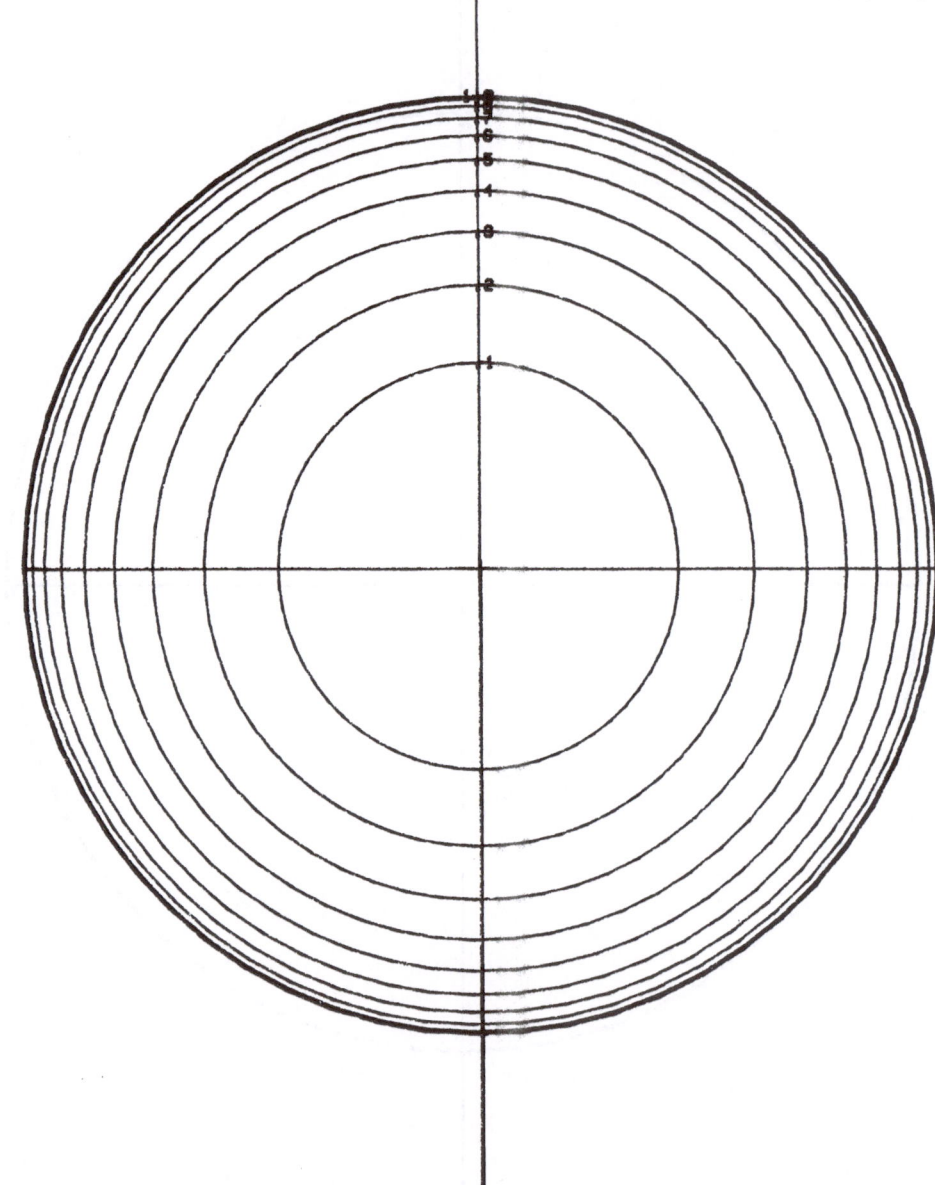

We observe that the projected distributions are much more
concentrated in the interior of the ball.

Let us insert (3.5.6) into (3.5.5). We want to replace $G_k(h)$ by $G_n(h) = G(h)$. Our aim is to apply the theory of Section 1 – Section 4 for $k \geq 3$.

$$G_k(h) = P_k(x^1 \leq h) = \int\limits_{-\infty}^{h} \int\limits_{-\infty}^{\infty} \cdots \int\limits_{-\infty}^{\infty} f^k(\eta^1, \ldots, \eta^k) d\eta^k \ldots d\eta^1$$

$$\text{(3.5.7)} \qquad = \int\limits_{-\infty}^{h} \int\limits_{-\infty}^{\infty} \cdots \int\limits_{-\infty}^{\infty} f^n(\eta^1, \ldots, \eta^k, \xi^{k+1}, \ldots, \xi^n) d\xi^n \ldots d\xi^{k+1} d\eta^k \ldots d\eta^1$$

$$= G_n(h) = P(x^1 \leq h)$$

$$= \int\limits_{-\infty}^{h} \int\limits_{-\infty}^{\infty} \cdots \int\limits_{-\infty}^{\infty} f^n(\xi^1, \ldots, \xi^n) d\xi^n \ldots d\xi^1 .$$

Here a replacement is possible. The quantities $|T - c_k^{k-1}|$ and $|T - c_k^{k-1}|^2$ do not cause problems. Only the factor $\dfrac{1}{\sqrt{(c_k^1)^2 + \ldots + (c_k^k)^2}}$ remains for our considerations. This term can be estimated from below by

$$\text{(3.5.8)} \qquad \begin{aligned} \frac{1}{r(c_k)} &= \frac{1}{\sqrt{(c_k^1)^2 + \ldots + (c_k^k)^2 + \ldots (c_k^n)^2}} \leq \\ &\leq \frac{1}{\sqrt{(c_k^1)^2 + \ldots + (c_k^k)^2}} . \end{aligned}$$

We conclude

$$\text{(3.5.9)} \qquad \begin{aligned} \frac{E_{m,n}^k(S)}{E_{m,n}^k(Z)} &\leq \frac{\lambda_k(\Omega_k) k^2 \int\limits_0^t G_n(h)^{m-k} T^{-1} \cdot}{\lambda_{k-1}(\Omega_{k-1}) \int\limits_0^t G_n(h)^{m-k} h(Tt)^{-1} \cdot} \\ &\qquad \frac{\cdot \int\limits_{\mathbb{R}^{n-1^\bullet}} |T - c_k^{k-1}| f^n(c_k) d\bar{c}_k^* dh}{\cdot \int\limits_{\mathbb{R}^{n-1^\bullet}} |T - c_k^{k-1}|^2 \dfrac{1}{\sqrt{(c_k^1)^2 + \ldots + (c_k^n)^2}} f^n(c_k) d\bar{c}_k^* dh} . \end{aligned}$$

Here $\mathbb{R}^{n-1^\bullet} = \{x \mid x \in \mathbb{R}^n, x^k = 0\}$ and $\bar{c}_k^* = (c_k^1, \ldots, c_k^{k-1}, c_k^{k+1}, \ldots, c_k^n)^T$.

The evaluation of this quotient can be done in accordance with the treatment of (3.2.6) in Sections 2 – 4. We use (3.2.18) and observe that l ranges from 0 to $m - k$ in our case. With the according definition for $Q_1(F)$ and $Q_2(F)$ we obtain

$$\begin{aligned} \frac{E_{m,n}^k(S)}{E_{m,n}^k(Z)} &\leq \frac{\lambda_k(\Omega_k) k^2 (n-1) Q_1(F) Q_2(F)}{\lambda_{k-1}(\Omega_{k-1})} \leq \\ &\leq \frac{\lambda_k(\Omega_k) k^2}{\lambda_{k-1}(\Omega_{k-1})} (n-1) \frac{n}{2} \frac{\lambda_{n-1}(\Omega_{n-1})}{\lambda_{n-2}(\Omega_{n-2})} \left(1 + \frac{e\pi}{2}\right) m^{1/(n-1)} \end{aligned}$$

as in (3.4.4) and using that $m - k + 2 \leq m$. This enables us to estimate

$$
\frac{E_{m,n}^k(S)}{E_{m,n}^k(Z)} \leq \sqrt{\frac{2\pi}{k}} \, k^2 \frac{1}{2} \, n \, \frac{(n-1)\lambda_{n-1}(\Omega_{n-1})}{\lambda_{n-2}(\Omega_{n-2})} \left(1 + \frac{e\pi}{2}\right) m^{1/(n-1)} \leq
$$

(3.5.10)

$$
\leq \sqrt{\frac{2\pi}{k}} \, k^2 \frac{1}{2} \, n(n-1) \sqrt{\frac{2\pi}{n-1}} \left(1 + \frac{e\pi}{2}\right) m^{1/(n-1)} \leq
$$

$$
\leq m^{1/(n-1)} \, n \, k^{3/2} \sqrt{n-1} \left(1 + \frac{e\pi}{2}\right) \pi.
$$

Because of $E_{m,n}^k(Z) \leq 1$ we know that for $n \geq 3$ and $k \geq 3$

(3.5.11)
$$
E_{m,n}^k(S) \leq m^{1/(n-1)} \, n \, k^{3/2} \sqrt{n-1} \left(1 + \frac{e\pi}{2}\right) \pi.
$$

Now consider the case $k = 2$ and $n \geq 3$. Recall formula (2.3.11) of Theorem 4.

(3.5.12)
$$
E_{m,n}^2(S) = \binom{m}{2} 2\lambda_1(\omega_2) \int_0^1 G_2(h)^{m-2} \int_R \int_R |\det B_2| \cdot
$$
$$
\cdot W(\Pi_2(b_1)) \, f^2(\Pi_2(b_1)) \, f^2(\Pi_2(b_2)) \, db_1^1 \, db_2^1 \, dh.
$$

In the case $k = 2$ we have $W(\Pi_2(b_1)) = \frac{1}{2}$ for every b_1 and $B_2 = \begin{bmatrix} b_1^1 & 1 \\ b_2^1 & 1 \end{bmatrix}$, $|\det B_2| = |b_1^1 - b_2^1|$. We conclude

(3.5.13)
$$
|\det B_2| \leq |b_1^1| + |b_2^1|
$$

and

(3.5.14)
$$
E_{m,n}^2 \leq 2 \binom{m}{2} \lambda_1(\omega_2) \int_0^1 G_2(h)^{m-2} \int_R \int_R |b_1^1| \cdot
$$
$$
\cdot f^2(\Pi_2(b_1)) f^2(\Pi_2(b_2)) db_1^1 \, db_2^1 \, dh.
$$

Now we apply formulae (3.5.6) and (3.5.7), which are valid also for $k = 2$.

(3.5.15)
$$
E_{m,n}^2 \leq 2 \binom{m}{2} 2\lambda_1(\omega_2) \int_0^1 G(h)^{m-2} \int_{R^{n-1*}} \int_{R^{n-1*}} |b_1^1|
$$
$$
f^n(b_1) f^n(b_2) d\overline{b}_1^* \, d\overline{b}_2^* \, dh,
$$

where $R^{n-1*} = \{x \mid x \in R^n, x^2 = 0\}$ and $\overline{b}^* = (b^1, b^3, \ldots, b^n)^T$.

This yields

$$E^2_{m,n} \leq 2 \binom{m}{2} \lambda_1(\omega_2) \int_0^1 G_2(h)^{m-2} g(h) \frac{2}{n-2} \frac{\lambda_{n-3}(\omega_{n-2})}{\lambda_{n-1}(\omega_n)} \cdot$$

$$\cdot \int_h^1 \frac{(r^2 - h^2)^{(n-2)/2}}{r^{n-2}} \, dF(r) dh.$$

(3.5.16)

(Recall that $n \geq 3$).

And we know that

$$\int_h^1 \frac{(r^2 - h^2)^{(n-2)/2}}{r^{n-2}} \, dF(r) \leq \left[\int_h^1 \frac{(r^2 - h^2)^{(n-1)/2}}{r^{n-1}} \, dF(r) \right]^{\frac{n-2}{n-1}}$$

and

$$\int_h^1 \frac{(r^2 - h^2)^{(n-1)/2}}{r^{n-1}} \, dF(r) = \int_h^1 \int_{\frac{h}{r}}^1 (1 - \sigma^2)^{(n-3)/2} \sigma d\sigma dF(r)(n-1) \leq$$

$$\leq \int_h^1 \int_{\frac{h}{r}}^1 (1 - \sigma^2)^{(n-3)/2} d\sigma dF(r)(n-1) \leq$$

$$\leq (1 - \widehat{G}_F(h)) \, (n-1) \frac{\lambda_{n-1}(\omega_n)}{2\lambda_{n-2}(\omega_{n-1})} \leq$$

$$\leq (1 - \widehat{G}_F(h)) \, (n-1) \frac{1}{2} \sqrt{\frac{2\pi}{n-2}} \leq (1 - \widehat{G}_F(h)) \, n \leq$$

$$\leq (1 - G_F(h)) \, 2n.$$

So we have

$$E^2_{m,n} \leq 4 \binom{m}{2} 2n\pi \int_0^1 G(h)^{m-2}(1 - G(h))^{(n-2)/(n-1)} g(h) dh \cdot$$

$$\cdot \frac{2}{(n-2)} \frac{\lambda_{n-3}(\omega_{n-2})}{\lambda_{n-1}(\omega_n)} \leq$$

(3.5.17)

$$\leq \binom{m}{2} \frac{16\pi n}{n-2} \frac{n-2}{2\pi} \int_0^1 x^{m-2}(1-x)^{1-1/(n-1)} dx \leq$$

$$\leq 4n \, m^{1/(n-1)} = m^{1/(n-1)} \, n \, 4.$$

Summing up the steps required for changing the dimension and the bounds given by (3.5.11) and (3.5.17), we obtain

$$E_{m,n}(s_t) \leq n + nm^{1/(n-1)}4 + \sum_{k=3}^{n} m^{1/(n-1)}nk^{3/2}\pi\left(1 + \frac{e\pi}{2}\right)\sqrt{n-1} \leq$$

(3.5.18)
$$\leq m^{1/(n-1)}\sqrt{n-1}\,\pi\left(1 + \frac{e\pi}{2}\right) n \int_{0}^{n+1} x^{3/2}dx$$

$$= m^{1/(n-1)}(n+1)^4\,\frac{2}{5}\,\pi\left(1 + \frac{e\pi}{2}\right).$$

So we have our main result.

Theorem 7

For all distributions satisfying our conditions the expected number of pivot steps required for our complete Simplex-Method can be estimated as follows:

(3.5.19)
$$E_{m,n}(s_t) \leq m^{1/(n-1)}(n+1)^4\,\frac{2}{5}\,\pi\left(1 + \frac{e\pi}{2}\right).$$

Chapter 4

ASYMPTOTIC RESULTS

4.1 AN ASYMPTOTIC UPPER BOUND IN INTEGRAL FORM

In the chapter before we have demonstrated a rather complicated and lengthy derivation of our main theorem, which states that our Simplex-Method is polynomial with respect to the expected number of pivot steps. Our upper bounds had the size

$$0(m^{1/(n-1)}n^3) \text{ for } E_{m,n}(S) \text{ and}$$
$$0(m^{1/(n-1)}n^4) \text{ for } E_{m,n}(s_t).$$

These bounds seem to exaggerate the n-term. This is due to the fact that we have combined several kinds of worst cases in our analysis of the average behaviour. The methods of Chapter 3 are not suitable for improving our results or for deriving lower bounds. Also it is not possible to derive detailed information on the average behaviour based on the use of special distributions.

But these whishes can be satisfied when we study the asymptotic average behaviour. In addition, proofs will become less complicated.

The term "asymptotic" will be used in the sense that

(4.1.1) $$m \to \infty \quad \text{and } n \text{ is fixed}.$$

Then we obtain a function of the parameter m for every fixed value of n. This function $E_{m,n}(S)$ gives the expected values of the number of shadow-vertices for the corresponding parameters.

We try to derive upper and lower bound-functions of m. We do not deal with $E_{m,n}(s_t)$ here, because it is in most cases directly deducible from the values of $E_{m,n}(S)$, and for the study of special distributions as in Sections 3 – 5 several different distributions had to be analyzed. This would complicate our considerations too much.

For abbreviation of proofs and formulae we shall use a special notation for asymptotic relations.

(4.1.2)   a)   Two functions Λ, Θ of (m, n) are **asymptotically** equal, if there is a function $\varepsilon(m, n)$ tending to 0 for $m \to \infty$ and fixed n, such that $[1-\varepsilon(m,n)]\Lambda(m,n) \le \Theta(m,n) \le \Lambda(m,n)[1+\varepsilon(m,n)]$. We denote this property by $\Theta(m, n) \to= \Lambda(m, n)$.

            b) Θ is asymptotically less or equal Λ if there is a function $\varepsilon(m, n)$ as above such that $\Theta(m, n) \le \Lambda(m, n)[1 + \varepsilon(m, n)]$, what is denoted by $\Theta(m, n) \to\le \Lambda(m, n)$.

Respective notations are used for asymptotically greater or equal $(\to\ge)$, less $(\to<)$ and greater $(\to>)$.

In this first section we are going to derive an asymptotic upper bound for $E_{m,n}(S)$ under the three well-known conditions on our distributions over $\mathbb{R}^n \setminus \{0\}$.

Studying asymptotic properties, we are not allowed to generalize the results for distributions with bounded support to the set of all distributions satisfying our three basic conditions. For **fixed m and n** this was possible (as in Chapter 3). But here, some important steps of the proofs depend on this property. We start with formula (2.3.11) for distributions with density functions

$$E_{m,n}(S) = \binom{m}{n} n\lambda_{n-1}(\omega_n) \int_0^\infty G(h)^{m-n}\Lambda_S(h)\, dh, \quad \text{where}$$

(4.1.3)

$$\Lambda_S(h) = \int_{\mathbb{R}^{n-1}} \cdots \int_{\mathbb{R}^{n-1}} |\det B|\, W(b_1,\ldots,b_{n-1})f(b_1)\ldots f(b_n)\, d\bar{b}_1 \ldots d\bar{b}_n .$$

The main difficulties result from this function $\Lambda_S(h)$, particularly from the spherical measures $W(b_1,\ldots,b_{n-1})$. A very simple upper bound can be obtained by defining

(4.1.4) $$\Lambda_R(h) := \int_{\mathbb{R}^{n-1}} \cdots \int_{\mathbb{R}^{n-1}} |\det B|\, f(b_1)\ldots f(b_n)\, d\bar{b}_1 \ldots d\bar{b}_n .$$

Since

(4.1.5)
$$W(b_1,\ldots,b_{n-1}) \le \frac{1}{2} \, ,$$

we know that

(4.1.6)
$$\Lambda_S(h) \le \frac{1}{2}\Lambda_R(h) \, .$$

Unfortunately this bound is effective only for very small values of h (see the corresponding figure). So it is necessary to use a more skilfull estimation for the case of greater h. Let us try with the inequality of Cauchy-Schwartz.

$$\Lambda_S(h) \le [\Lambda_1(h)\Lambda_2(h)]^{1/2} \, , \text{ where}$$

(4.1.7)
$$\Lambda_1(h) = \int\limits_{\mathbb{R}^{n-1}} \ldots \int\limits_{\mathbb{R}^{n-1}} |\det B|^2 \, f(b_1)\ldots f(b_n) \, d\bar{b}_1 \ldots d\bar{b}_n$$

$$\Lambda_2(h) = \int\limits_{\mathbb{R}^{n-1}} \ldots \int\limits_{\mathbb{R}^{n-1}} W(b_1,\ldots,b_{n-1})^2 f(b_1)\ldots f(b_n) \, d\bar{b}_1 \ldots d\bar{b}_n \, .$$

Here the existence of $\Lambda_1(h)$ is not guaranteed. The same difficulty may arise for the function

(4.1.8)
$$g_2(h) = g(h)E((x^1)^2 \mid x^n = h) =$$

$$= g(h) \int\limits_{-\infty}^{\infty} \ldots \int\limits_{-\infty}^{\infty} (x^1)^2 f(x^1,\ldots,x^{n-1},h) \, dx^1 \ldots dx^{n-1} \, .$$

(The second equation holds for distributions with density-functions).

The $n-1$-dimensional spherical angle $W(b_1,\ldots,b_{n-1})$ can be estimated from above by the formula

(4.1.9)
$$W(b_1,\ldots,b_{n-1}) \le \frac{1}{\lambda_{n-1}(\Omega_{n-1})}\lambda_{n-1}(\mathrm{CH}(0,\frac{1}{h}b_1,\ldots,\frac{1}{h}b_{n-1}))$$

$$= \frac{1}{\lambda_{n-1}(\Omega_{n-1})}\frac{1}{h^{n-1}}\lambda_{n-1}(\mathrm{CH}(0,b_1,\ldots,b_{n-1}))$$

Note that the factor $\frac{1}{h^{n-1}}$ results from normalizing the vectors generating the simplex in such a way that all points lie in the hyperplane $\{x \mid x^n = 1\}$. So it is guaranteed that the entire set $M_W := \mathrm{CC}(c_1,\ldots,c_{n-1}) \cap \Omega_n$ is contained in the extended simplex. Regarding the rotational symmetry of our distribution, we obtain

Figure 4.1a

Upper and lower estimations for $W(c_1, c_2)$

For deriving a lower bound we use $CH(0, c_1, c_2)$, for the upper bound we use the extended simplex, where $h=1$.

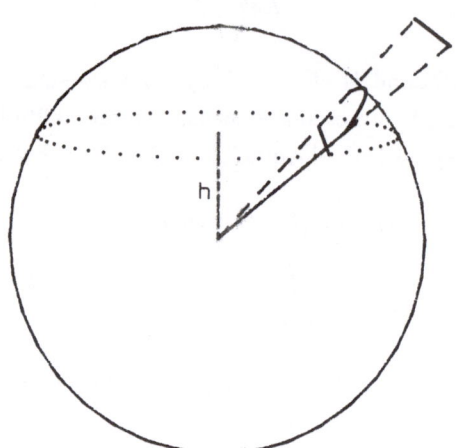

seen
from
the
side

h is small -relative to the radius of the unit ball- and the approximations or estimations are bad.

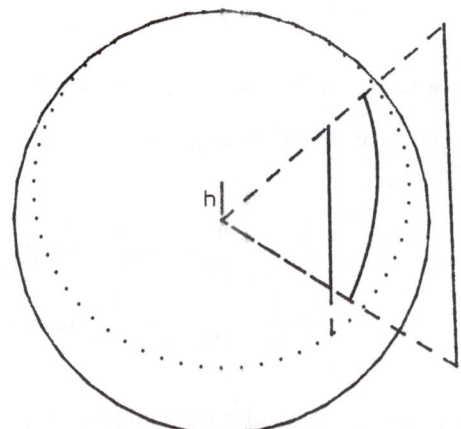

seen
from
above

Figure 4.1b

Upper and lower estimations for $W(c_1, c_2)$

Here h is great (close to 1) and the approximations are good.

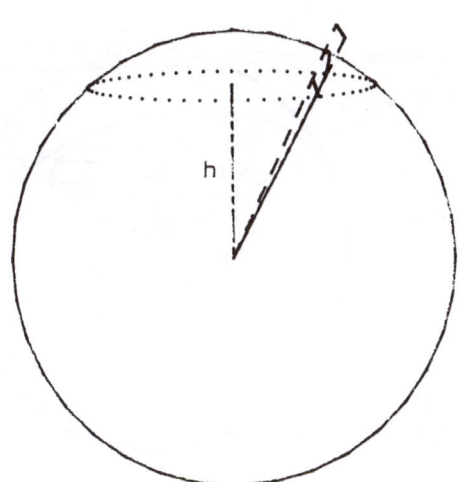

seen
from
the
side

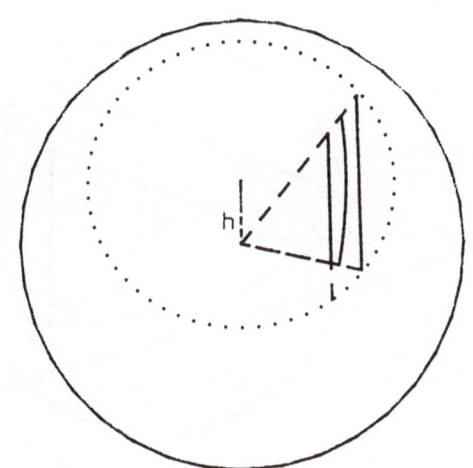

seen
from
avove

Figure 4.2a

Upper and lower bounds for $V(c_1,c_2,c_3)$

The lower bound comes from the volume of $CH(0,c_1,c_2,c_3)$, the upper bound from extending that simplex until $h=1$.

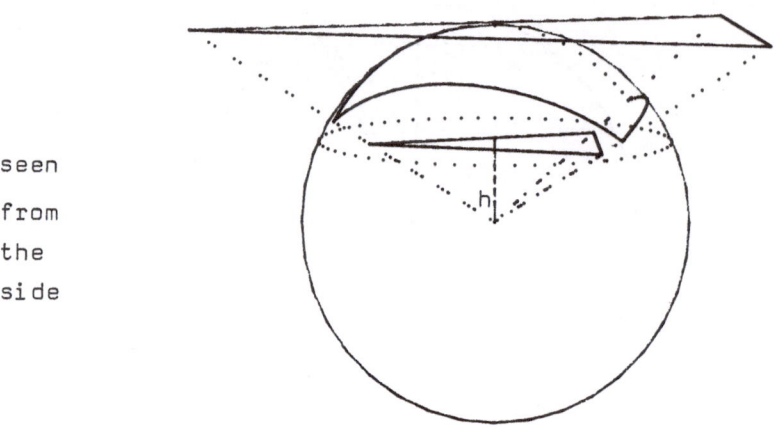

seen

from

the

side

Here h is rather small ($h \ll 1$) and the approximations are bad.

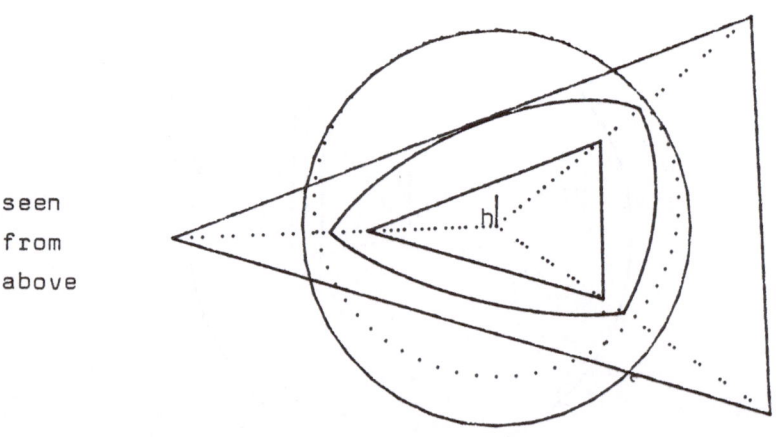

seen

from

above

Figure 4.2b

Upper and lower bounds for $V(c_1, c_2, c_3)$

Here h is great (close to 1) and the approximations
are good.

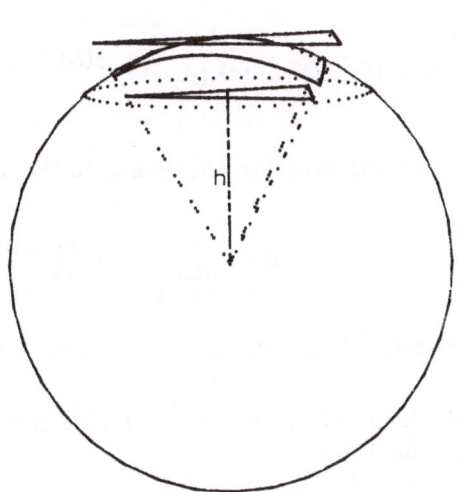

seen
from
the
side

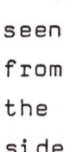

seen
from
above

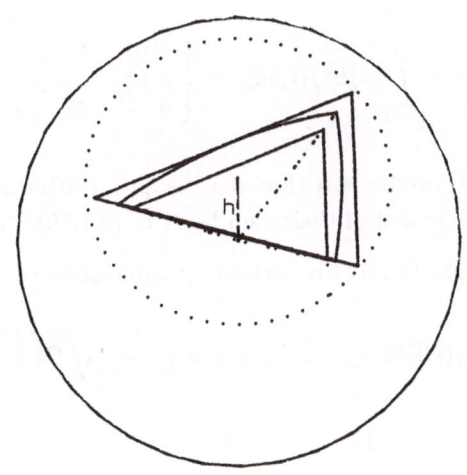

Lemma 4.1

If $g_2(h)$ exists for all $h \in (0, \infty)$, then

$$(4.1.10) \qquad \Lambda_1(h) = n! g_2(h)^{n-1} g(h)$$

$$(4.1.11) \quad \Lambda_2(h) \le \frac{1}{\lambda_{n-1}(\Omega_{n-1})^2} \frac{1}{h^{2n-2}} \frac{1}{(n-1)!} g_2(h)^{n-2} g(h)^2 \cdot \left(\frac{g_2(h)}{g(h)} + (n-1)h^2 \right).$$

Proof. First consider the equation. According to the definition of the matrix B we have

$$\det B = \sum_{\sigma \in S(n)} \text{sign}\,\sigma \prod_{i=1}^{n} \tilde{b}_i^{\sigma(i)}$$

where $S(n)$ is the group of permutations of n elements, and where

$$\tilde{b}_i^j := \begin{cases} b_i^j & \text{for } j = 1, \ldots, n-1 \quad \text{(compare definition of } B, (2.3.9)) \\ 1 & \text{for } j = n \end{cases}.$$

Hence

$$\Lambda_1(h) = \sum_{\sigma, \tau \in S(n)} \text{sign}(\sigma)\, \text{sign}(\tau) \left\{ \prod_{i=1}^{n} \int_{\mathbb{R}^{n-1}} \tilde{b}_i^{\sigma(i)} \tilde{b}_i^{\tau(i)} f(b_i)\,d\bar{b}_i \right\}.$$

The rotational symmetry of f guarantees that

$$\int_{\mathbb{R}^{n-1}} \tilde{b}_i^j \tilde{b}_i^k f(b_i)\,d\bar{b}_i = \begin{cases} g(h) & \text{for } j = k = n \\ g_2(h) & \text{for } j = k \le n-1 \\ 0 & \text{for } j \ne k \end{cases}$$

Consequently, all products with $\sigma \ne \tau$ disappear after integration. Evaluation of the remaining products and of their sum leads to (4.1.10).

The inequality (4.1.11) is derived by application of the well-known formula

$$\lambda_{n-1}(\text{CH}(0, b_1, \ldots, b_{n-1})) = \frac{1}{(n-1)!} \sqrt{D_1^2 + \ldots + D_{n-1}^2 + D_n^2}$$

with (for $i = 1, \ldots, n-1$)

$$D_i = \det \begin{bmatrix} b_1^1 & \cdots & b_1^{i-1} & b_1^{i+1} & \cdots & b_1^{n-1} & h \\ \vdots & & \vdots & \vdots & & \vdots & \vdots \\ b_{n-1}^1 & \cdots & b_{n-1}^{i-1} & b_{n-1}^{i+1} & \cdots & b_{n-1}^{n-1} & h \end{bmatrix}$$

and with

$$D_n = \det \begin{bmatrix} b_1^1 & \cdots & b_1^{n-1} \\ \vdots & & \vdots \\ b_{n-1}^1 & \cdots & b_{n-1}^{n-1} \end{bmatrix}.$$

We apply these formulae to the simplex $\lambda_{n-1}(\mathrm{CH}(0, \frac{1}{h}b_1, \ldots, \frac{1}{h}b_{n-1}))$. Using exactly the same methods as above, we obtain

$$\Lambda_2(h) \leq \frac{1}{\lambda_{n-1}(\Omega_{n-1})^2} \frac{1}{h^{2n-2}} \frac{1}{(n-1)!^2}$$
$$\cdot \{(n-1)h^2(n-1)!g_2(h)^{n-2}g(h)^2 + (n-1)!g_2(h)^{n-1}g(h)\}$$

Now (4.1.11) is at hand.

$\square$

Combining the two results of the lemma we have

$$\Lambda_S(h) \leq \frac{1}{\lambda_{n-1}(\Omega_{n-1})} \sqrt{n} \, \frac{1}{h^{n-1}} g_2(h)^{n-(3/2)} g(h)^{3/2} \cdot$$

(4.1.12)

$$\cdot \sqrt{\frac{g_2(h)}{g(h)} + (n-1)h^2}$$

(if $g_2(h)$ is finite).

In a theoretical sense (4.1.12) remains true even when $g_2(h)$ does not exist. Then the right side is to be regarded as ∞.

The disadvantage of this estimation for $\Lambda_S(h)$ is its bad behaviour for small values of h. For that reason we combine the two estimation results (4.1.6) and (4.1.12). We split the integration area $[0, \infty)$ into two parts $[0, q)$ and $[q, \infty)$ such that $0 < q < \infty$. In the first interval we apply (4.1.6) and in the second (4.1.12). So we obtain the following upper bound.

Theorem 8

Let q be a value of $(0, \infty)$ and let $g_2(h) < \infty$ for all $h \in [q, \infty)$. Then

$$E_{m,n}(S) \leq \binom{m}{n} \frac{n}{2} \lambda_{n-1}(\omega_n) \int_0^q G(h)^{m-n} \Lambda_R(h) \, dh +$$

(4.1.13)

$$+ \binom{m}{n} n^{3/2} \frac{\lambda_{n-1}(\omega_n)}{\lambda_{n-1}(\Omega_{n-1})} \int_q^\infty G(h)^{m-n} \frac{1}{h^{n-1}} \cdot$$

$$\cdot g_2(h)^{n-(3/2)} g(h)^{3/2} \sqrt{\frac{g_2(h)}{g(h)} + (n-1)h^2} \, dh \, .$$

$\square$

The condition of bounded support even enables us to derive a (general) formula for lower bounds of $E_{m,n}(S)$. Without loss of generality, we restrict the considerations to the case that

$$n \geq 3, \; F(1) = 1 \text{ and } F(r) < 1 \text{ for all } r < 1.$$

To begin with, it is clear that

(4.1.14) $$W(b_1, \ldots, b_{n-1}) \geq \frac{1}{\lambda_{n-1}(\Omega_{n-1})} \lambda_{n-1}(\mathrm{CH}(0, b_1, \ldots, b_{n-1})).$$

Now let $H(b_n)$ be the distance between b_n and the hyperplane through $0, b_1, \ldots, b_{n-1}$. Since the support is contained in Ω_n, we conclude

(4.1.15) $$H(b_n) \leq 2\sqrt{1 - h^2}.$$

And we know that

(4.1.16) $$\frac{H(b_n)}{n} \lambda_{n-1}(\mathrm{CH}(0, b_1, \ldots, b_{n-1})) = \lambda_n(\mathrm{CH}(0, b_1, \ldots, b_n)) = \frac{h}{n!} |\det B|.$$

So it is clear that

(4.1.17) $$W(b_1, \ldots, b_{n-1}) \geq \frac{h}{2\sqrt{1 - h^2}} \frac{1}{\lambda_{n-1}(\Omega_{n-1})(n - 1)!} |\det B|$$

and

$$\Lambda_S(h) \geq \frac{h}{2\sqrt{1 - h^2}} \frac{1}{\lambda_{n-1}(\Omega_{n-1})(n - 1)!}$$

$$\cdot \int_{\mathbb{R}^{n-1}} \cdots \int_{\mathbb{R}^{n-1}} |\det B|^2 \, f(b_1) \ldots f(b_n) \, d\bar{b}_1 \ldots d\bar{b}_n$$

Now we use (4.1.10) and obtain from (4.1.3)

Theorem 9

If the distribution has bounded support and if $n \geq 3$

(4.1.18) $$E_{m,n}(S) \geq \binom{m}{n} \frac{n^2}{2} \frac{\lambda_{n-1}(\omega_n)}{\lambda_{n-1}(\Omega_{n-1})} \int_0^1 G(h)^{m-n} g_2(h)^{n-1} g(h) \cdot \frac{h}{\sqrt{1 - h^2}} \, dh.$$

Remark. In the asymptotic case we are not forced to distinguish between $E_{m,n}(S)$ and the quotient $\frac{E_{m,n}(S)}{E_{m,n}(Z)}$, because $E_{m,n}(Z)$ tends to 1 for $m \to \infty$ and fixed n.

This can be proven as a

Lemma 4.2

For $m \to \infty$, n fixed, we know that

(4.1.19)
$$E_{m,n}(Z) \to 1.$$

Proof. $E_{m,n}(Z)$ gives the probability that $u \in CC(a_1, \ldots, a_m)$. If there is any such vector w not belonging to $CC(a_1, \ldots, a_m)$, then there is also a vector $\overline{w}$ such that $\overline{w}^T a_i < 0$ for all $i = 1, \ldots, m$ (Separation Theorem). This random event would be impossible, if every orthant of the 2^n contained at least one of the vectors a_i in its interior. Consider for that purpose the orthant containing $\overline{w}$.

But this probability can be estimated as follows:

$$P(\text{there exists a } \overline{w} \text{ such that } \overline{w}^T a_1 < 0, \ldots, \overline{w}^T a_m < 0)$$

$$\leq P(\text{at least one of the orthants is ``empty''})$$

$$\leq \sum_{j=1}^{2^n} P(\text{orthant } j \text{ is empty})$$

$$= \sum_{j=1}^{2^n} \left(1 - \frac{1}{2^n}\right)^m$$

$$= 2^n \left(1 - \frac{1}{2^n}\right)^m \to 0 \text{ for } m \to \infty, \, n \text{ fixed.}$$

So $P\,(\text{``}Z = 1\text{''})$ tends to 1 for $m \to \infty$, n fixed. Since Z can attain only the values $0, 1$ or higher integers (the latter only on a null set), this means that $E_{m,n}(Z) \to 1$ asympotically.

□

4.2 ASYMPTOTIC RESULTS FOR CERTAIN CLASSES OF DISTRIBUTIONS

An asymptotic bound for $E_{m,n}(S)$ can be deduced if we restrict our considerations to distributions with bounded support. Then we are allowed to assume that Ω_n is the smallest ball containing the support. Recall the general formulae for G, g and g_2 and $n \geq 3$.

(4.2.1)
$$G(h) = 1 - \frac{\lambda_{n-2}(\omega_{n-1})}{\lambda_{n-1}(\omega_n)} \int_h^1 \int_{\frac{h}{r}}^1 (1 - \sigma^2)^{(n-3)/2} \, d\sigma \, dF(r)$$

$$(4.2.2) \qquad g(h) = \frac{\lambda_{n-2}(\omega_{n-1})}{\lambda_{n-1}(\omega_n)} \int_h^1 \frac{(r^2 - h^2)^{(n-3)/2}}{r^{n-2}} \, dF(r)$$

$$(4.2.3) \qquad g_2(h) = \frac{\lambda_{n-2}(\omega_{n-1})}{(n-1)\lambda_{n-1}(\omega_n)} \int_h^1 \frac{(r^2 - h^2)^{(n-1)/2}}{r^{n-2}} \, dF(r)$$

For h approximating 1 we observe the following relations between these three functions. Let us use $\Phi(h)$ as an abbreviation for $1 - G(h)$. Notice that under our condition $g_2(h)$ exists in any case.

Proposition 1.

For $n \geq 3$ we have a function $\alpha : [0, 1] \to \mathbb{R}$

$$(4.2.4) \qquad g_2(h) = \Phi(h)(1 + \alpha(h))$$

such that $\alpha(h) \to 0$ for $h \to 1$.

$\square$

Proof. $g_2(h)$ and $\Phi(h)$ are differentiable for $n \geq 3$ at $h > 0$. Hence we can apply the rule of l'Hospital. It yields

$$\lim_{h \to 1} \frac{g_2(h)}{\Phi(h)} = \lim_{h \to 1} \frac{\int_h^1 \frac{(r^2 - h^2)^{(n-3)/2}}{r^{n-2}} \, dF(r)}{\int_h^1 \frac{(r^2 - h^2)^{(n-3)/2}}{r^{n-2}} \, dF(r)} = 1.$$

Proposition 2.

For $n \geq 3$ we have a function $\beta : [0, 1] \to \mathbb{R}$ such that $\beta(h) \to 0$ for $h \to 1$ and

$$(4.2.5) \quad g(h) \leq \Phi(h)^{(n-3)/(n-1)} (n-1)^{(n-3)/(n-1)} \left[\frac{\lambda_{n-2}(\omega_{n-1})}{\lambda_{n-1}(\omega_n)} \right]^{2/(n-1)} (1 + \beta(h)).$$

Proof. We exploit the concavity of the function $f(x) = x^{(n-3)/(n-1)}$. Here we obtain

$$g(h) = \frac{\lambda_{n-2}(\omega_{n-1})}{\lambda_{n-1}(\omega_n)} \int_h^1 \frac{(r^2 - h^2)^{(n-3)/2}}{r^{n-2}} \, dF(r) \le$$

$$\le \frac{\lambda_{n-2}(\omega_{n-1})}{\lambda_{n-1}(\omega_n)} \left[\int_h^1 \frac{(r^2 - h^2)^{(n-1)/2}}{r^{n-2}} \, dF(r) \right]^{(n-3)/(n-1)} .$$

$$\cdot \left[\int_h^1 \frac{1}{r^{n-2}} \, dF(r) \right]^{2/(n-1)} =$$

$$= \left[\frac{\lambda_{n-2}(\omega_{n-1})}{\lambda_{n-1}(\omega_n)(n-1)} \int_h^1 \frac{(r^2 - h^2)^{(n-1)/2}}{r^{n-2}} \, dF(r) \right]^{(n-3)/(n-1)} .$$

$$\cdot (n-1)^{(n-3)/(n-1)} \left[\frac{\lambda_{n-2}(\omega_{n-1})}{\lambda_{n-1}(\omega_n)} \right]^{2/(n-1)} \left[\int_h^1 \frac{1}{r^{n-2}} \, dF(r) \right]^{2/(n-1)} =$$

$$= g_2(h)^{(n-3)/(n-1)} (n-1)^{(n-3)/(n-1)} \left[\frac{\lambda_{n-2}(\omega_{n-1})}{\lambda_{n-1}(\omega_n)} \right]^{2/(n-1)} .$$

$$\cdot \left[\int_h^1 \frac{1}{r^{n-2}} \, dF(r) \right]^{2/(n-1)} .$$

We take into regard that the last factor satisfies

$$\left[\int_h^1 \frac{1}{r^{n-2}} \, dF(r) \right]^{2/(n-1)} \le 1 + \gamma(h)$$

for a function $\gamma(h)$ with $\gamma(h) \to 0$ when $h \to 1$. This results from $\frac{1}{h^{n-2}} \to 1$ for $h \to 1$ and from $\int_h^1 dF(r) \le 1$. Proposition 1 delivered a function $\alpha(h)$ disappearing for $h \to 1$, which allows to estimate

$$g(h) \le \Phi(h)^{(n-3)/(n-1)} [1 + \alpha(h)]^{(n-3)/(n-1)} (n-1)^{(n-3)/(n-1)}.$$

$$\cdot \left[\frac{\lambda_{n-2}(\omega_{n-1})}{\lambda_{n-1}(\omega_n)} \right]^{2/(n-1)} (1 + \gamma(h)).$$

Setting $1 + \beta(h) := (1 + \alpha(h))^{(n-3)/(n-1)} (1 + \gamma(h))$ yields the desired function $\beta(h)$. Now we are prepared to prove

Theorem 10

Let $n \geq 2$ and let $F(\bar{r}) = 1$ for a $\bar{r} \in \mathbb{R}$. Then there is a function $\varepsilon_F(m, n)$, depending on F, such that $\varepsilon_F(m, n) \to 0$ for $m \to \infty$, n fixed, and

$$(4.2.6) \qquad E_{m,n}(S) \leq m^{1/(n-1)} n^2 \sqrt{2\pi}(1 + \varepsilon_F(m, n)).$$

Proof. The claim is trivial for $n = 2$, because a two-dimensional polyhedron cannot have more than m boundary simplices.

For $n \geq 3$ we apply (4.1.13) and we restrict our discussion to distributions with $F(1) = 1$ and $F(r) < 1$ for all $r < 1$. For every q in $(0, 1)$ we have $E_{m,n}(S) \leq I_1 + I_2$, where

$$I_1 := \binom{m}{n} \frac{n}{2} \lambda_{n-1}(\omega_n) \int_0^q G(h)^{m-n} \Lambda_R(h) \, dh$$

$$I_2 := \binom{m}{n} n^{3/2} \frac{\lambda_{n-1}(\omega_n)}{\lambda_{n-1}(\Omega_{n-1})} \cdot$$

$$\cdot \int_q^1 G(h)^{m-n} \frac{1}{h^{n-1}} \, g_2(h)^{n-(3/2)} \, g(h)^{3/2} \sqrt{\frac{g_2(h)}{g(h)} + (n-1)h^2} \, dh \, .$$

Here I_1 disappears asymptotically, because

$$I_1 \leq \binom{m}{n} \frac{n}{2} \lambda_{n-1}(\omega_n) G(q)^{m-n} \int_0^q \Lambda_R(h) \, dh \leq C(n) \, m^n \, G(q)^{m-n}$$

and because $G(q) < 1$ for $q < 1$.

For the asymptotic evaluation of I_2 we exploit that $\frac{g_2(h)}{g(h)} \leq 1$ and that $h^2 \leq 1$ in any case. Hence

$$\sqrt{\frac{g_2(h)}{g(h)} + (n-1)h^2} \leq \sqrt{n} \, .$$

So we have

$$I_2 \leq \binom{m}{n} n^2 \frac{\lambda_{n-1}(\omega_n)}{\lambda_{n-1}(\Omega_{n-1})} \frac{1}{q^{n-1}} \int_q^1 G(h)^{m-n} \, g_2(h)^{n-(3/2)} \, g(h)^{3/2} \, dh \, .$$

We substitute

$$G(h) = 1 - \Phi(h) \, .$$

For given $\delta > 0$ (dependent on F), we can choose $q > 1$ sufficiently close to 1, such that for all $h \in [q, 1]$ the following properties are valid

1) $g_2(h) \leq (1 + \delta)\Phi(h)$

2) $g(h) \leq \Phi(h)^{(n-3)/(n-1)}(n - 1)^{(n-3)/(n-1)} \left[\dfrac{\lambda_{n-2}(\omega_{n-1})}{\lambda_{n-1}(\omega_n)}\right]^{2/(n-1)} (1 + \delta)$.

3) $\dfrac{1}{q^{n-1}} \leq \dfrac{1}{h^{n-1}} \leq 1 + \delta$.

Taking such a q, we get

$$I_2 \leq \binom{m}{n} n^2 \frac{\lambda_{n-1}(\omega_n)}{\lambda_{n-1}(\Omega_{n-1})} \frac{1}{q^{n-1}} \int_q^1 (1 - \Phi(h))^{m-n}\Phi(h)^{n-(3/2)}g(h)^{3/2}dh\,(1 + \delta)^{n-(3/2)}$$

$$\leq \binom{m}{n} n^2 \frac{\lambda_{n-1}(\omega_n)}{\lambda_{n-1}(\Omega_{n-1})} \int_0^1 (1 - \Phi)^{m-n}\Phi^{n-(3/2)}\Phi^{1/2-1/(n-1)}d\Phi.$$

$$\cdot (n - 1)^{(n-3)/2(n-1)} \left[\frac{\lambda_{n-2}(\omega_{n-1})}{\lambda_{n-1}(\omega_n)}\right]^{1/(n-1)} (1 + \delta)^n\ .$$

Now we apply

$$\frac{\lambda_{n-1}(\omega_n)}{\lambda_{n-1}(\Omega_{n-1})} \leq \sqrt{2\pi n} \quad \text{(Appendix 2.28), and}$$

$$\frac{\lambda_{n-2}(\omega_{n-1})}{\lambda_{n-1}(\omega_n)} \leq \sqrt{\frac{n-1}{2\pi}} \quad \text{(Appendix 2.26)}.$$

We obtain

$$I_2 \leq \binom{m}{n} n^2 \sqrt{2\pi n}(n - 1)^{1/2-1/(n-1)} \left[\frac{n-1}{2\pi}\right]^{1/2(n-1)}(1 + \delta)^n$$

$$\cdot \int_0^1 (1 - \Phi)^{m-n}\Phi^{n-1-(1/(n-1))}d\Phi\ .$$

According to (Appendix 1.15) the integral can be estimated as follows

$$\binom{m}{n} \int_0^1 (1 - \Phi)^{m-n}\Phi^{n-1-(1/(n-1))}d\Phi \leq \frac{1}{n}m^{1/(n-1)}.$$

This yields

$$I_2 \leq m^{1/(n-1)}n^2\sqrt{2\pi}(1 + \delta)^n\ .$$

For m large enough, we have $I_1 \leq I_2\delta$. Hence for any given $\varepsilon > 0$ we can choose $\delta > 0$ sufficiently small such that $(1 + \delta)^{n+1} < \varepsilon$. Since ε depends on m, n and F, our claim is true.

$\square$

The methods used in this section can be exploited to obtain distributions with arbitrary small size of growth.

Theorem 11

For every $\varepsilon > 0$ there is a distribution with bounded support, such that

(4.2.7) $E_{m,n}(S) = 0(m^\varepsilon)$ *for $m \to \infty$ and fixed $n \geq 3$.*

Proof. For given $\varepsilon > 0$ we construct a feasible RDF satisfying (4.2.7)

$$F(r) := \begin{cases} \dfrac{\int_0^r (1-\tau^2)^k \tau^{n-1} d\tau}{\int_0^1 (1-\tau^2)^k \tau^{n-1} d\tau} & \text{for } r \leq 1 \\[4mm] 1 & \text{for } r \geq 1 \end{cases}$$

Here k is chosen large enough such that $\frac{k+1}{k+2} > 1 - 2\varepsilon$.

Then we obtain the representations

$$g(h) = C_1(n,k) \int_h^1 (r^2 - h^2)^{(n-3)/2} (1-r^2)^k r \, dr$$

$$= C_2(n,k)(1-h^2)^{(n-3)/2+k+1} \int_0^1 (1-\sigma)^{(n-3)/2} \sigma^k d\sigma =$$

$$= C_3(n,k)(1-h^2)^{(n-3)/2+k+1}$$

and

$$\Phi(h) = \int_h^1 g(\sigma) d\sigma = C_4(n,k) \int_h^1 (1-\sigma^2)^{(n-3)/2+k+1} d\sigma \geq$$

$$\geq C_4(n,k) \int_h^1 (1-\sigma^2)^{(n-3)/2+k+1} \sigma d\sigma =$$

$$= C_5(n,k)(1-h^2)^{(n-3)/2+k+2} .$$

So it is clear that a "constant" $C_6(n,k)$ exists such that

$$g(h) \leq C_6(n,k)\Phi(h)^{\frac{n-3+2(k+1)}{n-3+2(k+2)}} \leq C_6(n,k)\Phi(h)^{\frac{k+1}{k+2}}$$

because of $\Phi \leq 1$. Hence

$$g(h) \leq C_6(n,k) \, \Phi(h)^{1-2\varepsilon}.$$

We insert and substitute

$$E_{m,n}(S) \rightarrow\leq \binom{m}{n} C_7(n,k) \int_0^1 (1-\Phi)^{m-n} \Phi^{n-(3/2)} \Phi^{(1/2)-\varepsilon} d\Phi.$$

Finally, we have

$$E_{m,n}(S) \rightarrow\leq C_8(n,k) m^\varepsilon.$$

$\square$

W. Schmidt (1968), Lindberg (1981) and Borgwardt (1980) have shown that it is possible to construct a distribution yielding a growth of only $0(1)$ for $E_{m,n}(V)$, the expected number of boundary simplices, if it is regarded as a function of m for fixed n. They used the RDF

(4.2.8)
$$F(r) := \begin{cases} 0 & r < 1 \\ 1 - \frac{1}{r} & r \geq 1 \end{cases}$$

and proved that

(4.2.9) $E_{m,n}(V) \leq C(n)$ for $m \rightarrow \infty$ and fixed n, with unknown $C(n)$.

(Recall that $E_{m,n}(S)$ is bounded from above by $E_{m,n}(V)$.)

We could generalize this result to a greater class of distributions in Borgwardt (1980). The proof seems to be of minor interest here. But it seems to be important that we are able to give explicit upper bounds for $E_{m,n}(S)$ for a subset of those distributions, involved in Borgwardt (1980). This derivation can be done by the aid of Theorem 8 and formula (4.1.13). Note that our method is not applicable to Schmidt's distribution (4.2.8). So we intend to prove

Figure 4.3

Illustration of W. Schmidt's distribution

F(r) = 0   for r <1   The distribution has

F(r) = 1 - $\frac{1}{r}$ for r $\geqq$1   unbounded support.

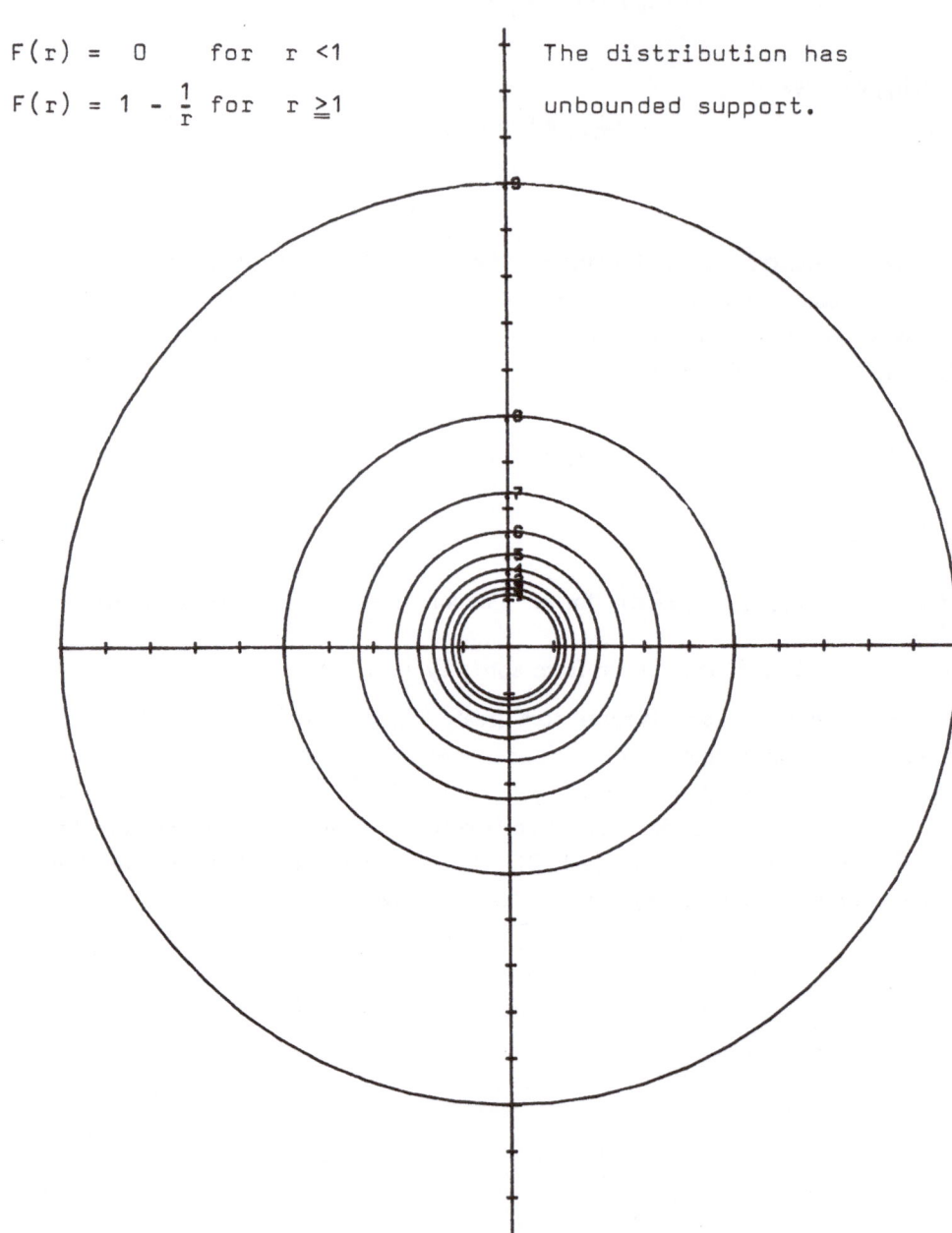

The numbers at the circles tell the probability that a
randomly chosen point lies inside.

Theorem 12

If there are real values $k > 0$, $l > 1$, such that

(4.2.10)
$$\lim_{r \to \infty} r^l \cdot [1 - F(r)] = k, \text{ then we know that}$$

(4.2.11)
$$\lim_{m \to \infty} \sup E_{m,n}(S) \le n\sqrt{2\pi} \left(\frac{n+l}{l-2}\right)^{(n-1)/2} \cdot \sqrt{1 + (n-1)(l-1)}$$

$\square$

Proof of the Theorem.   We first formulate an auxiliary result. Let $\delta > 0$ and let $J(r)$ be a monotone function of r. If the condition (4.2.10) holds, then there is an F such that for $h > \bar{r}$

$$\int_h^\infty J(r)dF(r) \le \int_h^\infty J(r)\frac{kl}{r^{l+1}}dr\,(1+\delta)$$

and

$$\int_h^\infty J(r)dF(r) \ge \int_h^\infty J(r)\frac{kl}{r^{l+1}}dr\,(1-\delta)\,.$$

(In the proof in Borgwardt (1980) (only in the proof) we made a little mistake by replacing $dF(r)$ by $\frac{1}{r^l}dr$ instead of $\frac{1}{r^{l+1}}dr$. With the correct replacement the proof works.)

The proof for these inequalities follows immediately from the elementary theory of Lebesgue integration. Before applying the upper bound in (4.1.13) we prove two estimations which hold asymptotically for $\bar{r} \to \infty$ and $h > r$.

1) $\dfrac{g_2(h)}{(1 - G(h))h} \to\le \dfrac{\Gamma\left(\frac{l-1}{2}\right)\Gamma\left(\frac{n+l+1}{2}\right)}{\Gamma\left(\frac{l}{2}\right)\Gamma\left(\frac{n+l}{2}\right)} \le \dfrac{\left(\frac{n+l}{2}\right)^{1/2}}{\left(\frac{l}{2} - 1\right)^{1/2}}$ (Appendix 1.11).

This follows from

$$\frac{g_2(h)}{(1-G(h))h} = \frac{1}{n-1} \frac{\int\limits_{h}^{\infty} \frac{(r^2-h^2)^{(n-1)/2}}{r^{n-2}} dF(r)}{h \int\limits_{h}^{\infty} \int\limits_{\frac{h}{r}}^{1} (1-\sigma^2)^{(n-3)/2} d\sigma dF(r)} \to \le$$

$$\to \le \frac{1}{n-1} \frac{\int\limits_{h}^{\infty} \frac{(r^2-h^2)^{(n-1)/2}}{r^{n-2+l+1}} dr}{h \int\limits_{h}^{\infty} \int\limits_{\frac{h}{r}}^{1} (1-\sigma^2)^{(n-3)/2} \sigma d\sigma \frac{1}{r^{l+1}} dr} =$$

$$= \frac{(n-1)}{(n-1)} \frac{\int\limits_{h}^{\infty} \frac{(r^2-h^2)^{(n-1)/2}}{r^{n+l-1}} dr}{\int\limits_{h}^{\infty} \frac{(r^2-h^2)^{(n-1)/2}}{r^{n+l}} dr\, h} =$$

$$= \frac{\frac{1}{h^{n+l-1-n}} \int\limits_{0}^{1} (1-x^2)^{(n-1)/2} x^{l-2} dx}{h \frac{1}{h^{n+l-n}} \int\limits_{0}^{1} (1-x^2)^{(n-1)/2} x^{l-1} dx} =$$

$$= \frac{\int\limits_{0}^{1} (1-x^2)^{(n-1)/2} x^{l-2} dx}{\int\limits_{0}^{1} (1-x^2)^{(n-1)/2} x^{l-1} dx} = \frac{\Gamma\left(\frac{l-1}{2}\right)\Gamma\left(\frac{n+l+1}{2}\right) k_2}{\Gamma\left(\frac{l}{2}\right)\Gamma\left(\frac{n+l}{2}\right) k_1}.$$

Here we have used that for $p+1>0$, $m+1>0$.

$$\int\limits_{0}^{1} x^m (1-x^2)^p dx = \frac{\Gamma(p+1)\Gamma\left(\frac{m+1}{2}\right)}{2\Gamma\left(\frac{m+3}{2}+p\right)}.$$

2) For $h > \bar{r}$ we have also

$$\frac{g(h)\left(\frac{g_2(h)}{g(h)} + (n-1)h^2\right)}{g_2(h)} \to \le 1 + (n-1)^2 \frac{\Gamma\left(\frac{n+1}{2}\right)\Gamma\left(\frac{l-1}{2}\right)}{\Gamma\left(\frac{n-1}{2}\right)\Gamma\left(\frac{l+1}{2}\right)} =$$

$$= 1 + (n-1)^2 \frac{l-1}{n-1}.$$

This follows from

$$\frac{g_2(h) + (n-1)h^2 g(h)}{g_2(h)} = 1 + (n-1)^2 \frac{h^2 \int\limits_h^\infty \frac{(r^2-h^2)^{(n-3)/2}}{r^{n-2}} dF(r)}{\int\limits_h^\infty \frac{(r^2-h^2)^{(n-1)/2}}{r^{n-2}} dF(r)}$$

$$\to \, \le 1 + (n-1)^2 \frac{h^2 \int\limits_h^\infty \frac{(r^2-h^2)^{(n-3)/2}}{r^{n-2+l+1}} dr}{\int\limits_h^\infty \frac{(r^2-h^2)^{(n-1)/2}}{r^{n-2+l+1}} d(r)} =$$

$$= 1 + (n-1)^2 \frac{h^2 \frac{1}{h^{l+1}} \int\limits_0^\infty (1-x^2)^{(n-3)/2} x^l \, dx}{\frac{1}{h^{l-1}} \int\limits_0^\infty (1-x^2)^{(n-1)/2} x^{l-2} \, dx}$$

$$= 1 + (n-1)^2 \frac{\Gamma\left(\frac{n-1}{2}\right)\Gamma\left(\frac{l+1}{2}\right)}{\Gamma\left(\frac{n+1}{2}\right)\Gamma\left(\frac{l-1}{2}\right)}$$

$$= 1 + (n-1)^2 \frac{l-1}{n-1} \, .$$

Note that the integral-function in the numerator of the first line is not monotone, but has two monotone pieces. So we can apply our method to the two pieces. Now we are allowed to replace $g_2(h)^{n-1}$ by using claim 1) and $\sqrt{g_2(h) + (n-1)h^2 g(h)}$ by using claim 2). The first term in the sum of (4.1.13) is asymptotically irrelevant. Consequently the asymptotic estimation is

$$E_{m,n}(S) \le \binom{m}{n} n^{3/2} \frac{\lambda_{n-1}(\omega_n)}{\lambda_{n-1}(\Omega_{n-1})} \left(\frac{n+l}{l-2}\right)^{(n-1)/2}$$

$$\sqrt{1 + (n-1)^2 \frac{l-1}{n-1}} \int\limits_0^1 (1-\Phi)^{m-n} \Phi^{n-1} d\Phi \, .$$

We use that the integral is $\frac{1}{n} \frac{1}{\binom{m}{n}}$ (Appendix 1.13) and that $\frac{\lambda_{n-1}(\omega_n)}{\lambda_{n-1}(\Omega_{n-1})} \le \sqrt{2\pi n}$. So we obtain the inequality.

$\square$

The result of this theorem leads immediately to the

Corollary

If $l = n^2$, we have the asymptotic relation

(4.2.12) $\lim\limits_{m\to\infty} \sup E_{m,n}(S) \le n^{5/2} C$, where $C \in \mathbb{R}$.

$\square$

Now we want to prove an additional Theorem, which states that such examples are impossible when the support is bounded.

Theorem 13

If $n \geq 3$ and $F(1) = 1$, $F(r) < 1$ for all $r < 1$, then

$$(4.2.13) \qquad\qquad E_{m,n}(S) \to \infty \text{ for } m \to \infty, \ n \text{ fixed.}$$

Proof. We apply (4.1.18) and show that our lower bound tends to infinity in the asymptotic case. Let $\delta > 0$ be given. If we choose $q < 1$ large enough, we can guarantee that

$$E_{m,n}(S) \geq \binom{m}{n} \frac{n^2}{2} \frac{\lambda_{n-1}(\omega_n)}{\lambda_{n-1}(\Omega_n)} \frac{q}{\sqrt{1-q^2}} \int_0^{\Phi(q)} (1-\Phi)^{m-n} \Phi^{n-1} d\Phi (1-\delta).$$

Here we have applied (4.2.4) and our substitution, as well as the inequality

$$\frac{h}{\sqrt{1-h^2}} \geq \frac{q}{\sqrt{1-q^2}} \quad \text{for } h \geq q.$$

The corresponding term

$$\binom{m}{n} \frac{n^2}{2} \frac{\lambda_{n-1}(\omega_n)}{\lambda_{n-1}(\Omega_{n-1})} \frac{q}{\sqrt{1-q^2}} \int_{\Phi(q)}^{1} (1-\Phi)^{m-n} \Phi^{n-1} d\Phi (1-\delta)$$

disappears asymptotically as shown in the proof of (4.2.6). Hence we obtain for given $\varepsilon > 0$ and $m(q)$, q great enough

$$E_{m,n}(S) \geq \binom{m}{n} \frac{n^2}{2} \frac{\lambda_{n-1}(\omega_n)}{\lambda_{n-1}(\Omega_{n-1})} \frac{q}{\sqrt{1-q^2}} \int_0^{1} (1-\Phi)^{m-n} \Phi^{n-1} d\Phi (1-\varepsilon).$$

Application of (Appendix 1.13) yields

$$E_{m,n}(S) \geq \frac{n}{2} \frac{\lambda_{n-1}(\omega_n)}{\lambda_{n-1}(\Omega_{n-1})} \frac{q}{\sqrt{1-q^2}} (1-\varepsilon)$$

for m chosen great enough. Since q can be chosen arbitrarily close to 1, we know that $E_{m,n}(S)$ tends to infinity. So all distributions with bounded support yield unbounded growth of $E_{m,n}(S)$ to infinity.

$\square$

4.3 SPECIAL DISTRIBUTIONS WITH BOUNDED SUPPORT

In this section we shall derive asymptotic upper and lower bounds for $E_{m,n}(S)$ under two special distributions:

1) Uniform distribution on ω_n.

2) Uniform distribution on Ω_n.

Both distributions have some nice features, which are very useful for our considerations. The most important property is that "the conditional distributions in the different hyperplanes $\{x \mid x^n = h_1\}$, $\{x \mid x^n = h_2\}$... are equivalent (with the exception of multiplication with a constant factor depending on h)".

Until now we have studied only upper bounds. Consequently, we do not have a good feeling about the sharpness, the efficiency and the quality of these upper bounds. This causes a great desire for lower bounds.

But the derivation of lower (asymptotic) bounds makes sense only for special distributions, because the order of growth can become arbitrarily small (as mentioned in the section before).

The treatment of lower bounds requires some preliminary considerations. For simplification we shall assume that the distributions have a density function. To begin with, we know that

$$(4.3.1) \qquad E_{m,n}(S) = \frac{E_{m,n}(S)}{E_{m,n}(Z)} E_{m,n}(Z) \, .$$

Now we turn to the quotient $\frac{E_{m,n}(S)}{E_{m,n}(Z)}$. Here we use our twofold integral transformation

$$(4.3.2)$$
$$\frac{E_{m,n}(S)}{E_{m,n}(Z)} = \frac{\binom{m}{n} \, [(n-2)!]^2 \, n\lambda_{n-1}(\omega_n) \, \lambda_{n-2}(\omega_{n-1})}{\binom{m}{n} \, [(n-2)!]^2 \, \lambda_{n-1}(\omega_n) \, \lambda_{n-2}(\omega_{n-1})} \cdot$$

$$\cdot \frac{\displaystyle\int_0^1 G(h)^{m-n} \int_{\mathbb{R}^{n-1}} \int_0^{\sqrt{1-h^2}} |\theta - c_n^{n-1}| \int_{\mathbb{R}^{n-2}} \cdots \int_{\mathbb{R}^{n-2}} \lambda_{n-2}(\mathrm{CH}(c_1,\ldots,c_{n-1}))^2}{\displaystyle\int_0^1 G(h)^{m-n} \int_{\mathbb{R}^{n-1}} \int_0^{\sqrt{1-h^2}} |\theta - c_n^{n-1}| \int_{\mathbb{R}^{n-2}} \cdots \int_{\mathbb{R}^{n-2}} \lambda_{n-2}(\mathrm{CH}(c_1,\ldots,c_{n-1}))^2}$$

$$\cdot \frac{W(c_1,\ldots,c_{n-1})f(c_1)\ldots f(c_{n-1})d\bar{c}_1\ldots d\bar{c}_{n-1}d\theta\, f(c_n)d\bar{c}_n\, dh}{V(c_1,\ldots,c_n)f(c_1)\ldots f(c_{n-1})d\bar{c}_1\ldots d\bar{c}_{n-1}d\theta\, f(c_n)d\bar{c}_n\, dh} \, .$$

Our asymptotic considerations rely on the fact that for growing m and fixed n the integral parts $[0, q]$ (for $q < 1$) become more and more irrelevant. Only the parts with

h close to 1 remain important. For h sufficiently great we can estimate

(4.3.3)
$$V(c_1,\ldots,c_n) = \frac{h}{n}\frac{\lambda_{n-1}(\mathrm{CH}(c_1,\ldots,c_n))}{\lambda_n(\Omega_n)}(1+\gamma(h)) =$$
$$= \frac{h}{n}\frac{|\theta - c_n^{n-1}|}{(n-1)}\frac{\lambda_{n-2}(\mathrm{CH}(c_1,\ldots,c_{n-1}))}{\lambda_n(\Omega_n)}(1+\gamma(h))$$

with $\gamma(h) \to 1$ for $h \to 1$.

And we know that

(4.3.4)
$$W(c_1,\ldots,c_{n-1}) = \frac{\sqrt{h^2+\theta^2}}{(n-1)}\frac{\lambda_{n-2}(\mathrm{CH}(c_1,\ldots,c_{n-1}))}{\lambda_{n-1}(\Omega_{n-1})}(1+\eta_1(h)) =$$
$$= \frac{h}{n-1}\frac{\lambda_{n-2}(\mathrm{CH}(c_1,\ldots,c_{n-1}))}{\lambda_{n-1}(\Omega_{n-1})}(1+\eta_2(h))$$
$$\text{with } \eta_1(h),\ \eta_2(h) \to 1 \quad \text{for } h \to 1.$$

After some reductions we obtain

Theorem 14

(4.3.5)
$$\frac{E_{m,n}(S)}{E_{m,n}(Z)} \to = \frac{n^2\lambda_n(\Omega_n)}{\lambda_{n-1}(\Omega_{n-1})}\cdot$$

$$\cdot\frac{\int\limits_0^1 G(h)^{m-n}h \int\limits_{\mathbb{R}^{n-1}} \int\limits_0^{\sqrt{1-h^2}} |\theta - c_n^{n-1}| \int\limits_{\mathbb{R}^{n-2}}\cdots\int\limits_{\mathbb{R}^{n-2}} \lambda_{n-2}(\mathrm{CH}(c_1,\ldots,c_{n-1}))^3}{\int\limits_0^1 G(h)^{m-n}h \int\limits_{\mathbb{R}^{n-1}} \int\limits_0^{\sqrt{1-h^2}} |\theta - c_n^{n-1}|^2 \int\limits_{\mathbb{R}^{n-2}}\cdots\int\limits_{\mathbb{R}^{n-2}} \lambda_{n-2}(\mathrm{CH}(c_1,\ldots,c_{n-1}))^3}$$

$$\frac{\cdot f(c_1)\ldots f(c_{n-1})d\bar{\bar{c}}_1\ldots d\bar{\bar{c}}_{n-1}d\theta f(c_n)d\bar{c}_n\,dh}{\cdot f(c_1)\ldots f(c_{n-1})d\bar{\bar{c}}_1\ldots d\bar{\bar{c}}_{n-1}d\theta f(c_n)d\bar{c}_n\,dh}.$$

$\square$

4.4 ASYMPTOTIC BOUNDS UNDER UNIFORM DISTRIBUTIONS

Now we want to give asymptotic bounds for the two special cases by insertion of the according terms into (4.3.5). We start with uniform distribution on ω_n and restrict our analysis on $n \geq 3$. In this special case we have

(4.4.1)
$$G(h) = 1 - \frac{\lambda_{n-2}(\omega_{n-1})}{\lambda_{n-1}(\omega_n)} \int_h^1 (1 - \sigma^2)^{(n-3)/2} d\sigma$$

(4.4.2)
$$g(h) = \frac{\lambda_{n-2}(\omega_{n-1})}{\lambda_{n-1}(\omega_n)} (1 - h^2)^{(n-3)/2}$$

(4.4.3)
$$g_2(h) = \frac{\lambda_{n-2}(\omega_{n-1})}{(n-1)\lambda_{n-1}(\omega_n)} (1 - h^2)^{(n-1)/2}.$$

We already know that $E_{m,n}(Z) \to = 1$. So we concentrate on (4.3.5) and we use a fictive density function (see Chapter 2, Section 4). The special property of the special distribution mentioned at the beginning of Section 3 yields

(4.4.4)
$$\frac{E_{m,n}(S)}{E_{m,n}(Z)} \to = \frac{n^2 \lambda_n(\Omega_n)}{\lambda_{n-1}(\Omega_{n-1})} \frac{\int_0^1 G(h)^{m-n}h \int_{R^{n-1}}}{\int_0^1 G(h)^{m-n}h \int_{R^{n-1}}}$$

$$\frac{\int_0^{\sqrt{1-h^2}} |\theta - c_n^{n-1}|\sqrt{1-h^2-\theta^2}^{3(n-2)+(n-4)(n-1)} d\theta f(c_n) d\bar{c}_n dh}{\int_0^{\sqrt{1-h^2}} |\theta - c_n^{n-1}|^2\sqrt{1-h^2-\theta^2}^{3(n-2)+(n-4)(n-1)} d\theta f(c_n) d\bar{c}_n dh} =$$

(4.4.5)
$$= \frac{n^2 \lambda_n(\Omega_n)}{\lambda_{n-1}(\Omega_{n-1})} \frac{\int_0^1 G(h)^{m-n}h \int_{R^{n-1}} \int_0^1 |\tau\sqrt{1-h^2} - c_n^{n-1}|\cdot}{\int_0^1 G(h)^{m-n}h \int_{R^{n-1}} \int_0^1 |\tau\sqrt{1-h^2} - c_n^{n-1}|^2\cdot}$$

$$\frac{\cdot|(1-h^2)(1-\tau^2)|^{(n^2-2n-2)/2}(1-h^2)^{1/2} d\tau f(c_n) d\bar{c}_n dh}{\cdot|(1-h^2)(1-\tau^2)|^{(n^2-2n-2)/2}(1-h^2)^{1/2} d\tau f(c_n) d\bar{c}_n dh}$$

For the evaluation of the numerator integral we take into regard that

a) $|\tau\sqrt{1-h^2} - c_n^{n-1}| \leq \tau\sqrt{1-h^2} + |c_n^{n-1}|$

b) $\int_{R^{n-1}} f(c_n) d\bar{c}_n = g(h)$

c) $\int_{R^{n-1}} |c_n^{n-1}| f(c_n) d\bar{c}_n = g(h) \frac{2\lambda_{n-3}(\omega_{n-2})}{(n-2)\lambda_{n-2}(\omega_{n-1})} (1 - h^2)^{1/2}$

d) $\displaystyle\int_0^1 \tau |1-\tau^2|^{(n^2-2n-2)/2} d\tau = \frac{1}{n^2-2n}$

e) $\displaystyle\int_{\mathbb{R}^{n-1}} \int_0^1 |\tau\sqrt{1-h^2} - c_n^{n-1}|(1-\tau^2)^{(n^2-2n-2)/2} d\tau f(c_n) d\bar{c}_n \geq$

$\displaystyle\geq \int_{\mathbb{R}^{n-1}} \int_0^1 |c_n^{n-1}|(1-\tau^2)^{(n^2-2n-2)/2} d\tau f(c_n) d\bar{c}_n \, .$

for symmetry reasons.

So we obtain the following upper and lower asymptotic bounds for $\frac{E_{m,n}(S)}{E_{m,n}(Z)}$:

$$\frac{n^2\lambda_n(\Omega_n)}{\lambda_{n-1}(\Omega_{n-1})} \frac{\int_0^1 G(h)^{m-n} h g(h)(1-h^2)^{(n^2-2n)/2} dh}{\int_0^1 G(h)^{m-n} h g(h)(1-h^2)^{(n^2-2n+1)/2} dh}$$

$$\frac{\left\{ \frac{1}{n^2-2n} + \frac{2\lambda_{n-3}(\omega_{n-2})}{(n-2)\lambda_{n-2}(\omega_{n-1})} \int_0^1 |1-\tau^2|^{(n-2n-2)/2} d\tau \right\}}{\left(\frac{1}{n^2-2n+1} + \frac{1}{n-1} \right) \int_0^1 (1-\tau^2)^{(n^2-2n-2)/2} d\tau} \to \leq$$

$(4.4.6)$ $\qquad\qquad \to \leq \dfrac{E_{m,n}(S)}{E_{m,n}(Z)} \to \leq$

$$\to \leq \frac{n^2\lambda_n(\Omega_n)}{\lambda_{n-1}(\Omega_{n-1})} \frac{\int_0^1 G(h)^{m-n} h g(h)(1-h^2)^{(n^2-2n)/2} dh}{\int_0^1 G(h)^{m-n} h g(h)(1-h^2)^{(n^2-2n+1)/2} dh} \, .$$

$$\cdot \frac{\left\{ \frac{1}{n^2-2n} + \frac{2\lambda_{n-3}(\omega_{n-2})}{(n-2)\lambda_{n-2}(\omega_{n-1})} \int_0^1 (1-\tau^2)^{(n^2-2n-2)/2} d\tau \right\}}{\left\{ \left(\frac{1}{n^2-2n+1} + \frac{1}{n-1} \right) \int_0^1 (1-\tau^2)^{(n^2-2n-2)/2} d\tau \right\}}$$

In both inequalities we find the quotient of integrals

$$\frac{\int_0^1 G(h)^{m-n} \, h \, g(h) \, (1-h^2)^{(n^2-2n)/2} dh}{\int_0^1 G(h)^{m-n} \, h \, g(h) \, (1-h^2)^{(n^2-2n+1)/2} dh} \, .$$

We prove that this quotient **asymptotically** has the upper bound

$(4.4.7)$ $\qquad\qquad m^{1/(n-1)} \left[\dfrac{\lambda_{n-2}(\omega_{n-1})}{(n-1)\lambda_{n-1}(\omega_n)} \right]^{1/(n-1)} \leq m^{1/(n-1)}$

and the lower bound

$$(4.4.8) \qquad m^{1/(n-1)} \left\{ \frac{\lambda_{n-2}(\omega_{n-1})}{n(n-1)\lambda_{n-1}(\omega_n)} \right\}^{1/(n-1)}$$

by use of (4.2.4), (4.4.3), the substitution $\Phi = 1 - G(h)$ and (Appendix 1.15).

The terms not depending on m in the upper bound of (4.4.6) are

$$(4.4.9) \qquad \frac{n^2 \lambda_n(\Omega_n)}{\lambda_{n-1}(\Omega_{n-1})} \cdot \frac{\frac{1}{n(n-2)} + \left\{ \frac{2\lambda_{n-3}(\omega_{n-2})}{(n-2)\lambda_{n-2}(\omega_{n-1})} \int_0^1 (1-\tau^2)^{(n^2-2n-2)/2} d\tau \right\}}{\left\{ \left(\frac{1}{(n-1)^2} + \frac{1}{n-1} \right) \right\} \int_0^1 (1-\tau^2)^{(n^2-2n-2)/2} d\tau}.$$

The corresponding quotient for the lower bound can be obtained by dropping the term $\frac{1}{n(n-2)}$ in the numerator of (4.4.9).

Evaluation of (4.4.9) for the case $n = 3$ (Appendix 2.25 – 2.27) leads to the value $\frac{160}{3\pi} \le 18 = n^2 \cdot 2$ (using the left side of (4.4.7)).

For $n \ge 4$ we make use of

$$\int_0^1 (1-\tau^2)^{(n^2-2n-2)/2} d\tau \ge \frac{1}{2} \sqrt{\frac{2\pi}{n(n-2)}}$$

and of

$$\frac{\lambda_{n-2}(\omega_{n-1})}{\lambda_{(n-3)}(\omega_{n-2})} \le \sqrt{\frac{2\pi}{n-3}} \quad \text{(Appendix 2.25),}$$

as well as of

$$\frac{\lambda_{n-1}(\omega_n)\lambda_{n-3}(\omega_{n-2})}{\lambda_{n-2}(\omega_{n-1})\lambda_{n-2}(\omega_{n-1})} \le 1 \quad \text{(Appendix 2.26).}$$

The completing factor in (4.4.7) can be estimated from above by 1. By this way we obtain as an upper bound for (4.4.9)

$$n^2 2 \left(1 + \sqrt{\frac{n-2}{n(n-3)}} \right).$$

The factor $(1 + \sqrt{\frac{n-2}{n(n-3)}})$ attains its maximum for $n = 4$. Here it satisfies

$$\left(1 + \sqrt{\frac{2}{4}} \right) = 1 + \sqrt{\frac{1}{2}}.$$

For the derivation of an asymptotical lower bound calculate the value for $n = 3$

$$18 \frac{16}{9\pi} \ge n^2 2 \frac{1}{2}.$$

For $n \geq 4$ we use that

$$\frac{\lambda_{n-1}(\omega_n)}{\lambda_{n-2}(\omega_{n-1})} \frac{\lambda_{n-3}(\omega_{n-2})}{\lambda_{n-2}(\omega_{n-1})} \geq 1 \sqrt{\frac{n-3}{n-1}}$$

and come to the lower bound

$$n^2 \, 2 \sqrt{\frac{n-3}{n-1}} \left(\frac{(n-1)^3}{n^2(n-2)}\right)$$

To complete the estimation for $\frac{E_{m,n}(S)}{E_{m,n}(Z)}$, we must not forget the factor arising in (4.4.8)

$$\left[\frac{\lambda_{n-2}(\omega_{n-1})}{n(n-1)\lambda_{n-1}(\omega_n)}\right]^{1/(n-1)} \, .$$

This factor is greater than $\frac{2}{7}$ if $n = 3$. So the complete estimation from below is

$$\frac{E_{m,n}(S)}{E_{m,n}(Z)} \geq m^{1/(n-1)} \, n^2 \, 2 \left(\frac{1}{n(n-1)} \sqrt{\frac{n-2}{2\pi}}\right)^{1/(n-1)} \cdot \left(\frac{(n-1)^3}{n^2(n-2)}\right) \frac{n-3}{n-1} \, .$$

All the factors in brackets attain their minimum in the set $n \geq 4$ exactly at $n = 4$. So we conclude that their product for $n \geq 4$ is bounded from below by $\frac{1}{6}$. So we have our

Theorem 15

For uniform distribution on ω_n we have for $n \geq 3$

1) an upper bound (asymptotic $m \to \infty$, n fixed)

(4.4.10) $$E_{m,n}(S) \to \leq m^{1/(n-1)} n^2 2\gamma(n)$$

where

$$\gamma(n) = \begin{cases} 1 & \text{for } n = 3 \\ 1 + \sqrt{\frac{n-2}{n(n-3)}} & \text{for } n \geq 4 \, . \end{cases}$$

So $\gamma(n) \leq 1 + \sqrt{\frac{1}{2}}$ for $n \geq 3$ and $\gamma(n) \to 1$ for $n \to \infty$.

2) a lower bound (asymptotic $m \to \infty$, n fixed)

(4.4.11) $$E_{m,n}(S) \to \geq m^{1/(n-1)} \, n^2 \, 2\eta(n)$$

where

$$\eta(n) = \begin{cases} \frac{1}{7} & \text{for } n = 3 \\ \left(\frac{1}{n(n-1)} \sqrt{\frac{n-2}{2\pi}}\right)^{1/(n-1)} \frac{(n-1)^3}{n^2(n-2)} \sqrt{\frac{n-3}{n-1}} & \text{for } n \geq 4 \, . \end{cases}$$

So $\eta(n) \geq \frac{1}{7}$ for $n \geq 3$ and $\eta(n) \to 1$ for $n \to \infty$.

$\square$

In the same manner we can derive bounds on $E_{m,n}(S)$ based on uniform distribution on Ω_n. For $n \geq 3$ we have here

$$(4.4.12) \qquad G(h) = 1 - \frac{\lambda_{n-1}(\Omega_{n-1})}{\lambda_n(\Omega_n)} \int_h^1 (1 - \sigma^2)^{(n-1)/2} d\sigma$$

$$(4.4.13) \qquad g(h) = \frac{\lambda_{n-1}(\Omega_{n-1})}{\lambda_n(\Omega_n)} (1 - h^2)^{(n-1)/2}$$

$$(4.4.14) \qquad g_2(h) = \frac{\lambda_{n-1}(\Omega_{n-1})}{(n+1)\lambda_n(\Omega_n)} (1 - h^2)^{(n+1)/2}.$$

The quotient corresponding to (4.3.5) gets the form

$$\frac{E_{m,n}(S)}{E_{m,n}(Z)} \rightarrow = \frac{n^2 \lambda_n(\Omega_n)}{\lambda_{n-1}(\Omega_{n-1})} \frac{\int_0^1 G(h)^{m-n} h \int_{\mathbb{R}^{n-1}}}{\int_0^1 G(h)^{m-n} \int_{\mathbb{R}^{n-1}}}$$

$$(4.4.15)$$

$$\frac{\int_0^{\sqrt{1-h^2}} |\theta - c_n^{n-1}| \sqrt{1 - h^2 - \theta^2}^{3(n-2)+(n-2)(n-1)} d\theta f(c_n) d\bar{c}_n \, dh}{\int_0^{\sqrt{1-h^2}} |\theta - c_n^{n-1}|^2 \sqrt{1 - h^2 - \theta^2}^{3(n-2)+(n-2)(n-1)} d\theta f(c_n) d\bar{c}_n \, dh}.$$

Again, we make use of

a) $|\tau\sqrt{1 - h^2} - c_n^{n-1}| \leq \tau\sqrt{1 - h^2} + |c_n^{n-1}|$

b) $\displaystyle\int_{\mathbb{R}^{n-1}} f(c_n) d\bar{c}_n = g(h)$

c) $\displaystyle\int_{\mathbb{R}^{n-1}} |c_n^{n-1}| f(c_n) d\bar{c}_n = g(h) \frac{2\lambda_{n-2}(\Omega_{n-2})}{n\lambda_{n-1}(\Omega_{n-1})} (1 - h^2)^{1/2}$

d) $\displaystyle\int_0^1 \tau(1 - \tau^2)^{(n^2-4)/2} d\tau = \frac{1}{n^2 - 2}$

e) $\displaystyle\int_{\mathbb{R}^{n-1}} \int_0^1 |\tau\sqrt{1 - h^2} - c_n^{n-1}|(1 - \tau^2)^{(n^2-4)/2} d\tau f(c_n) d\bar{c}_n \geq$

$$\geq \int_{\mathbb{R}^{n-1}} \int_0^1 |c_n^{n-1}|(1 - \tau^2)^{(n^2-4)/2} d\tau f(c_n) d\bar{c}_n.$$

So we obtain the two inequalities

$$
\frac{n^2 \lambda_n(\Omega_n)}{\lambda_{n-1}(\Omega_{n-1})} \frac{\int_0^1 G(h)^{m-n} hg(h)(1-h^2)^{(n^2-2)/2} dh \cdot}{\int_0^1 G(h)^{m-n} hg(h)(1-h^2)^{(n^2-1)/2} dh \cdot}
$$

$$
\frac{\cdot \frac{2\lambda_{n-2}(\Omega_{n-2})}{n\lambda_{n-1}(\Omega_{n-1})} \int_0^1 (1-\tau^2)^{(n^2-4)/2} d\tau}{\cdot \left(\frac{1}{n^2-1} + \frac{1}{n+1}\right) \int_0^1 (1-\tau^2)^{(n^2-4)/2} d\tau} \to \leq
$$

(4.4.16)
$$
\to \leq \frac{E_{m,n}(S)}{E_{m,n}(Z)} \to \leq
$$

$$
\to \leq \frac{n^2 \lambda_n(\Omega_n) \int_0^1 G(h)^{m-n} hg(h)(1-h^2)^{(n^2-2)/2} dh \cdot}{\lambda_{n-1}(\Omega_{n-1}) \int_0^1 G(h)^{m-n} hg(h)(1-h^2)^{(n^2-1)/2} dh \cdot}
$$

$$
\frac{\cdot \left\{\frac{1}{n^2-2} + \frac{2\lambda_{n-2}(\Omega_{n-2})}{n\lambda_{n-1}(\Omega_{n-1})} \int_0^1 (1-\tau^2)^{(n^2-4)/2} d\tau \right\}}{\cdot \left\{\left(\frac{1}{n^2-1} + \frac{1}{n+1}\right) \int_0^1 (1-\tau^2)^{(n^2-4)/2} d\tau \right\}}
$$

The integral quotient depending on h yields

$$
\frac{\int_0^1 G(h)^{m-n} hg(h)(1-h^2)^{(n^2-2)/2} dh}{\int_0^1 G(h)^{m-n} hg(h)(1-h^2)^{(n^2-1)/2} dh}
$$

and has asymptotically the upper bound

(4.4.17)
$$
m^{1/(n+1)} \left[\frac{\lambda_{n-1}(\Omega_{n-1})}{(n+1)\lambda_n(\Omega_n)}\right]^{1/(n+1)}
$$

and the asymptotical lower bound

(4.4.18)
$$
m^{1/(n+1)} \left[\frac{\lambda_{n-1}(\Omega_{n-1})}{n(n+1)\lambda_n(\Omega_n)}\right]^{1/(n+1)} .
$$

The remaining terms, independent of h resp. m, give for an upper bound and $n = 3$
$9 \cdot 2, 7 = n^2 \cdot 2 \cdot 1,35$; for an upper bound while $n \geq 4$

$$
n^2 \cdot 2 \cdot \left(1 + \frac{n}{\sqrt{n^2 - 2}\sqrt{n-1}}\right) \frac{n^2-1}{n^2} \leq n^2 \cdot 2 \cdot \left(1 + \frac{1}{\sqrt{2}}\right).
$$

Similarly we get results on the lower bound. For $n = 3$ the evaluation of the lower bound leads to

$$18\frac{64}{27\pi} \geq n^2 \cdot 2 \cdot \frac{3}{4} \,.$$

The completing factor in (4.4.18) is $\geq \frac{1}{2}$.

For $n \geq 4$ we have the lower bound

$$n^2 \cdot 2 \cdot \frac{\sqrt{n^2 - 1}(n - 1)}{n^2} \,.$$

For the complete lower bound we have to include

$$\left[\frac{\lambda_{n-1}(\Omega_{n-1})}{n(n+1)\lambda_n(\Omega_n)}\right]^{1/(n+1)} \,,$$

For $n = 3$ this term is

$$\left(\frac{\pi}{3 \cdot 4 \cdot \frac{4}{3} \cdot \pi}\right) = \left(\frac{1}{16}\right)^{1/4} = \frac{1}{2} \,.$$

Finally we obtain the

Theorem 16

For uniform distribution on Ω_n we have an asymptotic (m tending to infinity and $n \geq 3$ fixed) upper bound

(4.4.19) $$E_{m,n}(S) \rightarrow \leq m^{1/(n+1)} \, n^2 \, 2\gamma(n)$$

where

$$\gamma(n) = \begin{cases} 1,35 & \text{for } n = 3 \\ 1 + \dfrac{n}{\sqrt{n^2 - 2}\sqrt{n-1}} \, \dfrac{n^2-1}{n^2} & \text{for } n \geq 4 \,. \end{cases}$$

So $\gamma(n) \leq 1 + \sqrt{\frac{1}{2}}$ for $n \geq 3$ and $\gamma(n) \rightarrow 1$ for $n \rightarrow \infty$.

Also we have the lower bound

(4.4.20) $$E_{m,n}(S) \rightarrow \geq m^{1/(n+1)} \, n^2 \, 2\eta(n)$$

where

$$\eta(n) = \begin{cases} \dfrac{3}{8} & \text{for } n = 3 \\ \left(\dfrac{1}{(n+1)\sqrt{2\pi n}}\right)^{1/(n+1)} \dfrac{\sqrt{n^2-1}(n-1)}{n^2} & \text{for } n \geq 4 \,. \end{cases}$$

So $\eta(n) \geq \frac{1}{3}$ and $\eta(n) \rightarrow 1$ for $n \rightarrow \infty$.

$\square$

4.5 ASYMPTOTIC BOUNDS
UNDER GAUSSIAN DISTRIBUTION

We are now going to deal with a distribution with unbounded support, the Gaussian distribution. It has some very nice properties. The most important one is the "independence" of components, which can be recognized by regarding the density function on $\mathbb{R}^n$

$$(4.5.1) \qquad f(x) = \pi^{-\frac{n}{2}} e^{-(x^1)^2 - \ldots - (x^n)^2} \quad (x \in \mathbb{R}^n).$$

A consequence of this independence is that in the sets $\{x \mid x^n = h\}$ the "inner distributions" are equal for every h (independent of h). The radial distribution function is

$$(4.5.2) \qquad F(r) = \pi^{-\frac{n}{2}} n \lambda_{n-1}(\omega_n) \int_0^r e^{-\tau^2} \tau^{n-1} d\tau.$$

The marginal functions are

$$(4.5.3) \qquad G(h) = \frac{1}{\sqrt{\pi}} \int_{-\infty}^h e^{-\tau^2} d\tau$$

$$(4.5.4) \qquad g(h) = \frac{1}{\sqrt{\pi}} e^{-h^2}$$

$$(4.5.5) \qquad g_2(h) = \frac{1}{\sqrt{\pi}} \frac{1}{2} e^{-h^2}.$$

Again, we try to estimate $E_{m,n}(S)$ by calculation and estimation of the quotient mentioned above. But here the "spherical measures" will cause more trouble than in Section 4. First of all we apply our well-known coordinate transformations and obtain

$$E_{m,n}(S) = \binom{m}{n} n(n-2)! \lambda_{n-1}(\omega_n) \lambda_{n-2}(\omega_{n-1}) \cdot$$

$$(4.5.6) \qquad \cdot \int_0^\infty G(h)^{m-n} \int_{\mathbb{R}^{n-1}} \int_0^\infty |\theta - c_n^{n-1}| \int_{\mathbb{R}^{n-2}} \cdots \int_{\mathbb{R}^{n-2}}$$

$$\{\lambda_{n-2}(\mathrm{CH}(c_1, \ldots, c_{n-1}))\}^2 W(c_1, \ldots, c_{n-1}) f(c_1) \ldots f(c_{n-1})$$

$$d\bar{c}_1 \ldots d\bar{c}_{n-1} d\theta f(c_n) d\bar{c}_n dh,$$

as well as

$$E_{m,n}(Z) = \binom{m}{n}(n-2)!\lambda_{n-1}(\omega_n)\lambda_{n-2}(\omega_{n-1})$$

(4.5.7)
$$\int_0^\infty G(h)^{m-n} \int_{\mathbb{R}^{n-1}} \int_0^\infty |\theta - c_n^{n-1}| \int_{\mathbb{R}^{n-2}} \cdots \int_{\mathbb{R}^{n-2}}$$

$$\{\lambda_{n-2}(CH(c_1,\ldots,c_{n-1}))\}^2 V(c_1,\ldots,c_n) f(c_1)\ldots f(c_{n-1})$$

$$d\bar{c}_1 \ldots d\bar{c}_{n-1} d\theta\, f(c_n) d\bar{c}_n\, dh\,.$$

If $h \gg \|\bar{c}_i\|_\infty := \max(|c_i^1|,\ldots,|c_i^{n-1}|)$ for $i = 1,\ldots,n$, the spherical measures W and V could be calculated easily. It is interesting and helpful that in the asymptotic case this condition is satisfied with extremely high probability. So it will be possible to concentrate on the subset of $\mathbb{R}^n$ where this condition is satisfied. The following arguments show that such a restriction is allowed for the asymptotic case.

1) In (4.5.6) and (4.5.7) we find integrals of the form $\int_0^\infty G(h)^{m-n}\ldots dh$.

As we have seen in this chapter before, the values of such integrals are asymptotically approximated by integrals of the type $\int_q^\infty G(h)^{m-n}\ldots dh$ for every positive q. More formal: We know that

$$\lim_{\substack{m \to \infty \\ n \text{ fixed}}} \frac{\int_q^\infty G(h)^{m-n}\ldots dh}{\int_0^\infty G(h)^{m-n}\ldots dh} = 1\,.$$

2) $E_{m,n}(S)$ and $E_{m,n}(Z)$ are bounded. So we know that

$$\int_0^\infty G(h)^{m-n} \int_{\mathbb{R}^{n-1}} \int_0^\infty |\theta - c_n^{n-1}| \int_{\mathbb{R}^{n-2}} \cdots \int_{\mathbb{R}^{n-2}} \cdot$$

$$\cdot (\lambda_{n-2}\{CH(c_1,\ldots,c_{n-1})\})^2\, W(\ldots)_{\text{resp.}V}\, f(c_1)\ldots f(c_{n-1}) \cdot$$

(4.5.8)
$$\cdot\, d\bar{c}_1 \ldots d\bar{c}_{n-1} d\theta\, f(c_n) d\bar{c}_n\, dh \to =$$

$$\to = \int_q^\infty G(h)^{m-n} \int_{\overline{\nabla}(\sqrt{h})} \int_0^{\sqrt{h}} |\theta - c_n^{n-1}| \int_{\overline{\overline{\nabla}}(\sqrt{h})} \cdots \int_{\overline{\nabla}(\sqrt{h})} \cdots$$

$$f(c_1)\ldots f(c_{n-1}) d\bar{c}_1 \ldots d\bar{c}_{n-1} d\theta\, f(c_n) d\bar{c}_n\, dh\,.$$

Here

$$\nabla(\sqrt{h}) = \{x \mid x \in \mathbb{R}^n, \|\bar{x}\|_\infty \leq \sqrt{h}\}$$

$$\overline{\nabla}(\sqrt{h}) = \{x \mid x \in \mathbb{R}^{n-1}, \|x\|_\infty \leq \sqrt{h}\}$$

$$\overline{\overline{\nabla}}(\sqrt{h}) = \{x \mid x \in \mathbb{R}^{n-2}, \|x\|_\infty \leq \sqrt{h}\}$$

So we are allowed to concentrate on the restricted area.

3) In this region the evaluation of W and V becomes quite simple, because we know that

$$W(c_1, \ldots, c_{n-1}) = \frac{\lambda_{n-1}(CC(c_1, \ldots, c_{n-1}) \cap (\Omega_n)}{\lambda_{n-1}(\Omega_{n-1})},$$

$$\lambda_{n-1}(CC\{c_1, \ldots, c_{n-1}\} \cap \Omega_n) \leq \lambda_{n-1}(CH\{0, \frac{1}{h}c_1, \ldots, \frac{1}{h}c_{n-1}\}) =$$

$$= \lambda_{n-1}(CH\{0, c_1, \ldots, c_{n-1}\})\frac{1}{h^{n-1}}$$

and

$$\lambda_{n-1}(CC\{c_1, \ldots, c_{n-1}\} \cap \Omega_n) \leq \lambda_{n-1}(CH\{0, \frac{1}{\Psi}c_1, \ldots, \frac{1}{\Psi}c_{n-1}\}) =$$

$$= \lambda_{n-1}(CH\{0, c_1, \ldots, c_{n-1}\})\left(\frac{1}{\Psi}\right)^{n-1},$$

where $\Psi = (\max\{\|c_1\|_2, \ldots, \|c_{n-1}\|_2\})$. In the restricted integration area we have $h \leq \Psi \leq \sqrt{h^2 + (n-1)h}$. Consequently, $\frac{\Psi}{h}$ tends to 1 while $h \to \infty$. So we are allowed to replace

$$W(c_1, \ldots, c_{n-1}) \text{ by } \frac{\lambda_{n-1}(CH\{0, c_1, \ldots, c_{n-1}\})}{h^{n-1} \cdot \lambda_{n-1}(\Omega_{n-1})}$$

and

$$V(c_1, \ldots, c_n) \text{ by } \frac{\lambda_{n-1}(CH\{0, c_1, \ldots, c_n\})}{h^n \lambda_n(\Omega_n)}.$$

But

$$\lambda_{n-1}(CH\{0, c_1, \ldots, c_{n-1}\}) \geq \frac{h}{n-1} \lambda_{n-2}(CH\{c_1, \ldots, c_{n-1}\})$$

and

$$\lambda_{n-1}(CH\{0, c_1, \ldots, c_{n-1}\}) \leq \frac{\sqrt{h^2 + h}}{n-1} \lambda_{n-2}(CH\{c_1, \ldots, c_{n-1}\}).$$

The quotient $\frac{h}{\sqrt{h^2 + h}}$ again tends to 1 for $h \to \infty$. So $\lambda_{n-1}(CH\{0, c_1, \ldots, c_{n-1}\})$ can be replaced by $\frac{h}{n-1}\lambda_{n-2}(CH\{c_1, \ldots, c_{n-1}\})$ in the asymptotic case. In addition, we make use of the fact that

$$\lambda_n(CH\{0, c_1, \ldots, c_n\}) = \frac{h}{n}\lambda_{n-1}(CH\{c_1, \ldots, c_n\}) =$$

$$= \frac{h}{n}|\theta - c_n^{n-1}|\frac{1}{n-1}\lambda_{n-2}(CH\{c_1, \ldots, c_{n-1}\}).$$

Figure 4.4

Illustration of the restricted area $\nabla(\sqrt{h}\,)$ in $\mathbf{R}^3$
and of $\bar{\nabla}(\sqrt{h}\,)$

seen
from
the
side

different
values
of h

seen
from
above

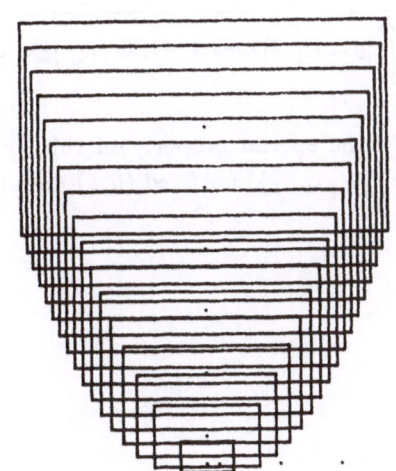

The result of all these considerations is the asymptotic equation

(4.5.9)

$$\frac{E_{m,n}(S)}{E_{m,n}(Z)} \to= \frac{n^2 \lambda_n(\Omega_n) \int\limits_q^\infty G(h)^{m-n}.}{\lambda_{n-1}(\Omega_{n-1}) \int\limits_q^\infty G(h)^{m-n}.}$$

$$\cdot \frac{\int\limits_{\overline{\nabla}(\sqrt{h})} \int\limits_0^{\sqrt{h}} |\theta - c_n^{n-1}| \int\limits_{\overline{\overline{\nabla}}(\sqrt{h})} \cdots \int\limits_{\overline{\overline{\nabla}}(\sqrt{h})} \lambda_{n-2}(CH\{c_1,\ldots,c_{n-1}\})^3.}{\int\limits_{\overline{\nabla}(\sqrt{h})} \int\limits_0^{\sqrt{h}} |\theta - c_n^{n-1}|^2 \int\limits_{\overline{\overline{\nabla}}(\sqrt{h})} \cdots \int\limits_{\overline{\overline{\nabla}}(\sqrt{h})} \lambda_{n-2}(CH\{c_1,\ldots,c_{n-1}\})^3.}$$

$$\cdot \frac{\cdot h^{-n+2} f(c_1) \ldots f(c_{n-1}) d\overline{\overline{c}}_1 \ldots d\overline{\overline{c}}_{n-1} d\theta f(c_n) d\overline{c}_n dh}{\cdot h^{-n+1} f(c_1) \ldots f(c_{n-1}) d\overline{\overline{c}}_1 \ldots d\overline{\overline{c}}_{n-1} d\theta f(c_n) d\overline{c}_n dh}.$$

Evaluation of the inner integrals yields

(4.5.10)

$$\frac{E_{m,n}(S)}{E_{m,n}(Z)} = \frac{n^2 \lambda_n(\Omega_n)}{\lambda_{n-1}(\Omega_{n-1})} \frac{\int\limits_q^\infty G(h)^{m-n} h^{-n+2} e^{-h^2(n-1)}.}{\int\limits_q^\infty G(h)^{m-n} h^{-n+1} e^{-h^2(n-1)}.}$$

$$\cdot \frac{\int\limits_{\overline{\nabla}(\sqrt{h})} \int\limits_0^{\sqrt{h}} |\theta - c_n^{n-1}| e^{-\theta^2(n-1)} d\theta f(c_n) d\overline{c}_n dh}{\int\limits_{\overline{\nabla}(\sqrt{h})} \int\limits_0^{\sqrt{h}} |\theta - c_n^{n-1}|^2 e^{-\theta^2(n-1)} d\theta f(c_n) d\overline{c}_n dh}.$$

This simplification is due to the special property of this distribution, because the average value of $\lambda_{n-2}(CH\{c_1,\ldots,c_{n-1}\})$ for fixed (h,θ) does not depend upon (h,θ).

For deriving an **upper bound** we estimate $|\theta - c_n^{n-1}| \leq |\theta| + |c_n^{n-1}|$ and obtain

$$\frac{E_{m,n}(S)}{E_{m,n}(Z)} \to\leq \frac{n^2\lambda_n(\Omega_n)}{\lambda_{n-1}(\Omega_{n-1})} \frac{\int\limits_q^\infty G(h)^{m-n} h^{-n+2} e^{-h^2(n-1)}}{\int\limits_q^\infty G(h)^{m-n} h^{-n+1} e^{-h^2(n-1)}}$$

(4.5.11)

$$\frac{\left[\int\limits_{\overline{\nabla}(\sqrt{h})} f(c_n)d\bar{c}_n \int\limits_0^h \theta e^{-\theta^2(n-1)}d\theta +\right.}{\left[\int\limits_{\overline{\nabla}(\sqrt{h})} f(c_n)d\bar{c}_n \int\limits_0^h \theta^2 e^{-\theta^2(n-1)}d\theta +\right.}$$

$$\left. + \int\limits_{\overline{\nabla}(\sqrt{h})} |c_n^{n-1}| f(c_n)d\bar{c}_n \int\limits_0^h e^{-\theta^2(n-1)}d\theta\right] dh$$

$$\left. + \int\limits_{\overline{\nabla}(\sqrt{h})} (c_n^{n-1})^2 f(c_n)d\bar{c}_n \int\limits_0^h e^{-\theta^2(n-1)}d\theta\right] dh$$

$.$

All the inner integrals over $[0, \sqrt{h}]$ and $\overline{\nabla}(\sqrt{h})$ are asymptotic approximations for the corresponding integrals over $[0, \infty)$, resp. $\mathbb{R}^{n-1}$. Evaluation of those "complete" integrals yields

(4.5.12)

$$\frac{E_{m,n}(S)}{E_{m,n}(Z)} \to\leq \frac{n^2\lambda_n(\Omega_n)\int\limits_0^\infty G(h)^{m-n} h^{-n+2} e^{-h^2(n-1)}}{\lambda_{n-1}(\Omega_{n-1})\int\limits_0^\infty G(h)^{m-n} h^{-n+1} e^{-h^2(n-1)}}$$

$$\frac{g(h)dh \cdot 2[(n-1) + \sqrt{n-1}]}{g(h)dh \cdot \sqrt{\pi} \cdot n} \; .$$

For the derivation of the corresponding **lower bound** we use the fact that (as a result of symmetry) the replacement of $|\theta - c_n^{n-1}|$ by c_n^{n-1} in the integral of (4.5.10) decreases the numerator. So we get — following the same way as above —

(4.5.13)

$$\frac{E_{m,n}(S)}{E_{m,n}(Z)} \to\geq \frac{n^2\lambda_n(\Omega_n)\int\limits_0^\infty G(h)^{m-n} h^{-n+2} e^{-h^2(n-1)}g(h)dh}{\lambda_{n-1}(\Omega_{n-1})\int\limits_0^\infty G(h)^{m-n} h^{-n+1} e^{-h^2(n-1)}g(h)dh} \; .$$

$$\cdot \frac{2(n-1)}{\sqrt{\pi}n} \; .$$

For the evaluation of the quotient of integrals appearing in (4.5.12) and (4.5.13) we apply the formula (see RENYI, p. 135)

$$(4.5.14) \qquad G(h) = \frac{1}{\sqrt{\pi}} \int_{-\infty}^{h} e^{-\tau^2} d\tau = 1 - \frac{1}{\sqrt{\pi}} \frac{e^{-h^2}}{2h\left(1 + \frac{\Theta(h)}{h^2}\right)}$$

$$\text{where } 0 \le \Theta(h) \le 1 \quad \text{for } h \ge 0.$$

We introduce $\Phi = 1 - G(h)$ as a new integration variable with

$$\Phi(h) = \frac{1}{\sqrt{\pi}} \frac{e^{-h^2}}{2h\left(1 + \frac{\Theta(h)}{h^2}\right)} .$$

This equation shows that

$$\lim_{h \to \infty} \frac{\Phi(h)}{\frac{e^{-h^2}}{h}} = \frac{1}{2\sqrt{\pi}}$$

and that

$$\lim_{h \to \infty} \frac{\sqrt{\ln \frac{1}{\Phi(h)}}}{h} = 1 .$$

So we replace $\frac{e^{-h^2}}{h}$ by $2\sqrt{\pi}\Phi$, and h by $\sqrt{\ln \frac{1}{\Phi(h)}}$ and arrive at

$$(4.5.15) \qquad \frac{\int_{0}^{\frac{1}{2}} (1 - \Phi)^{m-n} \Phi^{n-1} \cdot \sqrt{\frac{1}{\ln \Phi}} d\Phi}{\int_{0}^{\frac{1}{2}} (1 - \Phi)^{m-n} \Phi^{n-1} d\Phi} \to =$$

$$\to = \frac{\int_{0}^{1} (1 - \Phi)^{m-n} \Phi^{n-1} \cdot \sqrt{\frac{1}{\ln \Phi}} d\Phi}{\int_{0}^{1} (1 - \Phi)^{m-n} \Phi^{n-1} d\Phi}$$

This quotient is estimated in (Appendix 1.18–1.20) and has

$$(4.5.16) \qquad \text{the upper bound} \sqrt{\ln \frac{m}{n-1}}$$

$$(4.5.17) \qquad \text{the lower bound} \sqrt{\ln \frac{m+1}{n}} \quad \text{(asymptotically)}.$$

Finally we take into regard that

$$(4.5.18) \qquad n^2 \sqrt{\frac{2\pi}{n+1}} \le \frac{n^2 \lambda_n(\Omega_n)}{\lambda_{n-1}(\Omega_{n-1})} \le \sqrt{\frac{2\pi}{n}} n^2 \quad \text{and}$$

$$\frac{n^2 \lambda_n(\Omega_n)}{\lambda_{n-1}(\Omega_{n-1})} = 9 \cdot \frac{4}{3} = 12 \quad \text{for } n = 3.$$

So we arrive at the inequalities

$$n^2 \frac{2}{\sqrt{\pi}} \frac{(n-1)}{n} \sqrt{\frac{2\pi}{n+1}} \sqrt{\ln \frac{m+1}{n}} \to\le \frac{E_{m,n}(S)}{E_{m,n}(Z)} \to\le$$

$$\to\le \frac{2}{\sqrt{\pi}} \frac{(n-1)+\sqrt{n-1}}{n} \sqrt{\frac{2\pi}{n}} \, n^2 \sqrt{\ln \frac{m}{n-1}} \quad \text{for } n \ge 4 \text{ and}$$

(4.5.19)

$$n^2 \frac{2}{\sqrt{\pi}} \frac{(n-1)}{n} \frac{4}{3} \sqrt{\ln \frac{m+1}{n}} \to\le \frac{E_{m,n}(S)}{E_{m,n}(Z)} \to\le$$

$$\to\le \frac{2}{\sqrt{\pi}} \frac{(n-1)+\sqrt{n-1}}{n} \frac{4}{3} \, n^2 \sqrt{\ln \frac{m}{n-1}} \quad \text{for } n = 3.$$

Now we use that

$$\lim_{\substack{m\to\infty \\ n \text{ fixed}}} \frac{\ln \frac{m+1}{n}}{\ln m} = \lim_{\substack{m\to\infty \\ n \text{ fixed}}} \frac{\ln \frac{m}{n-1}}{\ln m} = 1$$

and obtain for $n \ge 4$

$$n^{\frac{3}{2}} 2^{\frac{3}{2}} \frac{(n-1)}{\sqrt{n(n+1)}} \sqrt{\ln m} \to\le \frac{E_{m,n}(S)}{E_{m,n}(Z)} \to\le$$

(4.5.20)

$$\to\le n^{\frac{3}{2}} 2^{\frac{3}{2}} \frac{(n-1)+\sqrt{n-1}}{n} \sqrt{\ln m} .$$

The term $\frac{(n-1)}{\sqrt{n(n+1)}}$ attains its minimum for $n \ge 4$ exactly at $n = 4$ with $\frac{3}{2\sqrt{5}} \ge \frac{2}{3}$.

The term $(n-1)+\sqrt{n-1}$ can be estimated by

$$\frac{(n-1)+\sqrt{n-1}}{n} \le \frac{n+\sqrt{n}}{n} \le 1 + \frac{1}{\sqrt{n}} \le 1 + \frac{1}{2} \quad \text{for } n \ge 4 .$$

For $n = 3$ the lower bound of (4.5.19) gives

$$n^2 \frac{2}{\sqrt{\pi}} \frac{2}{3} \frac{4}{3} \sqrt{\ln m} \ge n^{\frac{3}{2}} 2^{\frac{3}{2}} \frac{\sqrt{2}}{\sqrt{\pi}\sqrt{3}} \frac{4}{3} \sqrt{\ln m} \ge n^{\frac{3}{2}} 2^{\frac{3}{2}} \frac{9}{20} \frac{4}{3} \sqrt{\ln m} = n^{\frac{3}{2}} 2^{\frac{3}{2}} \frac{3}{5} \sqrt{\ln m}.$$

Here the upper bound satisfies

$$n^2 \frac{2}{\sqrt{\pi}} \frac{2+\sqrt{2}}{3} \frac{4}{3} \sqrt{\ln m} \le n^{\frac{3}{2}} 2^{\frac{3}{2}} \frac{8(1+\sqrt{2})}{\sqrt{\pi}\sqrt{3}\,3} \sqrt{\ln m} \le n^{\frac{3}{2}} 2^{\frac{3}{2}} \frac{8}{3} \frac{5}{6} \sqrt{\ln m} \le n^{\frac{3}{2}} 2^{\frac{3}{2}} \frac{20}{9} \sqrt{\ln m}.$$

And we have our final

Theorem 17

For Gaussian distribution on $\mathbb{R}^n$ we have asymptotic bounds on $E_{m,n}(S)$ of the following kind:

$$(m \to \infty, n \geq 3 \text{ fixed}).$$

1) Upper bound

(4.5.21)

$$E_{m,n}(S) \to \leq \sqrt{\ln m}\ n^{\frac{3}{2}}\ 2^{\frac{3}{2}}\ \gamma(n)$$

$$\text{where } \gamma(n) = \begin{cases} \frac{20}{9} & \text{for } n = 3 \\ \frac{(n-1)+\sqrt{n-1}}{n} & \text{for } n \geq 4 \end{cases}.$$

So $\gamma(n) \to 1$ for $n \to \infty$, and $\gamma(n) \leq \frac{20}{9}$ for $n \geq 3$.

2) Lower bound

(4.5.22)

$$E_{m,n}(S) \to \geq \sqrt{\ln m}\ n^{\frac{3}{2}}\ 2^{\frac{3}{2}}\ \eta(n)$$

$$\text{where } \eta(n) = \begin{cases} \frac{3}{5} & \text{for } n = 3 \\ \frac{(n-1)}{\sqrt{n}(n+1)} & \text{for } n \geq 4 \end{cases}.$$

So $\eta(n) \to 1$ for $n \to \infty$, and $\eta(n) \geq \frac{3}{5}$ for $n \geq 3$.

$\square$

Chapter 5

PROBLEMS WITH NONNEGATIVITY CONSTRAINTS

5.1 THE GEOMETRY

Now we deal with linear programming problems of the type

$$(5.1.1) \quad \begin{array}{ll} \text{Maximize} & v^T x \\ \text{subject to} & a_1^T x \leq 1, \ldots, a_m^T x \leq 1, x \geq 0 \\ \text{where} & x, a_1, \ldots, a_m, v \in \mathbb{R}^n. \end{array}$$

Here the feasibility region will be denoted by $\hat{X}$. The origin is — in any case — a vertex of $\hat{X}$ and can be used as an initial vertex. Hence Phase I is superfluous.

The shadow-vertex-algorithm is applicable here, too. The strategy for solving the problem will be quite similar to the method described in (1.4.6) and (1.4.7).

To begin with, we need some definitions and notation. The nonnegativity constraints $x^j \geq 0$ for $j = 1, \ldots, n$ will be formulated as follows

$$(5.1.2) \quad \begin{array}{l} \rho(-e_j)^T x \leq 1 \quad \text{for all } \rho > 0, \rho \in \mathbb{R}, \\ \text{where } e_j \text{ is the } j\text{-th unit vector of } \mathbb{R}^n. \end{array}$$

Let e be the vector $(1, \ldots, 1)^T \in \mathbb{R}^n$. The dual polyhedron $\hat{Y} \subset \mathbb{R}^n$ (dual relative to $\hat{X} \subset \mathbb{R}^n$) is here

$$(5.1.3) \quad \hat{Y} := \{y \mid y^T x \leq 1 \text{ for all } x \in \hat{X}\}$$

Lemma 5.1

(5.1.4) $$\hat{Y} = CH(0, a_1, \ldots, a_m) + CC(-e_1, \ldots, -e_n).$$

Proof. At first we show $\supset$.

Let $y \in CH(0, a_1, \ldots, a_m) + CC(-e_1, \ldots, -e_n)$ or equivalently

$$y = \lambda_1 a_1 + \ldots + \lambda_m a_m - \rho_1 e_1 - \ldots - \rho_n e_n$$

with $\lambda_1, \ldots, \lambda_m, \rho_1, \ldots, \rho_n \geq 0$ and $\lambda_1 + \ldots + \lambda_m \leq 1$. For an arbitrary $x \in \hat{X}$ we conclude that $y^T x \leq \lambda_1 + \ldots + \lambda_m \leq 1$. Hence y belongs to $\hat{Y}$.

For the proof of $\subset$ let $y \in \hat{Y}$.

Assume that y is not an element of the set on the right hand side of the claim. Since this set is convex, there is a vector z such that $y^T z > 1$ and $\bar{y}^T z \leq 1$ for all $\bar{y} \in CH(\ldots) + CC(\ldots)$. So z itself is an element of $\hat{X}$. But now y cannot belong to $\hat{Y}$ because of (5.1.3) and we have a contradiction.

$\square$

This time, our condition of **nondegeneracy** will be

(5.1.5)   *Every set of n vectors out of $\{a_1, \ldots, a_m, -e_1, \ldots, -e_n, v\}$*

is linearly independent and every set of $n + 1$ vectors

out of $\{a_1, \ldots, a_m\}$ is in general position.

Let $\Delta = \{\Delta^1, \ldots, \Delta^n\} \subset \{1, \ldots, m + n\}$ denote an n-element index-set such that $\Delta^i < \Delta^{i+1}$ for $i = 1, \ldots, n - 1$.

For simplification and abbreviation we set $a_{m+j} := -e_j$ and we assign the restriction $x^j \geq 0$ with the index $m + j$. Now we define a "number of normal restrictions" for every Δ by setting

(5.1.6) $l(\Delta) := \sharp(\Delta \cap \{1, \ldots, m\})$ $\sharp$ stands for cardinality of a set.

Here $a_{\Delta^1}, \ldots, a_{\Delta^l}$ are the "normal" and $a_{\Delta^{l+1}}, \ldots, a_{\Delta^n}$ are the "standard" or non-negativity restrictions. Consider a vertex x_Δ of $\hat{X}$. In x_Δ there are exactly n active constraints. Corresponding to that vertex we have the dual region $\{y \mid y^T x_\Delta = 1\} \cap \hat{Y}$. x_Δ satisfies the following n equations

(5.1.7)

$$
\begin{array}{cc}
a_{\Delta^1}^T x = 1 & x^{(\Delta^{l+1}-m)} = 0 \\
\vdots & \vdots \\
a_{\Delta^l}^T x = 1 & x^{(\Delta^n - m)} = 0 \\
(\sharp = l) & (\sharp = n - l)
\end{array}
$$

We define

(5.1.8) $$\hat{\Sigma}(\Delta) = \begin{cases} CH(a_{\Delta^1}, \ldots, a_{\Delta^l}) + CC(a_{\Delta^{l+1}}, \ldots, a_{\Delta^n}) & \text{for } l > 0 \\ CC(-e_1, \ldots, -e_n) = CC(a_{m+1}, \ldots, a_{m+n}) & \text{for } l = 0 \end{cases}$$

and obtain the

Figure 5.1

<u>The intersection of span(u,v) with $\hat{Y}$</u>

We observe that the following boundary simplex-cones
are intersected:
background:
$CH(a_5)+CC(-e_1,-e_2)$; $CH(a_5,a_9)+CC(-e_1)$;

foreground
$CH(a_5,a_8,a_9)$; $CH(a_4,a_8,a_9)$; $CH(a_4,a_8)+CC(-e_2)$;

$CH(a_4)+CC(-e_2,-e_3)$

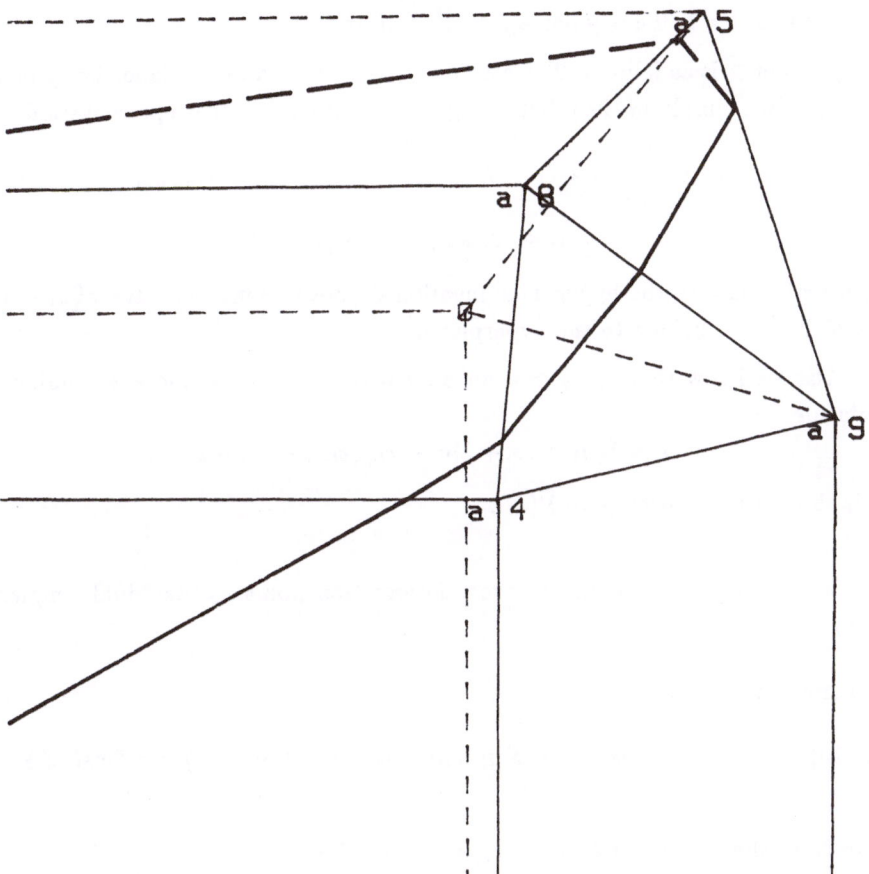

Notice that 0 is not an intersection of rays - the perspec-
tivic   view causes this wrong impression.

Lemma 5.2 *If $l > 0$, then*

$$(5.1.9) \qquad\qquad \widehat{\Sigma}(\Delta) = \{y \mid y^T x_\Delta = 1\} \cap \widehat{Y}.$$

Proof. Without loss of generality we set

$$\Delta^1 = 1, \ldots, \Delta^l = l, \ \Delta^{l+1} = m + l + 1, \ldots, \Delta^m = m + n$$

in this proof. First we prove $\subset$.

Let $y \in \widehat{\Sigma}(\Delta)$. Then

$$y = \lambda_1 a_1 + \ldots + \lambda_l a_l - \rho_{l+1} e_{l+1} - \ldots - \rho_n e_n$$

with $\lambda_1 + \ldots + \lambda_l = 1$; $\lambda_i, \rho_j \geq 0$ for $i = 1, \ldots, l$; $j = l+1, \ldots, n$. Because of (5.1.7) we have $y^T x_\Delta = 1$. That y belongs to $\widehat{Y}$ is clear.

The proof of $\supset$ is as follows. The set $\{y \mid y^T x_\Delta = 1\}$ is a hyperplane. Let y be a point of that plane, simultaneously belonging to $\widehat{Y}$. Then y has the representation

$$y = \lambda_1 a_1 + \ldots + \lambda_m a_m - \rho_{m+1} e_1 - \ldots - \rho_{m+n} e_n$$

$$\lambda_1 + \ldots + \lambda_m = 1, \lambda_i, \rho_j \geq 0.$$

x_Δ is orthogonal to the hyperplane mentioned above. Hence we have $x_\Delta^T a_i < 1$ for all a_i which do not belong to the hyperplane.

Also we know that $e_j^T x_\Delta > 0$ for all j with $x_\Delta^j > 0$. So there is a unique representation

$$y = \lambda_1 a_1 + \ldots + \lambda_l a_l - \rho_{l+1} e_{l+1} - \ldots - \rho_n e_n.$$

This shows that y belongs to $\widehat{\Sigma}(\Delta)$.

$\qquad\qquad\qquad\qquad\qquad\qquad\qquad\qquad\qquad\qquad\qquad\qquad\qquad\qquad$ □

So we are able to assign to every intersection point x_Δ its "dual simplex-cone" $\widehat{\Sigma}(\Delta)$.

Lemma 5.3

$$(5.1.10) \qquad x_\Delta \text{ is a vertex of } X \text{ if and only if } l = 0 \text{ or } \widehat{\Sigma}(\Delta) \text{ is a facet of } \widehat{Y}.$$

Proof. In the case $l = 0$ we have $x_\Delta = 0$ and all is clear.

Now let $l > 0$. Let x_Δ be a vertex of $\widehat{X}$. x_Δ is orthogonal to $\widehat{\Sigma}(\Delta)$. Then an arbitrary element of $\widehat{\Sigma}(\Delta)$ is on the boundary of $\widehat{Y}$, because $(1 + \delta)y^T x_\Delta > 1$ for all $\delta > 0$. This shows that $(1 + \delta)y \notin \widehat{Y}$ for all $\delta > 0$.

For the opposite direction let $\widehat{\Sigma}(\Delta)$ be contained in a boundary hyperplane of $\widehat{Y}$. Then x_Δ is a normal vector to that plane. Let $y \in \widehat{\Sigma}(\Delta)$. We assume that x_Δ is not feasible. Then there is an a_i such that $a_i^T x_\Delta > 1$ or an e_i with $-e_i^T x_\Delta > 0$. In the first case we have $y + (a_i - y)\frac{1}{2} \in \widehat{Y}$, but $\{y + (a_i - y)\frac{1}{2}\}^T x_\Delta > 1$. Hence $\widehat{\Sigma}(\Delta)$ is not contained in a supporting hyperplane for $\widehat{Y}$ with normal vector x_Δ. In the second case we have $y - \rho e_i \in \widehat{Y}$, but $x_\Delta^T(y - \rho e_i) > 1$. This leads to a contradiction, too. So we know that x_Δ is feasible and satisfies at least n of the restrictions exactly (as equations). Hence it is a vertex of $\widehat{X}$.

As before in Chapter I, Section 2 (1.2.8), the Lemma of Farkas enables us to show the

Lemma 5.4 *Let x_Δ be a vertex of $\widehat{X}$, $w \in \mathbb{R}^n$, $w \neq 0$.*

(5.1.11) *x_Δ is maximal with respect to $w^T x$ on $\widehat{X}$, if and only if $\widehat{\Sigma}(\Delta) \cap \mathbb{R}^+ x \neq \emptyset$.*

$\square$

The proof proceeds exactly in the same way as the proof of (1.2.8).

So we are again interested in those Δ's, whose $\widehat{\Sigma}(\Delta)$'s meet the boundary-condition and are simultaneously intersected by a certain two-dimensional plane. Again we have such shadow-vertices and are able to apply the shadow-vertex-algorithm.

Figure 5.2a

The polyhedra $\hat{Y}$ and $\hat{X}$ seen from the side

The following boundary simplex-cones are intersected:

$CC(-e_1,-e_2,-e_3)$; $CH(a_9)+CC(-e_1,-e_3)$; $CH(a_5,a_9)+CC(-e_1)$;

$CH(a_5,a_8,a_9)$; $CH(a_5,a_8)+CC(-e_2)$; $CH(a_4,a_8)+CC(-e_2)$;

$$CH(a_4)+CC(-e_2,-e_3) \ .$$

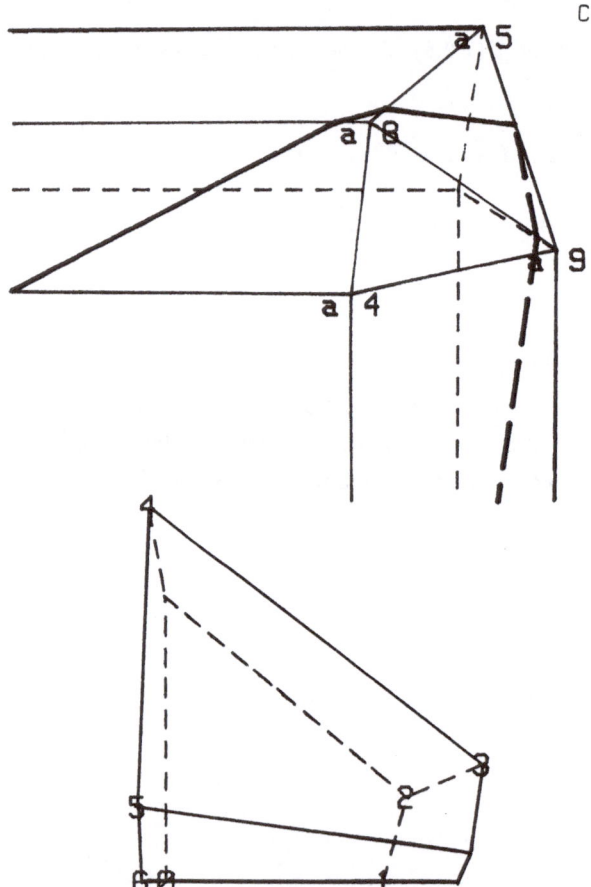

The corresponding sequence of vertices is $x_0,x_1,x_2,x_3,x_4,$ x_5,x_6.

Figure 5.2b

The polyhedra $\hat{Y}$ and $\hat{X}$ seen from above

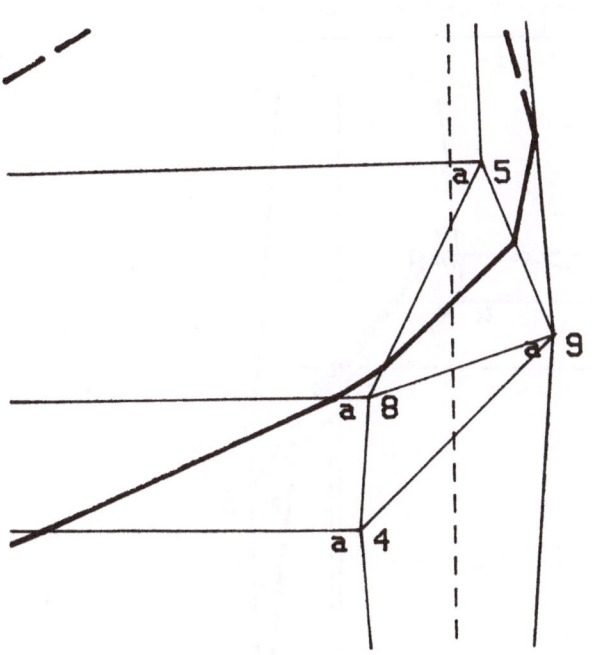

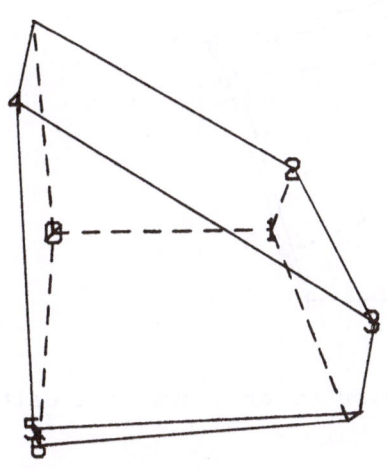

Figure 5.2c

The polyhedra Ŷ and X̂   seen from below

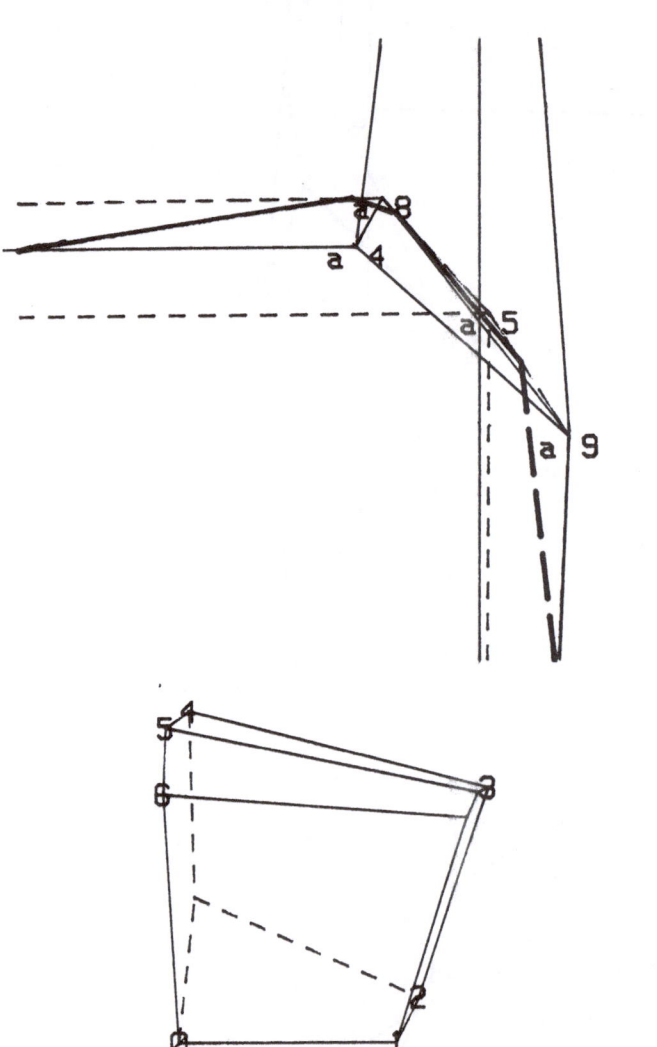

Under this perspective we observe that the path actually follows the shadow.

5.2 THE COMPLETE SOLUTION METHOD

Now we want to introduce a method for solving the complete problem. This method shows slight differences to the method described in Chapter I, Section 4 (1.4.6, 1.4.7), but is closely related. Let again Π_k be the projection $\Pi_k : \mathbb{R}^n \to \mathbb{R}^k$ $(k = 1, \ldots, n)$ with

(5.2.1)
$$\Pi_k \begin{bmatrix} x^1 \\ \vdots \\ x^n \end{bmatrix} = \begin{bmatrix} x^1 \\ \vdots \\ x^k \end{bmatrix}$$

and let I_k be the problem

(5.2.2)
$$\begin{aligned} \text{Maximize} \quad & \Pi_k(v)^T \Pi_k(x) \\ \text{subject to} \quad & \Pi_k(a_1)^T \Pi_k(x) \leq 1 \\ & \qquad \vdots \\ & \Pi_k(a_m)^T \Pi_k(x) \leq 1 \\ & x^1 \geq 0 \\ & \quad \vdots \\ & x^k \geq 0. \end{aligned}$$

Let $\hat{Y}_k$ denote the polyhedron

(5.2.3) $\quad \mathrm{CH}(0, \Pi_k(a_1), \ldots, \Pi_k(a_m)) + \mathrm{CC}(-\Pi_k(e_1), \ldots, -\Pi_k(e_k)).$

Now our strategy can be described as follows.

(5.2.4)

1) First we set $k = 2$ and start in $\hat{Y}_2$ at the "boundary simplex-cone" $\mathrm{CC}(-\Pi_2(e_1), -\Pi_2(e_2))$ with optimization direction $(-1, -1)^T$. We try to find that boundary simplex-cone of $\hat{Y}_2$, which is intersected by $\mathbb{R}^+ \Pi_2(v)$.
If we do not find a solution, then there is no solution for the complete problem either and we go to 5).

2) If $k < n$ then we set $k = k + 1$. Else we go to 6).

3) Starting from the boundary simplex-cone of $\hat{Y}_{k-1}$ just found
$\mathrm{CH}(\Pi_{k-1}(a_{\Delta^1}), \ldots, \Pi_{k-1}(a_{\Delta^l})) + \mathrm{CC}(\Pi_{k-1}(a_{\Delta^{l+1}}), \ldots, \Pi_{k-1}(a_{\Delta^{k-1}}))$
we find (in one pivot step) a vector a_i ($i \notin \Delta^1, \ldots, \Delta^{k-1}$ and $i \leq m+k$) such that
if $i > m$
$\mathrm{CH}(\Pi_k(a_{\Delta^1}), \ldots, \Pi_k(a_{\Delta^l})) + \mathrm{CC}(\Pi_k(a_{\Delta^{l+1}}), \ldots, \Pi_k(a_{\Delta^{k-1}}), \Pi_k(a_i))$
if $i \leq m$

$CH(\Pi_k(a_{\Delta^1}), \ldots, \Pi_k(a_{\Delta^l}), \Pi_k(a_i)) + CC(\Pi_k(a_{\Delta^{l+1}}), \ldots, \Pi_k(a_{\Delta^{k-1}}))$
is a boundary simplex-cone of $\hat{Y}_k$. This facet of $\hat{Y}_k$ is intersected by
$\text{span}\left(\left[\begin{array}{c}\Pi_{k-1}(v)\\0\end{array}\right], \Pi_k(e_k)\right).$

4)   Now we apply in $\hat{Y}_k$ the shadow-vertex algorithm using the two-dimensional plane $\text{span}(\Pi_k(v), \Pi_k(e_k), e_k)$. By this way we are searching for that boundary simplex-cone
$CH(\Pi_k(a_{\Delta^1}), \ldots, \Pi_k(a_{\Delta^l})) + CC(\Pi_k(a_{\Delta^{l+1}}), \ldots, \Pi_k(a_{\Delta^k})),$
which is intersected by $\mathbb{R}^+(\Pi_k(v))$. If we do not find such a solution, then the global problem has none and we go to 5). Else we return to 2).

5)   Print: Problem unsolvable. Go to 7).

6)   Print the data of the last boundary simplex-cone (the solution).

7)   STOP.

So we reach the solution (if one exists) after $(n-1)$ applications (runs) of the shadow-vertex algorithm, using the respective planes

(5.2.5) $\text{span}(\Pi_k(v), \Pi_k(e_k)).$

The changes of dimensions (stages) in 3) require at most $(n-2)$ pivot steps in total. Initialization does not require any step here. Now we are interested in the number of facets of $\hat{Y}_k$ (boundary simplex-cones) in the single stages k, which are intersected by the corresponding two-dimensional plane. Summation over all stages $k = 2, \ldots, n$ gives the number of pivot steps for the complete solution.

5.3 A SIMPLIFICATION OF THE BOUNDARY-CONDITION

We are now going to analyze the application of the shadow-vertex algorithm for an arbitrary stage k. As before, a simplex-cone has to be counted, if it satisfies the boundary- and the intersection-condition. In this section we want to deal with the boundary-condition. We try to understand exactly the conditions under which $\hat{\Sigma}_k(\Delta) = \Pi_k(CH(a_{\Delta^1}, \ldots, a_{\Delta^l}) + CC(a_{\Delta^{l+1}}, \ldots, a_{\Delta^k}))$ is a boundary simplex-cone of $\Pi_k(\hat{Y}) = \hat{Y}_k$.

Let again $l > 0$ and $\Delta^1 < \ldots < \Delta^l \leq m < \Delta^{l+1} < \ldots < \Delta^k \leq m + k$. (Here Δ shall denote an index set of k elements). And let Π_l^Δ be a special projection of $\mathbb{R}^k$ into $\mathbb{R}^l$, namely

$\Pi_l^\Delta(x) = (x^{i_1}, \ldots, x^{i_l})^T$ where $i_1 < \ldots < i_l \leq k$ and
                                    where $m + i_1 \notin \Delta, \ldots, m + i_l \notin \Delta$

(in other words: all components of x with $x^j = 0$, where $m + j \in \Delta$, are dropped). In the same way we define $Y_i^\Delta := \Pi_i^\Delta(Y_k) = \Pi_i^\Delta(\Pi_k(CH(0, a_1, \ldots, a_m)))$.

Lemma 5.5

(5.3.1) $\widehat{\Sigma}_k(\Delta)$ is a facet of $\widehat{Y}_k$ if and only if

a) $\Pi_i^\Delta(\Pi_k(CH(a_{\Delta^1}, \ldots, a_{\Delta^l})))$ is a boundary simplex of Y_i^Δ, and

b)   the corresponding normal vector in $\mathbb{R}^l$ is strictly positive.

Proof. For simplicity and without loss of generality assume that $k = n$. First let $\widehat{\Sigma}(\Delta)$ be a facet of $\widehat{Y}$. Then there is a vector $x_\Delta \in \mathbb{R}^n$ such that

$$
\begin{aligned}
x_\Delta^T a_i &= 1 & &\text{for } i \in \{\Delta^1, \ldots, \Delta^l\} \\
x_\Delta^T a_i &< 1 & &\text{for } i \notin \{\Delta^1, \ldots, \Delta^l\} \text{ and } i \leq m \\
x_\Delta^T a_i &= -x_\Delta^T e_{i-m} = 0 & &\text{for } i \in \{\Delta^{l+1}, \ldots, \Delta^m\} \\
x_\Delta^T a_i &= -x_\Delta^T e_{i-m} < 0 & &\text{for } i \notin \{\Delta^{l+1}, \ldots, \Delta^m\} \text{ and } i > m.
\end{aligned}
$$

Now consider the vector $\Pi_i^\Delta(x_\Delta)$. We know that $\Pi_i^\Delta(x_\Delta) > 0$, because x_Δ is nonnegative and all zero-elements are dropped during the projection. So property b) is satisfied. For the proof of property a) recall that $\Pi_i^\Delta(x_\Delta)$ satisfies $\Pi_i^\Delta(x_\Delta)^T \Pi_i^\Delta(a_i) = x_\Delta^T a_i = 1$ for $i \in \{\Delta^1, \ldots, \Delta^l\}$ and $\Pi_i^\Delta(x_\Delta)^T \Pi_i^\Delta(a_i) = x_\Delta^T a_i < 1$ for $i \notin \{\Delta^1, \ldots, \Delta^l\}$, $i \leq m$.

Both properties hold, because only zero-components of x_Δ have been dropped. So we have a boundary simplex of Y_i^Δ indeed.

Now we turn to the opposite direction. Let $CH(\Pi_i^\Delta(a_{\Delta^1}), \ldots, \Pi_i^\Delta(a_{\Delta^l}))$ be a boundary simplex of Y_i^Δ with a positive normal vector $w \in \mathbb{R}^l$. Then define $\overline{w} \in \mathbb{R}^n$ such that $\overline{w}^i = w^i$ for $i \leq n$, $i + m \notin \Delta$ and $\overline{w}^i = 0$ else.

So we have

$$
\begin{aligned}
\overline{w}^T a_i &= w^T \Pi_i^\Delta(a_i) = 1 & &\text{for } i \in \{\Delta^1, \ldots, \Delta^l\} \text{ and} \\
\overline{w}^T a_i &= w^T \Pi_i^\Delta(a_i) < 1 & &\text{for } i \notin \{\Delta^1, \ldots, \Delta^l\}, \ i \leq m,
\end{aligned}
$$

because we have started from a boundary simplex of Y_i^Δ. For $i \leq n$, $i + m \in \Delta$, we have $-\rho \overline{w}^T e_i = 0 < 1$ for all $\rho \geq 0$, because $\overline{w}^i = 0$. For $i \leq n$, $i + m \in \Delta$, we know that $w^i > 0$ and consequently $-\overline{w}^T e_i < 0$. This confirms the boundary simplex-cone property.

5.4 EXPLICIT FORMULATION OF THE INTERSECTION-CONDITION

Now let us analyze the intersection-condition. The intersection plane in stage k is $\mathrm{span}(\Pi_k(v), \Pi_k(e_k))$. We consider an arbitrary facet of

$$
\widehat{Y}_k = \Pi_k(\widehat{Y}) : \widehat{\Sigma}(\Delta) = \Pi_k(CH(a_{\Delta^1}, \ldots, a_{\Delta^l}) + CC(a_{\Delta^{l+1}}, \ldots, a_{\Delta^k}))
$$

where $\Delta^1 < \Delta^2 < \ldots < \Delta^l \le m \le \Delta^{l+1} < \ldots < \Delta^k \le m + k$. Intersection with the plane mentioned above means that

$$\{0\} \ne \mathrm{span}(\Pi_k(v), \Pi_k(e_k) \cap \Pi_k(\mathrm{CC}(a_{\Delta^1}, \ldots, a_{\Delta^l}, a_{\Delta^{l+1}}, \ldots, a_{\Delta^k}))).$$

This is equivalent to the condition that

(5.4.1)   Either $\Pi_{k-1}(v)$ or $-\Pi_{k-1}(v)$ belongs to $\Pi_{k-1}(\mathrm{CC}(a_{\Delta^1}, \ldots, a_{\Delta^k}))$.

If this condition holds, then there is a "side simplex-cone" (even two) of the form

$$\Pi_{k-1}(\mathrm{CC}(a_{\Delta^1}, \ldots, a_{\Delta^{i-1}}, a_{\Delta^{i+1}}, \ldots, a_{\Delta^k}))$$

which contains $\Pi_{k-1}(v)$ or $-\Pi_{k-1}(v)$ (Theorem of Caratheodory).

Note that one vector a_{Δ^i} is omitted here and let us restrict our concentration to $\Pi_{k-1}(v)$. We try to make this condition evaluable. Unfortunately the input-vectors $a_{\Delta^{l+1}}, \ldots, a_{\Delta^k}$ spanning the cone or the side cone do not come from a rotation-invariant distribution. For that reason we have to separate these vectors from the others. We must distinguish between four cases:

$$
\begin{array}{cccc}
& m+k \in \Delta & & m+k \notin \Delta \\
\nearrow & \searrow & \nearrow & \searrow \\
\Delta^i < m+k & \Delta^i = m+k \quad & \Delta^i \le m & \Delta^i > m \\
\text{Case 1} & \text{Case 2} \quad & \text{Case 3} & \text{Case 4}
\end{array}
$$

Case 1

Here $\Delta^k = m + k \in \Delta$. So $\Pi_k(-e_k)$ belongs to the spanning vectors of the cone.

Claim 1

(5.4.2)   If the problem is nondegenerate, then Case 1 does not occur.

Proof. Since $a_{\Delta^k} = -e_k$, we have $\Pi_{k-1}(a_{\Delta^k}) = 0$. So the intersection condition is simplified to

$$\Pi_{k-1}(v) \in \Pi_{k-1}(\mathrm{CC}(a_{\Delta^1}, \ldots, a_{\Delta^{i-1}}, a_{\Delta^{i+1}}, \ldots, a_{\Delta^{k-1}})),$$

which leads to a system of $k - 1$ equations in $k - 2$ variables. In case of nondegeneracy the system does not have a solution.

Case 2

Now $\Delta^k = m + k \in \Delta$, but $\Pi_k(-e_k)$ is the "omitted direction" for generating the side-cone.

Claim 2

(5.4.3)   There is at most one simplex cone of $\hat{Y}_k$, where the situation of Case 2 can occur (summed up over all possible values of $l = 0, \ldots, k - 1$).

Proof. Let $\hat{\Sigma}_k(\Delta)$ and $\hat{\Sigma}_k(\overline{\Delta})$ be two facets of $\hat{Y}_k$ such that $m + k \in \Delta \cap \overline{\Delta}$. The "omitted direction" in both cases may be $\Delta^k = \overline{\Delta}^k = m + k$. Then the intersection conditions are:

There are $\lambda_1, \ldots, \lambda_k \geq 0$, $\rho \in \mathbb{R}$, such that

$$\Pi_k(v) = \Pi_k(\lambda_1 a_{\Delta^1} + \ldots + \lambda_{k-1} a_{\Delta^{k-1}} + \rho(-e_k))$$

and there are $\overline{\lambda}_1, \ldots, \overline{\lambda}_k \geq 0$, $\overline{\rho} \in \mathbb{R}$ such that

$$\Pi_k(v) = \Pi_k(\overline{\lambda}_1 a_{\overline{\Delta}^1} + \ldots + \overline{\lambda}_{k-1} a_{\overline{\Delta}^{k-1}} + \overline{\rho}(-e_k)).$$

So we know that η_1, η_2 and σ_1 exist such that

$$\eta_1[\Pi_k(v) + \sigma\Pi_k(e_k)] \in \hat{\Sigma}_k(\Delta) \quad \text{for all } \sigma \leq \sigma_1$$
$$\eta_2[\Pi_k(v) + \sigma\Pi_k(e_k)] \in \hat{\Sigma}_k(\overline{\Delta}) \quad \text{for all } \sigma \leq \sigma_1$$

Now $\eta_1[\Pi_k(v) + \sigma_1\Pi_k(e_k)]$ belongs to $\hat{\Sigma}_k(\Delta)$ and $\eta_2[\Pi_{k-1}(v) + \sigma_1\Pi_k(e_k)]$ belongs to $\hat{\Sigma}(\Delta)$. This shows that $\eta_1 = \eta_2$. Without loss of generality let $\eta_1 = \eta_2 = 1$.

Hence there is a point $y \in \hat{\Sigma}(\Delta) \cap \Sigma(\overline{\Delta})$ such that for all $\sigma < \min\{\sigma_1, 0\}$:

$$\Pi_k(y) + \sigma\Pi_k(e_k) \in \hat{\Sigma}(\Delta) \cap \hat{\Sigma}(\overline{\Delta}).$$

That ray $\{\Pi_k(y) + \sigma\Pi_k(e_k) \mid \rho < \min(\sigma_1, 0)\}$ is contained in a $k - 2$-dimensional side-simplex cone of $\hat{\Sigma}(\Delta)$ if Δ and $\overline{\Delta}$ are different. Then $\Pi_k(v)$ would be a linear combination of $k - 1$ generating input vectors including $\Pi_k(-e_k)$ and this would contradict nondegeneracy. Hence $\Delta = \overline{\Delta}$.

Case 3

$m + k \notin \Delta$, $\Delta^i \leq m$. Here the "omitted" direction belongs to the "normal" restriction vectors, which are distributed symmetrically under rotations. Intersection means that $\lambda_1, \ldots, \lambda_l \geq 0$ exist such that

(5.4.4) $v^j = \lambda_1 a_{\Delta^1}^j + \ldots + \lambda_{i-1} a_{\Delta^{i-1}}^j + \lambda_{i+1} a_{\Delta^{i+1}}^j + \ldots + \lambda_l a_{\Delta^l}^j$
for all $j \leq k - 1$, where $m + j \notin \Delta$ ($\# = l - 1$)

(5.4.5) $v^j \leq \lambda_1 a_{\Delta^1}^j + \ldots + \lambda_{i-1} a_{\Delta^{i-1}}^j + \lambda_{i+1} a_{\Delta^{i+1}}^j + \ldots + \lambda_l a_{\Delta^l}^j$
for all $j \leq k$, where $m + j \in \Delta$ ($\# = k - l$)

The second part is a system of $k - l$ inequalities in $l - 1$ variables, which are already determined by the system of equations.

Case 4

Again we assume that $m + k \notin \Delta$. But now $\Delta^i > m$. Here we have $\lambda_1, \ldots, \lambda_l \geq 0$ such that

$$(5.4.6) \qquad v^j = \lambda_1 a^j_{\Delta_1} + \ldots + \lambda_l a^j_{\Delta_l}$$

for all $j \leq k - 1$, with $m + j \notin \Delta$ ($\sharp = l - 1$)

$$(5.4.7) \qquad v^j = \lambda_1 a^j_{\Delta_1} + \ldots + \lambda_l a^j_{\Delta_l}$$

for $m + j = \Delta^i$ ($\sharp = 1$)

$$(5.4.8) \qquad v^j \leq \lambda_1 a^j_{\Delta_1} + \ldots + \lambda_l a^j_{\Delta_l}$$

for $j \leq k$ where $m + j \in \Delta$ and $m + j \neq \Delta^i$ ($\sharp = k - l - 1$)

(5.4.6) and (5.4.7) define a system of l equations in l variables. (5.4.8) is a system of $k - l - 1$ inequalities in l variables, which are already determined by (5.4.6) and (5.4.7).

Remark It is useful for the following steps that the system of equations consisting of (5.4.6) and (5.4.7) can be reduced to a system as in (5.4.4) (for Case 3) by weakening the intersection condition slightly. This is feasible when we want to derive upper bounds.

Lemma 5.6 *If (5.4.6) and (5.4.7) are satisfied simultaneously, then there are $q \in \{1, \ldots, l\}$ and $\eta_1, \ldots, \eta_{q-1}, \eta_{q+1}, \ldots, \eta_l \geq 0$ such that*

$$(5.4.9) \qquad v^j = \eta_1 a^j_{\Delta_1} + \ldots + \eta_{q-1} a^j_{\Delta_{q-1}} + \eta_{q+1} a^j_{\Delta_{q+1}} + \ldots + \eta_l a^j_{\Delta_l}$$

for all $j \leq k$, with $m + j \notin \Delta$ and $j \neq k$ ($\sharp = l - 1$)

Proof. First we drop equation (5.4.7) for $m + j = \Delta^i$. Then the rest (5.4.6) means that the vector generated by v and consisting of all components with $m + j \notin \Delta$ and $j \neq k$ is a positive combination of l vectors of dimension $l - 1$. According to the Theorem of Caratheodory there is also a positive combination of $l - 1$ of these vectors which already generates the desired vector.

Now Case 4 is tractable in a similar manner as Case 3.

5.5 COMPONENTWISE SIGN-INDEPENDENCE AND THE INTERSECTION CONDITION

Recall once more our stochastic assumptions

$$a_1, \ldots, a_m, v \quad \text{are distributed over } \mathbb{R}^n \setminus \{0\}$$

- identically

- independently

- symmetrically under rotations.

The conditions do not imply independence between the components of the single vectors a_i. But they yield a weak kind of independence, the so-called sign-independence of components.

Lemma 5.7 *If the random vector $a \in \mathbb{R}^n$ is distributed symmetrically under rotations over $\mathbb{R}^n \setminus \{0\}$, then the random variables*

(5.5.1) $$\text{sign}(a^1), \ldots, \text{sign}(a^n) \text{ are independent.}$$

$\Box$

We call this property componentwise sign-independence. This concept is in a certain sense similar to the column-wise sign-invariance of the model used independently in the papers of ADLER & MEGIDDO (1983) and TODD (1983).

Proof. The proof is immediate, because the density function and the radial distribution function only depend on the length of a, which is

$$\sqrt{(a^1)^2 + \ldots + (a^n)^2}.$$

The signs of the single components are irrelevant.

$\Box$

For the moment we forget about Cases 1 and 2, where the intersecting direction $\Pi_k(e_k)$ is one of the spanning vectors of $\widehat{\Sigma}(\Delta)$ (compare Section 4), because in every stage at most 1 such situation (positive intersection) can occur. So we take into account that we restrict to the case where $m + k \notin \Delta$.

Recall that an arbitrary simplex-cone (one of the candidates) has to be counted only if the following conditions are all satisfied

(5.5.2)   1) $\Pi_l^\Delta(\text{CH}(a_{\Delta^1}, \ldots, a_{\Delta^l}))$ is a boundary simplex of Y_l^Δ.

   2)   The normal vector w belonging to the boundary simplex of 1) is positive in all its l entries.

3) $\Pi_i^{\Delta}(\Pi_{k-1}(v))$ is a positive combination of at most $l-1$ of the vectors $\Pi_i^{\Delta}(\Pi_{k-1}(a_{\Delta^1})), \ldots, \Pi_i^{\Delta}(\Pi_{k-1}(a_{\Delta^l}))$.

4a)   System (5.4.5) has to be satisfied. It consists of $k-l$ inequalities in $l-1$ variables. (The values of the variables are already determined by (5.4.4) or

4b)   System (5.4.8) has to be satisfied. It consists of $k-l-1$ inequalities in l variables, which are all determined by the system of equations formed by (5.4.6) and (5.4.7).

So conditions 1), 2), 3) deal only with components $j \le k$, where $m+j \notin \Delta$, whereas 4a) resp. 4b) are concerned with components $j \le k$, $m+j \in \Delta$.

Lemma 5.8 *Satisfaction of 4a) or 4b) does not depend on the components with $m+j \notin \Delta$. It has probability $\frac{1}{2}^{n-1}$ (for 4a)) and $\frac{1}{2}^{n-l-1}$ (for 4b)).*

Proof. Consider 4a). We know that P (inequality j is satisfied) $= \frac{1}{2}$, because the variables are already determined and because the signs are arbitrarily invertible. The rotational symmetry and the componentwise sign-independence yield

$$P(\text{inequalities for } j \text{ where } m+j \in \Delta \text{ are satisfied}) = \frac{1}{2}^{k-l}.$$

In the same manner we obtain for 4b)

$$P(\text{inequalities are all satisfied}) = \frac{1}{2}^{k-l-1}.$$

So these requirements (4a) or 4b)) cause a reduction factor of at least $\frac{1}{2}^{k-l-1}$. Hence

(5.5.3) $P\left(\hat{\Sigma}(\Delta) \text{ is an intersected boundary simplex-cone}\right)$

$\le P$ (events 1), 2), 3) and (4a) or 4b)) are satisfied simultaneoulsy)

$\le P$ (events 1), 2), 3) are satisfied simultaneoulsy) $\cdot \frac{1}{2}^{k-l-1}$.

Let us analyze conditions 1) through 3) further. Let w be the normal vector mentioned in 2).

Lemma 5.9

(5.5.4)   *Condition 2) is independent of 1) resp. 3) and has probability $\left(\frac{1}{2}\right)^l$.*

Proof. That independence is an immediate consequence of rotational symmetry. Since $a_1, \ldots, a_l$ are distributed symmetrically under rotations, also the normal vector on the boundary simplex of Y_l^Δ is distributed in that manner. So the probability for nonnegativity is $(\frac{1}{2})^l$.

$\square$

So we arrive at

(5.5.5) $\qquad P\,(\widehat{\Sigma}(\Delta)$ satisfies the

boundary condition and the intersection condition) $\leq$

$$\leq \left(\frac{1}{2}\right)^{k-1} P\{\ 1)\text{ and }3)\text{ are satisfied }\}.$$

5.6 THE AVERAGE NUMBER OF PIVOT STEPS

We are going to evaluate the expected number of intersected boundary simplex-cones for stage k. First we consider the special cases. For initialization of stage k (change of dimensions) we count one step.

Case 1 ($m + k \in \Delta$ and $\Delta^i \neq m + k$) does not occur.

Case 2 ($m + k \in \Delta$ and $\Delta^i = m + k$) can occur only once for positive intersection and only once for negative intersection in stage k. So we count two steps.

For $l = 0$ we find only one $\widehat{\Sigma}_k(\Delta)$. So we count one step.

For $l = 1$ and $m + k \notin \Delta$ there is only one candidate, because $\Delta^2 = m + 1$, $\Delta^3 = m + 2, \ldots, \Delta^{k-1} = m + k - 1$ is demanded and only one such boundary simplex-cone is possible. Again we count one step.

The remaining cases ($m + k \notin \Delta$ and $l \geq 2$) can be treated by application of the methods and results of Chapter III, Section 5, particularly (3.5.10) and (3.5.16). We exploit that for $l \geq 2$

(5.6.1) $\qquad E_{m,n}^k(S_l) \leq 5+$ (number of candidates with l "normal" restrictions) $(\frac{1}{2})^{k-1}\,P\{\ 1)$ and 3) are satisfied$\}$

$= 5 + \binom{m}{l}\binom{k}{k-l}l \cdot \frac{1}{2}^{k-1} P[\Pi_l^\Delta(\text{CH}(a_{\Delta^1}, \ldots, a_{\Delta^l}))$ is a boundary
simplex of Y_l^Δ and $\Pi_l^\Delta(\text{CH}(a_{\Delta^1}, \ldots, a_{\Delta^{l-1}}))$ is intersected by
$\text{span}(\Pi_l^\Delta(v), \Pi_l^\Delta(e_k))]$.

Recall that (3.5.11) and (3.5.17) provide an upper bound in the following form for $n \geq 3, l \geq 2$

$\binom{m}{l}l\,P[\Pi_l^\Delta(\text{CH}(a_{\Delta^1}, \ldots, a_{\Delta^l}))$ is a boundary simplex of Y_l^Δ and

$\Pi_l^\Delta(\text{CH}(a_{\Delta^1}, \ldots, a_{\Delta^{l-1}}))$ is intersected by $\text{span}(\Pi_l^\Delta(v), \Pi_l^\Delta(e_k))] \leq$

$\leq m^{1/(n-1)}l^{3/2}n^{3/2}\pi(1 + \frac{e\pi}{2}).$

So we know that

$$(5.6.2) \qquad E_{m,n}^k(S_l) \leq 5 + (\tfrac{1}{2})^{k-1} \binom{k}{k-l} m^{1/(n-1)} l^{3/2} n^{3/2} \pi \left(1 + \frac{e\pi}{2}\right).$$

Summation over all values of l delivers

$$
\begin{aligned}
E_{m,n}^k(S) &\leq 5 + \sum_{l=2}^{k} (\tfrac{1}{2})^{k-1} \binom{k}{k-l} l^{3/2} m^{1/(n-1)} n^{3/2} \left(1 + \frac{e\pi}{2}\right)\pi \\
&\leq 5 + (\tfrac{1}{2})^{k-1} \sum_{r=0}^{k-1} \binom{k}{k-r-1} (r+1)^{3/2} \pi m^{1/(n-1)} n^{3/2} \cdot \left(1 + \frac{e\pi}{2}\right) \\
&\leq 5 + (\tfrac{1}{2})^{k-1} \sum_{r=0}^{k-1} \binom{k}{r+1} (r+1)^{3/2} \pi m^{1/(n-1)} n^{3/2} \cdot \left(1 + \frac{e\pi}{2}\right) \\
&\leq 5 + (\tfrac{1}{2})^{k-1} k^{3/2} \sum_{r=0}^{k-1} \binom{k-1}{r} \pi m^{1/(n-1)} n^{3/2} \cdot \left(1 + \frac{e\pi}{2}\right).
\end{aligned}
$$

So we have

$$(5.6.3) \qquad E_{m,n}^k(S_l) \leq 5 + k^{3/2} \pi\, m^{1/(n-1)}\, n^{3/2} \left(1 + \frac{e\pi}{2}\right).$$

At last we have to sum up over all stages $k = 2, \ldots, n$.

$$
\begin{aligned}
E_{m,n}^k(s_t) &\leq 5 + \sum_{k=2}^{n} k^{3/2}\, \pi\, m^{1/(n-1)}\, n^{3/2} \left(1 + \frac{e\pi}{2}\right) \\
(5.6.4) \qquad &\leq \int_{0}^{n+1} k^{3/2} dk\, \pi\, m^{1/(n-1)}\, n^{3/2} \left(1 + \frac{e\pi}{2}\right) \\
&= \frac{2}{5}(n+1)^{5/2}\, \pi\, m^{1/(n-1)}\, n^{3/2} \left(1 + \frac{e\pi}{2}\right) \\
&\leq \frac{2}{5}(n+1)^{4}\, m^{1/(n-1)}\, \pi \left(1 + \frac{e\pi}{2}\right)
\end{aligned}
$$

because of

$$\pi\left(1 + \frac{e\pi}{2}\right) \geq 5 \quad \text{and} \quad n^{3/2} \geq n.$$

We arrive at our final

Theorem 18 *Our algorithm in Chapter 5, Section 2 has an expected number of pivot steps not greater than*

$$(5.6.5) \qquad m^{1/(n-1)} (n+1)^4 \frac{2}{5} \pi \left(1 + \frac{e\pi}{2}\right)$$

for problems with m restrictions distributed according to the rotation-invariance model and n additional nonnegativity constraints.

$\square$

Chapter 6

APPENDIX

6.1 GAMMAFUNCTION AND BETAFUNCTION

The Gammafunction $\Gamma : (0, \infty) \to \mathbb{R}$ is defined by

$$(6.1.1) \qquad \Gamma(x) = \int_0^\infty e^{-t} t^{x-1} \, dt \text{ for } x > 0.$$

It has some interesting properties

$$(6.1.2) \qquad \Gamma(1) = 1$$

$$(6.1.3) \qquad \Gamma(x+1) = x\Gamma(x) \qquad \text{for } x > 0$$

$$(6.1.4) \qquad \Gamma(n+1) = n! \qquad \text{for } n = 0, 1, 2, \ldots$$

$$(6.1.5) \qquad \Gamma(\frac{1}{2}) = 2 \int_0^\infty e^{-t^2} \, dt = \sqrt{\pi}$$

$$(6.1.6) \qquad \Gamma\left(n + \frac{1}{2}\right) = \frac{(2n)! \sqrt{\pi}}{n! \, 2^{2n}}$$

$$(6.1.7) \qquad \Gamma(x) \text{ is a convex function for } x > 0,$$

because

$$\frac{\partial^2 \Gamma(x)}{\partial x^2} = \frac{\partial^2 \left[\int\limits_0^\infty e^{-t} e^{(\ln t)(x-1)} dt \right]}{\partial x^2} =$$

$$= \frac{\partial \left[\int\limits_0^\infty e^{-t} (\ln t) e^{(\ln t)(x-1)} dt \right]}{\partial x} =$$

$$= \int\limits_0^\infty e^{-t} (\ln t)^2 e^{(\ln t)(x-1)} dt > 0 \text{ for } x > 0.$$

(6.1.8) $\ln \Gamma(x)$ is also convex, because

$$\ln \Gamma(x) = \ln \int\limits_0^\infty e^{-t} e^{(\ln t)(x-1)} dt$$

and

$$\frac{\partial \ln \Gamma(x)}{\partial x} = \frac{1}{\Gamma(x)} \frac{\partial \Gamma(x)}{\partial x} = \frac{\int\limits_0^\infty e^{-t} (\ln t) t^{x-1} dt}{\int\limits_0^\infty e^{-t} t^{(x-1)} dt}$$

$$\frac{\partial^2 \ln \Gamma(x)}{x^2} = \frac{\int\limits_0^\infty e^{-t} (\ln t)^2 t^{(x-1)} dt \int\limits_0^\infty e^{-t} t^{(x-1)} dt}{(\Gamma(x))^2} -$$

$$- \frac{\int\limits_0^\infty e^{-t} ((\ln t) t^{x-1} dt \int\limits_0^\infty e^{-t} (\ln t) t^{(x-1)} dt}{(\Gamma(x))^2} \geq 0$$

because of the inequality of Cauchy-Schwarz. Hence $\ln \Gamma(x)$ is convex for $x > 0$. We conclude that for $\alpha \in (0,1)$ and $x \geq 1$

$$\ln \Gamma(x+\alpha) \leq \ln \Gamma(x) + \alpha[\ln \Gamma(x+1) - \ln \Gamma(x)] = \ln \Gamma(x) + \alpha \ln x$$

or equivalently

(6.1.9) $\frac{\Gamma(x+\alpha)}{\Gamma(x)} \leq x^\alpha$ for $\alpha \in (0,1)$ and $x \geq 1$.

For the derivation of a lower bound of this quotient it is useful to know that

$$\ln \Gamma(x+\alpha-1) \geq \ln \Gamma(x+\alpha) + \frac{1}{\alpha}(\ln \Gamma(x) - \ln \Gamma(x+\alpha))$$

and consequently

(6.1.10) $\frac{\Gamma(x+\alpha)}{\Gamma(x)} \geq (x+\alpha-1)^\alpha.$

So we have for $x \geq 1$ and $0 < \alpha < 1$

(6.1.11)
$$(x + \alpha - 1)^\alpha \leq \frac{\Gamma(x + \alpha)}{\Gamma(x)} \leq x^\alpha.$$

By such estimations of the Gammafunction the so-called Betafunction can be evaluated

(6.1.12)
$$B(k, l) = \int_0^1 x^{k-1}(1 - x)^{l-1} dx = \frac{\Gamma(k)\Gamma(l)}{\Gamma(k + l)} \text{ for } k > 0, l > 0.$$

Important for our purposes is the relation

(6.1.13)
$$\binom{m}{n} \int_0^1 x^{m-n}(1 - x)^{n-1} dx = \binom{m}{n} \frac{\Gamma(m - n + 1)\Gamma(n)}{\Gamma(m + 1)} =$$
$$= \frac{\Gamma(n)}{\Gamma(n + 1)} = \frac{1}{n} \text{ for } m, n \in \mathbb{N}.$$

In addition, we exploit equalities like

(6.1.14)
$$\binom{k}{l} \int_0^1 (1 - x)^{k-l} x^{l-1-\delta} dx = \frac{\Gamma(k + 1)\Gamma(l - \delta)}{\Gamma(k + 1 - \delta)\Gamma(l + 1)}$$

with $\delta > 0$ and $k \geq l \in \mathbb{N}$. In order to obtain an upper bound for that quotient, we estimate according to (6.1.11)

$$\frac{\Gamma(k + 1)}{\Gamma(k + 1 - \delta)} \leq (k + 1 - \delta)^\delta$$
$$\frac{\Gamma(l - \delta)}{\Gamma(l + 1)} \leq \frac{1}{l}\frac{\Gamma(l - \delta)}{\Gamma(l)} \leq \frac{1}{l}(l - 1)^\delta.$$

So we obtain

(6.1.15)
$$\binom{k}{l} \int_0^1 (1 - x)^{k-l} x^{l-1-\delta} dx \leq \frac{1}{l}\left(\frac{k + 1 - \delta}{l - 1}\right) \leq \frac{1}{l}k^\delta.$$

For the derivation of a lower bound we use

$$\frac{\Gamma(k + 1)}{\Gamma(k + 1 - \delta)} \geq k^\delta$$
$$\frac{\Gamma(l - \delta)}{\Gamma(l + 1)} \geq \frac{1}{l}\frac{\Gamma(l - \delta)}{\Gamma(l)} \geq \frac{1}{l}(l - \delta)^\delta$$

which leads to

$$(6.1.16) \qquad \binom{k}{l} \int_0^1 (1-x)^{k-l} x^{l-1-\delta}\, dx \geq \frac{1}{l}\Big(\frac{k}{l-\delta}\Big)^{\delta} \geq \frac{1}{l}\Big(\frac{k}{l}\Big)^{\delta}.$$

We summarize

$$(6.1.17) \qquad \frac{1}{l}\Big(\frac{k}{l}\Big)^{\delta} \leq \binom{k}{l} \int_0^1 (1-x)^{k-l} x^{l-1-\delta}\, dx \leq \frac{1}{l} k^{\delta}.$$

Further estimations are concerned with integrals of the kind

$$\int_0^1 (1-x)^{m-n} x^{n-1} \sqrt{\ln\frac{1}{x}}\, dx\,.$$

First we consider the integral

$$\int_0^1 (1-x)^{m-n} x^{n-1} \ln\frac{1}{x}\, dx\,.$$

It is identical with

$$\frac{-\partial \int_0^1 (1-x)^{m-n} x^l\, dx}{\partial l} \quad \text{at } l = n-1.$$

Here it is equal $-\frac{\partial B(m-n+1, l+1)}{\partial l} = B(m-n+1, l+1)\sum_{\nu=l+1}^{m-n+l+1}\frac{1}{\nu}$. We have

$$\sum_{\nu=n}^{m}\frac{1}{\nu} \leq \ln\Big(\frac{m}{n-1}\Big).$$

The Cauchy-Schwarz inequality yields

$$(6.1.18) \qquad \int_0^1 (1-x)^{m-n} x^{n-1} \sqrt{\ln\frac{1}{x}}\, dx \leq B(m-n+1, n)\sqrt{\ln\frac{m}{n-1}}$$

and

$$(6.1.19) \qquad \int_0^1 (1-x)^{m-n} x^{n-1} \ln\frac{1}{x}\, dx \leq B(m-n+1, n)\Big(\ln\frac{m}{n-1}\Big).$$

Finally, we are interested in a lower bound for that integral. We try to show that there exists a function $\varepsilon(m, n)$ such that

$$\int_0^1 (1-x)^{m-n} x^{n-1} \sqrt{\ln \frac{1}{x}} \, dx \, (1 + \varepsilon(m,n)) \geq$$

(6.1.20)

$$\geq \int_0^1 (1-x)^{m-n} x^{n-1} dx \sqrt{\ln \frac{m+1}{n}} = B(m-n+1, n)\sqrt{\ln \frac{m+1}{n}}$$

with $\varepsilon(m, n) \to 0$ for $m \to \infty$ and fixed n.

Proof. The function $\sqrt{\ln \frac{1}{x}}$ is certainly convex on $(0, 1)$, because

$$\frac{\partial \sqrt{\ln \frac{1}{x}}}{\partial x} = \frac{1}{2} \frac{1}{\sqrt{\ln \frac{1}{x}}} x \left(-\frac{1}{x^2}\right) = -\frac{1}{2x\sqrt{\ln \frac{1}{x}}}$$

and

$$\frac{\partial^2 \sqrt{\ln \frac{1}{x}}}{\partial x^2} = \frac{2\sqrt{\ln \frac{1}{x}} - \frac{1}{\sqrt{\ln \frac{1}{x}}}}{4x^2 (\ln \frac{1}{x})}.$$

This term is positive as long as

$$2\sqrt{\ln \frac{1}{x}} > \frac{1}{\sqrt{\ln \frac{1}{x}}} \text{ or } \ln \frac{1}{x} > \frac{1}{2}.$$

Equivalent is $x < e^{-\frac{1}{2}}$. The remaining interval $[e^{-\frac{1}{2}}, 1]$ looses its influence in the asymptotic case. Here the function in question can be continued as a constant function $f(x) = \sqrt{\frac{1}{2}}$ such that the complete function is convex. Then the mistake caused by that manipulation tends to 0 in the asymptotic case. But for convex functions f we know that $E[f(x)] \geq f(E[x])$. Here $E(x)$ stands for

$$\frac{\int_0^1 (1-x)^{m-n} x^n dx}{\int_0^1 (1-x)^{m-n} x^{n-1} dx} = \frac{\Gamma(n+1)\Gamma(m+1)\Gamma(m-n+1)}{\Gamma(n)\Gamma(m+2)\Gamma(m-n+1)} = \frac{n}{m+1}.$$

Consequently, the expectation value of $\sqrt{\ln \frac{1}{x}}$ is greater or equal (asymptotically)

$$\sqrt{\ln \frac{m+1}{n}} \geq \sqrt{\ln \frac{m}{n}}.$$

6.2 UNIT BALL AND UNIT SPHERE

The unit ball of $\mathbb{R}^n$ will be denoted by Ω_n; its surface, the unit sphere, by ω_n, and the Euclidean norm by $\|\ \|$. So

(6.2.1) $$\Omega_n = \{x \mid \|x\| \leq 1\} \quad \text{for } n \geq 1$$

(6.2.2) $$\omega_n = \{x \mid \|x\| = 1\} \quad \text{for } n \geq 2.$$

These bodies have the following Lebesgue measures

(6.2.3) $$\lambda_n(\Omega_n) = \frac{\pi^{\frac{n}{2}}}{\Gamma\left(\frac{n+2}{2}\right)}$$

(6.2.4) $$\lambda_{n-1}(\omega_n) = \frac{2\pi^{\frac{n}{2}}}{\Gamma\left(\frac{n}{2}\right)}.$$

We transfer formula (6.2.4) to $n = 1$ as a completion of our definition and obtain the general relation

$$n\lambda_n(\Omega_n) = \lambda_{n-1}(\omega_n).$$

For special cases this means

(6.2.5) $$\lambda_0(\omega_1) = \frac{2\pi^{\frac{1}{2}}}{\pi^{\frac{1}{2}}} = 2$$

(6.2.6) $$\lambda_1(\omega_2) = 2\pi$$

(6.2.7) $$\lambda_2(\omega_3) = 4\pi$$

(6.2.8) $$\lambda_3(\omega_4) = 2\pi^2$$

(6.2.9) $$\lambda_1(\Omega_1) = 2$$

(6.2.10) $$\lambda_2(\Omega_2) = \pi$$

(6.2.11) $$\lambda_3(\Omega_3) = \frac{4}{3}\pi$$

(6.2.12) $$\lambda_4(\Omega_4) = \frac{\pi^2}{2}.$$

These volumes can be derived (for $n \geq 2$) by consideration of the polar coordinates

$$\Psi_1 \in [0, 2\pi], \Psi_2 \in [0, \pi]; \Psi_3 \in [0, \pi]; \ldots; \Psi_{n-1} \in [0, \pi]$$

in the form

$$(6.2.13) \quad \omega_n \ni x = \begin{bmatrix} \sin \Psi_1 & \cdot & \sin \Psi_2 & \cdot & \sin \Psi_3 & \cdot & \dots & \cdot & \sin \Psi_{n-1} \\ \cos \Psi_1 & \cdot & \sin \Psi_2 & \cdot & \sin \Psi_3 & \cdot & \dots & \cdot & \sin \Psi_{n-1} \\ & & \cos \Psi_2 & \cdot & \sin \Psi_3 & \cdot & \dots & \cdot & \sin \Psi_{n-1} \\ & & & & \cos \Psi_3 & \cdot & \dots & \cdot & \sin \Psi_{n-1} \\ & & & & & & \cos \Psi_{n-2} & \cdot & \sin \Psi_{n-1} \\ & & & & & & & & \cos \Psi_{n-1} \end{bmatrix}.$$

Hence the surface-integration-element of ω_n is

$$1 \sin \Psi_2 (\sin \Psi_3)^2 (\sin \Psi_4)^3 \dots (\sin \Psi_{n-1})^{n-2}$$

and

$$(6.2.14) \quad \lambda_{n-1}(\omega_n) = \int_0^{2\pi} \int_0^\pi \dots \int_0^\pi \sin \Psi_2 (\sin \Psi_3)^2 \dots (\sin \Psi_{n-1})^{n-2} d\Psi_{n-1} \dots d\Psi_2 d\Psi_1.$$

The arbitrary point out of Ω_n has the polar-coordinate-representation
(6.2.15)

$$\Omega_n \ni x = \begin{bmatrix} r & \cdot & \sin \Psi_1 & \cdot & \sin \Psi_2 & \cdot & \sin \Psi_3 & \cdot & \dots & \cdot & \sin \Psi_{n-1} \\ r & \cdot & \cos \Psi_1 & \cdot & \sin \Psi_2 & \cdot & \sin \Psi_3 & \cdot & \dots & \cdot & \sin \Psi_{n-1} \\ r & \cdot & & & \cos \Psi_2 & \cdot & \sin \Psi_3 & \cdot & \dots & \cdot & \sin \Psi_{n-1} \\ r & \cdot & & & & & \cos \Psi_3 & \cdot & \dots & \cdot & \sin \Psi_{n-1} \\ \vdots & & & & & & & & & & \vdots \\ r & \cdot & & & & & & & \cos \Psi_{n-2} & \cdot & \sin \Psi_{n-1} \\ r & \cdot & & & & & & & & & \cos \Psi_{n-1} \end{bmatrix}.$$

The volume-integration element is

$$r^{n-1} \sin \Psi_2 \dots (\sin \Psi_{n-1})^{n-2}.$$

So we have

$$(6.2.16) \quad \lambda_n(\Omega_n) = \int_0^1 \int_0^{2\pi} \int_0^\pi \dots \int_0^\pi r^{n-1} \sin \Psi_2 \dots (\sin \Psi_{n-1})^{n-2} d\Psi_{n-1} \dots d\Psi_1 dr.$$

We observe the recursive relations

$$(6.2.17)$$
$$\lambda_{n-1}(\omega_n) = \lambda_{n-2}(\omega_{n-1}) 2 \int_0^{\pi/2} (\sin \Psi_{n-1})^{n-2} d\Psi_{n-1} =$$
$$= \lambda_{n-2}(\omega_{n-1}) \int_{-1}^1 \sqrt{1-h^2}^{n-3} dh$$

and

$$\lambda_n(\Omega_n) = \lambda_{n-1}(\Omega_{n-1}) 2 \int_0^{\pi/2} (\sin \Psi_{n-1})^n d\Psi_{n-1} =$$

(6.2.18)

$$= \lambda_{n-1}(\Omega_{n-1}) \int_{-1}^{+1} \sqrt{1-h^2}^{n-1} dh$$

Now we take into account that

(6.2.19) $$\int_0^1 \sqrt{1-h^2}^k dh = \frac{\Gamma(k+1)\Gamma(\frac{1}{2})}{2\Gamma(k+\frac{3}{2})} = \int_0^{\pi/2} (\sin \Psi)^{k+1} d\Psi.$$

And we know that for $k \in \mathbb{N}$

$$2\int_0^1 (1-h^2)^{\frac{2k-1}{2}} dh = 2 \int_0^{\pi/2} (\sin \Psi)^{2k} d\Psi = \frac{1 \cdot 3 \cdot 5 \cdot \ldots \cdot (2k-1) \cdot \pi}{2 \cdot 4 \cdot 6 \cdot \ldots \cdot (2k)}$$

$$2\int_0^1 (1-h^2)^{\frac{2k}{2}} dh = 2 \int_0^{\pi/2} (\sin \Psi)^{2k+1} d\Psi = \frac{2 \cdot 2 \cdot 4 \cdot 6 \cdot \ldots \cdot (2k)}{3 \cdot 5 \cdot 7 \cdot \ldots \cdot (2k+1)}.$$

So

$$2\int_0^{\pi/2} (\sin \Psi)^{2k} d\Psi \, 2 \int_0^{\pi/2} (\sin \Psi)^{2k+1} d\Psi = \pi \frac{2}{2k+1} \text{ and}$$

$$2\int_0^{\pi/2} (\sin \Psi)^{2k} d\Psi \, 2 \int_0^{\pi/2} (\sin \Psi)^{2k-1} d\Psi = \pi \frac{2}{2k}.$$

We conclude

$$\sqrt{\frac{2\pi}{2k}} \geq \int_{-1}^{+1} (1-h^2)^{\frac{2k-1}{2}} dh = 2 \int_0^{\pi/2} (\sin \Psi)^{2k} d\Psi \geq \sqrt{\frac{2\pi}{2k+1}}$$

and following the same arguments

$$\sqrt{\frac{2\pi}{2k+1}} \geq \int_{-1}^{+1} (1-h^2)^{\frac{2k}{2}} dh = 2 \int_0^{\pi/2} (\sin \Psi)^{2k+1} d\Psi \geq \sqrt{\frac{2}{2k+2}}$$

Consequently

$$(6.2.20) \qquad \sqrt{\frac{2\pi}{l-1}} \geq \int_{-1}^{+1} (1-h^2)^{\frac{l-1}{2}}\,dh = 2\int_{0}^{\pi/2} (\sin\Psi)^l\,d\Psi \geq \sqrt{\frac{2}{l+1}}$$

Now we know that

$$(6.2.21) \qquad \frac{\lambda_{n-1}(\omega_n)}{\lambda_{n-3}(\omega_{n-2})} = \frac{2\pi}{n-2} \qquad \text{for } n \geq 3$$

$$(6.2.22) \qquad \frac{\lambda_n(\Omega_n)}{\lambda_{n-2}(\Omega_{n-2})} = \frac{2\pi}{n} \qquad \text{for } n \geq 3.$$

From (6.2.20) we conclude that

$$(6.2.23) \qquad \sqrt{\frac{2\pi}{n+1}} \leq \frac{\lambda_n(\Omega_n)}{\lambda_{n-1}(\Omega_{n-1})} \leq \sqrt{\frac{2\pi}{n}} \qquad \text{for } n \geq 2$$

and equivalently

$$(6.2.24) \qquad \sqrt{\frac{n}{2\pi}} \leq \frac{\lambda_{n-1}(\Omega_{n-1})}{\lambda_n(\Omega_n)} \leq \sqrt{\frac{n+1}{2\pi}} \qquad \text{for } n \geq 2$$

as well as

$$(6.2.25) \qquad \sqrt{\frac{2\pi}{n-1}} \leq \frac{\lambda_{n-1}(\omega_n)}{\lambda_{n-2}(\omega_{n-1})} \leq \sqrt{\frac{2\pi}{n-2}}$$

or

$$(6.2.26) \qquad \sqrt{\frac{n-2}{2\pi}} \leq \frac{\lambda_{n-2}(\omega_{n-1})}{\lambda_{n-1}(\omega_n)} \leq \sqrt{\frac{n-1}{2\pi}}$$

(6.2.27)   In the special cases $n = 1, \ldots, 4$ we have

$$2 \int_0^{\pi/2} (\sin \Psi)^0 d\Psi = \frac{\lambda_1(\omega_2)}{\lambda_0(\omega_1)} = \pi = \int_{-1}^{+1} \sqrt{1-h^2}^{-1} dh$$

$$2 \int_0^{\pi/2} \sin \Psi d\Psi = \frac{\lambda_2(\omega_3)}{\lambda_1(\omega_2)} = 2 = \int_{-1}^{+1} \sqrt{1-h^2}^{0} dh$$

$$2 \int_0^{\pi/2} (\sin \Psi)^2 d\Psi = \frac{\lambda_3(\omega_4)}{\lambda_2(\omega_3)} = \frac{\pi}{2} = \int_{-1}^{+1} \sqrt{1-h^2} dh$$

$$2 \int_0^{\pi/2} (\sin \Psi)^2 d\Psi = \frac{\lambda_2(\Omega_2)}{\lambda_1(\Omega_1)} = \frac{\pi}{2} = \int_{-1}^{+1} \sqrt{1-h^2} dh$$

$$2 \int_0^{\pi/2} (\sin \Psi)^3 d\Psi = \frac{\lambda_3(\Omega_3)}{\lambda_2(\Omega_2)} = \frac{4}{3} = \int_{-1}^{+1} (1-h^2) dh$$

$$2 \int_0^{\pi/2} (\sin \Psi)^4 d\Psi = \frac{\lambda_4(\Omega_4)}{\lambda_3(\Omega_3)} = \frac{3}{8}\pi = \int_{-1}^{+1} \sqrt{1-h^2}^{3} dh.$$

In addition we need estimations like

(6.2.28) $$\sqrt{\frac{n}{n+1}} \cdot \sqrt{n2\pi} \leq \frac{\lambda_{n-1}(\omega_n)}{\lambda_{n-1}(\Omega_{n-1})} = n\frac{\lambda_n(\Omega_n)}{\lambda_{n-1}(\Omega_{n-1})} \leq \sqrt{n2\pi}.$$

For $n = 2, 3$ we have

(6.2.29) $$\frac{\lambda_1(\omega_2)}{\lambda_1(\Omega_1)} = 2\frac{\pi}{2} = \pi$$

(6.2.30) $$\frac{\lambda_2(\omega_3)}{\lambda_2(\Omega_2)} = 3\frac{4}{3} = 4.$$

At the end of this section we prove a lemma which tells some results on integration over the unit sphere of dimension $n - 1$. For this purpose let $d\omega$ be the surface-integration-element of ω_{n-1}.

Lemma 6.1

(6.2.31) $$\int_{\omega_{n-1}} d\omega(x) = \lambda_{n-2}(\omega_{n-1}) \quad \text{for } n \geq 2$$

(6.2.32) $$\int_{\omega_{n-1}} x^{n-1} d\omega(x) = 0 \quad \text{for } n \geq 2$$

(6.2.33)
$$\int_{\omega_{n-1}} |x^{n-1}| d\omega(x) = \frac{2\lambda_{n-3}(\omega_{n-2})}{(n-2)\lambda_{n-2}(\omega_{n-1})} \lambda_{n-2}(\omega_{n-1}) \quad \text{for } n \geq 3$$

$$= \lambda_0(\omega_1) = 2 \quad \text{for } n = 2$$

(6.2.34)
$$\int_{\omega_{n-1}} (x^{n-1})^2 d\omega(x) = \frac{1}{n-1} \lambda_{n-2}(\omega_{n-1}) \quad \text{for } n \geq 3$$

$$= \lambda_{n-2}(\omega_{n-1}) = 2 \quad \text{for } n = 2.$$

Proof. (6.2.31) Is equivalent to the definition of $d\omega(x)$.

(6.2.32) Is a result of symmetry.

(6.2.33) The case $n = 2$ is immediate. For $n \geq 3$ we have

$$\int_{\omega_{n-1}} |x^{n-1}| d\omega(x) = 2 \int_0^{\frac{\pi}{2}} (\sin \Psi)^{n-3} \cos \Psi d\Psi \cdot \lambda_{n-3}(\omega_{n-2}) = 2 \frac{1}{n-2} \lambda_{n-3}(\omega_{n-2}).$$

(6.2.34) Again, the case $n = 2$ is immediate. For $n \geq 3$

$$\int_{\omega_{n-1}} (x^{n-1})^2 d\omega(x) = 2 \int_0^{\frac{\pi}{2}} (\cos \Psi)^2 (\sin \Psi)^{n-3} d\Psi \, \lambda_{n-3}(\omega_{n-2}) =$$

$$= 2\lambda_{n-3}(\omega_{n-2}) \frac{1}{n-1} \int_0^{\frac{\pi}{2}} (\sin \Psi)^{n-3} d\Psi = \frac{1}{n-1} \lambda_{n-2}(\omega_{n-1}).$$

6.3 ESTIMATIONS UNDER VARIATION OF THE WEIGHTS

In this section we prove Lemma 3.3 of Chapter III, Section 3.

Lemma 6.2

Let A, B, C, be functions from $[a, b]$ into $[0, \infty)$, and let E be an arbitrary distribution function on $[a, b]$. For an arbitrarily chosen point $\overline{x} \in [a, b]$ define the new distribution function

$$\overline{E}(x) := \begin{cases} 0 & x < \overline{x} \\ E(x) & x \geq \overline{x} \end{cases}$$

Then the following results hold in any case where the denominator-integrals are positive.

1) Let A be continuous and let $\bar{x}$ be the greatest value in $[a,b]$ where $A(x)$ attains its minimum, i. e.

$$A(\bar{x}) = \min_{x \in [a,b]} A(x).$$

If B is monotonically increasing, then we have

(6.3.1)
$$\frac{\int\limits_a^b B(x)dE(x)}{\int\limits_a^b B(x)A(x)dE(x)} \leq \frac{\int\limits_a^b B(x)d\overline{E}(x)}{\int\limits_a^b B(x)A(x)d\overline{E}(x)}.$$

2) If A is monotonically decreasing on $[a,b]$ and if B decreases on $[a,\bar{x}]$, then

(6.3.2)
$$\frac{\int\limits_a^b B(x)A(x)dE(x)}{\int\limits_a^b B(x)dE(x)} \geq \frac{\int\limits_a^b B(x)A(x)d\overline{E}(x)}{\int\limits_a^b B(x)d\overline{E}(x)}.$$

3) If A and B are both increasing (resp. both decreasing) monotonically then

(6.3.3)
$$\frac{\int\limits_a^b C(x)dx}{\int\limits_a^b C(x)A(x)dx} \geq \frac{\int\limits_a^b C(x)B(x)dx}{\int\limits_a^b C(x)B(x)A(x)dx}.$$

Proof. If $\bar{x} < b$, we divide the interval $[a,b]$ into $[a,\bar{x}]$ and $(\bar{x},b]$. The cases $\bar{x} = a$ or $\bar{x} = b$ are trivial. Then $A(x) \geq \overline{A}(x)$ for all $x \in [a,b]$, B is monotonically increasing and

$$\frac{\int\limits_a^b B(x)dE(x)}{\int\limits_a^b B(x)A(x)dE(x)} \leq \frac{\int\limits_a^{\bar{x}} B(x)dE(x) + \int\limits_{\bar{x}+}^b B(x)dE(x)}{\int\limits_a^{\bar{x}} B(x)A(\bar{x})dE(x) + \int\limits_{\bar{x}+}^b B(x)A(x)dE(x)} =$$

$$= \frac{1}{A(\bar{x})}\frac{Q+R}{Q+\Psi R} \leq \frac{|q+R|}{A(\bar{x})|q+\Psi R|}, \quad \text{where}$$

$$\Psi \geq 1, \quad Q = \int\limits_a^{\bar{x}} B(x)dE(x) \leq \int\limits_a^{\bar{x}} dE(x)B(\bar{x}) =: q,$$

$$R = \int\limits_{\bar{x}+}^b B(x)dE(x).$$

But the last term is exactly

$$\frac{\int\limits_a^b B(x)d\overline{E}(x)}{\int\limits_a^b A(x)B(x)d\overline{E}(x)}.$$

So the claim is proven.

Proof of 2)

We know that

$$\frac{\int\limits_{\overline{x}+}^b B(x)A(x)dE(x)}{\int\limits_{\overline{x}+}^b B(x)dE(x)} \le A(\overline{x}) \le \frac{\int\limits_a^{\overline{x}} B(x)A(x)dE(x)}{\int\limits_a^{\overline{x}} B(x)dE(x)}.$$

So we obtain

$$\frac{\int\limits_a^b B(x)A(x)dE(x)}{\int\limits_a^b B(x)dE(x)} \ge \frac{\int\limits_a^{\overline{x}} B(x)A(\overline{x})dE(x) + \int\limits_{\overline{x}+}^b B(x)A(x)dE(x)}{\int\limits_a^{\overline{x}} B(x)dE(x) + \int\limits_{\overline{x}+}^b B(x)dE(x)} \ge$$

$$\ge \frac{\int\limits_a^{\overline{x}} B(x)A(x)d\overline{E}(x) + \int\limits_{\overline{x}}^b B(x)A(x)dE(x)}{\int\limits_a^{\overline{x}} B(x)d\overline{E}(x) + \int\limits_{\overline{x}}^b B(x)dE(x)} = \frac{\int\limits_a^b B(x)A(x)d\overline{E}(x)}{\int\limits_a^b B(x)d\overline{E}(x)}$$

Proof of 3)

Let A and B be increasing monotonically. For arbitrary $\eta \in [a, b]$ we have

$$\frac{\int\limits_\eta^b C(x)dx}{\int\limits_a^b C(x)dx} \le \frac{\int\limits_\eta^b C(x)B(x)dx}{\int\limits_a^b C(x)B(x)dx}$$

(the denominator-integrals cannot be 0, because $\int\limits_a^b C(x)B(x)A(x)dx = 0$), since B is increasing monotonically. By weighting with $C(x)B(x)$ instead of $C(x)$ we emphasize (put more weight on) the higher values of A.

The theory of Lebesgue integration yields

$$\frac{\int\limits_a^b C(x)A(x)dx}{\int\limits_a^b C(x)dx} \leq \frac{\int\limits_a^b C(x)B(x)A(x)dx}{\int\limits_a^b C(x)B(x)dx},$$

which is equivalent to the claim.

If A and B are decreasing functions, we have the opposite relation $\geq$ in the first inequality. Hence we emphasize the regions $[a, \eta]$, where A has its higher values. This leads to the same relation ($\leq$) in the second inequality.

References

Adler, I., Karp, R. & Shamir, R. [1983a]: *A Family of Simplex Variants Solving an* $m \times d$ *Linear Program in Expected Number of Pivot Steps Depending on* d *Only*, University of California, Computer Science Division, Berkeley, December 1983.

Adler, I., Karp, R. & Shamir, R. [1983b]: *A Simplex Variant Solving an* $m \times d$ *Linear Program in* $0(\min(m^2, d^2))$ *Expected Number of Pivot Steps*, University of California, Computer Science Division, Berkeley, December 1983.

Adler, I. & Meggido, N. [1983]: *A Simplex Algorithm whose Average Number of Steps is Bounded Between two Quadratic Functions of the Smaller Dimension*, Department of Industrial Engineering and Operations Research, University of California, Berkeley, California, December 1983.

Avis, D. & Chvatal, V. [1978]: *Notes on Bland's Pivoting Rule*, Mathematical Programming Study 8 (1978), 24–34.

Barnette, D. [1971]: *The Minimum Number of Vertices of a Simple Polytope*, Israel Journal of Mathematics 10 (1971), 121–125.

Barnette, D. [1974]: *An Upper Bound for the Diameter of a Polytope*, Discrete Mathematics 10 (1974), 9–13.

Bauer, H. [1974]: *Wahrscheinlichkeitstheorie und Grundzüge der Maßtheorie*, De Gruyter, Berlin, 1974.

Berenguer, S. E. & Smith, R. L. [1983]: *The Expected Number of Extreme Points of a Random Linear Program*, Technical Report No. 83-17, Dept. of Industrial and Operations Engineering, College of Engineering, University of Michigan, Ann Arbor, September 1983.

Blair, C. [1983]: *Random Linear Programs with Many Variables and Few Constraints*, College of Commerce and Business Administration, University of Illinois at Urbana-Champaign, April 1983.

Bland, R. [1977]: *New Finite Pivoting Rules for the Simplex-Method*, Mathematics of Operations Research 2 (1977), 103–107.

Bland, R. & Goldfarb, D. & Todd, M. [1981]: *The Ellipsoid Method: A Survey*, Operations Research 29 (1981), 1039–1091.

Borgwardt, K. H. [1977]: *Untersuchungen zur Asymptotik der mittleren Schrittzahl von Simplexverfahren in der linearen Optimierung*, Dissertation Universität Kaiserslautern.

Borgwardt, K. H. [1978]: *Untersuchungen zur Asymptotik der mittleren Schrittzahl von Simplexverfahren in der linearen Optimierung*, Operations Research Verfahren 28 (1978), 332–345.

Borgwardt, K. H. [1979]: *Zum Rechenaufwand von Simplexverfahren*, Operations Research Verfahren 31 (1979), 83–97.

Borgwardt, K. H. [1980]: *Die asymptotische Ordnung der mittleren Schrittzahl von Simplexverfahren*, Methods of Operations Research 37 (1980), 81–95.

Borgwardt, K. H. [1981]: *The Expected Number of Pivot Steps Required by a Certain Variant of the Simplex Method is Polynomial*, Methods of Operations Research 43 (1981), 35–41.

Borgwardt, K. H. [1982a]: *Some Distribution-Independent Results About the Asymptotic Order of the Average Number of Pivot Steps of the Simplex Method*, Mathematics of Operations Research 7 (1982), 441–462.

Borgwardt, K. H. [1982b]: *The Average Number of Pivot Steps Required by the Simplex-Method is Polynomial*, Zeitschrift für Operations Research 26 (1982), 157–177.

Borgwardt, K. H. [1985]: *Der durchschnittliche Rechenaufwand beim Simplexverfahren*, Operations Research Proceedings 1984, DGOR-Tagung, St. Gallen, 1984, 647–660.

Bronstein, I. [1970]: *Taschenbuch der Mathematik*, Harri Deutsch Verlag, Zürich, 1970.

Buck, R. C. [1943]: *Partition of Space*, American Mathematical Monthly 50 (1943), 541–544.

Carnal, H. [1970]: *Die konvexe Hülle von n rotationssymmetrisch verteilten Punkten*, Zeitschrift für Wahrscheinlichkeitsrechnung und verwandte Gebiete 15 (1970), 168–176.

Charnes, A. & Cooper, W. W. [1962]: *Programming with Linear Fractional Functionals*, Naval Research Logistics Quarterly 9 (1962), 181–186.

Collatz, L. & Wetterling, W. [1971]: *Optimierungsaufgaben*, Springer Verlag, Berlin, 1971.

Cottle, R. W. & Dantzig, G. B. [1968]: *Complementary Pivot Theory of Mathematical Programming*, in: G. Dantzig & A. Veinott, Jr. (eds.) Mathematics of the Decision Sciences, American Mathematical Society, Providence, R. I., 1968, 115–136.

Courant, R. [1972]: *Vorlesungen über Differential- und Integralrechnung 2*, Springer Verlag, Berlin 1972.

Dantzig, G. B. [1963]: *Linear Programming and Extensions*, Princeton University Press, Princeton, 1963.

Dantzig, G. B. [1966]: *Lineare Programmierung und Erweiterungen*, Springer Verlag, Berlin, 1966.

Dantzig, G. B. [1980]: *Expected Number of Steps of the Simplex Method for a Linear Program with a Convexity Constraint*, Dept. of Operations Research, Stanford University, Technical Report SOL 80-3R, 1980.

Dinkelbach, W. [1969]: *Sensitivitätsanalyse und parametrische Programmierung*, Springer Verlag, Berlin 1969.

Dwight, H. B. [1961]: *Tables of Integrals and other Mathematical Data*, Macmillan Publishing Co. (1961).

Eaves, C. & Scarf, H. [1976]: *The Solution of Systems of Piecewise Linear Equations*, Mathematics of Opertions Research 1 (1976), 1–27.

Efron, B. [1965]: *The Convex Hull of a Random Set of Points*, Biometrica 52 (3) and (4) (1965), 331–345.

Erwe, F. [1973]: *Differential- und Integralrechnung 2*, Bibliographisches Institut, Mannheim, 1973.

Gács, P. & Lovász, L. [1979]: *Khachiyan's Algorithm for Linear Programming*, Computer Science Department, Stanford University, 1979.

Gale, D. [1969]: *How to Solve Linear Inequalities*, American Mathematical Monthly 76 (1969), 589–599.

Gass, S. & Saaty, Th. [1955]: *The Computational Algorithm for the Parametric Objective Function*, Naval Research Logistics Quarterly 2 (1955), 39–45.

Geffroy, J. [1961]: *Localisation Asymptotique du Polyèdre d'Appui d'un Echantillon Laplacien à k Dimensions*, Publ. Inst. Stat. Univ. Paris 10 (1961), 212–228.

Goldfarb, D. [1983]: *Worst Case Complexity of the Shadow Vertex Simplex Algorithm*, Columbia University, Department of Industrial Engineering and Operations Research, May 1983.

Goldfarb, G. & Sit, W. J. [1979]: *Worst Case Behavior of the Steepest Edge Simplex Method*, Discrete Applied Mathematics 1 (1979), 277–285.

Gould, F. J. & Tolle, J. W. [1974]: *A Unified Approach to Complementarity in Optimization*, Discrete Mathematics 7 (1974), 225–271.

Grötschel, M. & Lovász, L. & Schrijver, A. [1981]: *The Ellipsoid Method and its Consequences in Combinatorial Optimization*, Combinatorica 1 (1981), 169–197.

Grünbaum, B. [1966]: *Convex Polytopes*, Wiley, New York, 1966.

Gradshteyn, I. S. & Ryzlik, I. M. [1965]: *Table of Integrals, Series and Products*, Academic Press, New York, 1965.

Haimovich, M. [1983]: *The Simplex Algorithm is Very Good! — On The Expected Number of Pivot Steps and Related Properties of Random Linear Programs*, 415 Uris Hall, Columbia University, New York, April 1983.

Haimovich, M. [1984a]: *A Note on Random Linear Complementarity Problems*, Columbia University, New York, May 1984.

Haimovich, M. [1984b]: *A Short Proof of Results on the Expected Number of Steps in Dantzig's Self Dual Algorithm*, Columbia University, New York, May 1984.

Haimovich, M. [1984c]: *On the Expected Behavior of Variable Dimension Simplex Algorithms*, Columbia University, New York, May 1984.

Hirche, J. [1979]: *Basishäufigkeit bei linearen Restriktionen*, Math. Operationsforschung und Statistik, Ser. Optimization 10 (1979), 27–37.

Hirsch, M. & Smale, S. [1979]: *On Algorithms for Solving $f(x) = 0$*, Communications on Pure and Applied Mathematics 32 (1979), 281–312.

Howe, R. [1983]: *Linear Complementarity and the Average Volume of Simplicial Cones*, Discussion Paper No. 670, Cowles Foundation for Research in Economics, Yale University, New Haven, Connecticut, June 1983.

Jeroslow, R. G. [1973]: *The Simplex Algorithm with the Pivot-Rule of Maximizing Criterion Improvement*, Discrete Mathematics (1973), 367–377.

Karmarkar, N. [1984]: *A New Polynomial-Time Algorithm for Linear Programming*, AT & T Bell Laboratories, Murray Hill, 1984.

Kelly, D. G. [1981]: *Some Results on Random Linear Programs*, Department of Statistics, University of North Carolina, 1981.

Kelly, D. G. & Tolle, J. W. [1979]: *Expected Number of Vertices of a Random Convex Polytope*, University of North Carolina at Chapel Hill, 1979.

Khachiyan, L. G. [1979]: *A Polynomial Algorithm in Linear Programming*, Doklady Akademia Nauk, UdSSR 244 (1979), 1093–1096. Translated in Sov. Math. Dokl. 20 (1979), 191–194.

Klee, V. [1964a]: *A Property of d-Polyhedral Graphs*, Journal of Mathematics and Mechanics 13 (1964), 1039–1042.

Klee, V. [1964b]: *On the Number of Vertices of a Convex Polytope*, Can. J. of Math. 16 (1964), 701–720.

Klee, V. [1964c]: *Diameters of Polyhedral Graphs*, Can. J. of Math. 16 (1964), 602–614.

Klee, V. [1965a]: *Paths on Polyhedra I*, J. Soc. Indust. Appl. Math. 13 (1965), 946–956.

Klee, V. [1965b]: *Heights of Convex Polytopes*, Journal Math. Anal. Appl. 11 (1965), 176–190.

Klee, V. [1965c]: *A Class of Linear Programming Problems Requiring a Large Number of Iterations*, Numer. Math. 7 (1965), 313–321.

Klee, V. [1966a]: *A Comparison of Primal and Dual Methods of Linear Programming*, Num. Math. 9 (1966), 227–235.

Klee, V. [1966b]: *Paths on Polyhedra II*, Pacific J. of Math. 17 (1966), 249–262.

Klee, V. & Kleinschmidt, P. [1985]: *The d-Step Conjecture and its Relatives*, University of Washington and University of Bochum, 1985.

Klee, V. & Minty, G. [1972]: *How Good is the Simplex-Algorithm?*, Inequalities III, O. Shisha (ed.), Academic Press, New York, 1972, 159–175.

Klee, V. & Walkup, D. [1967]: *The d-Step-Conjecture for Polyhedra of Dimension $d < 6$*, Acta Mathematica 117 (1967), 53–78.

Knuth, D. [1969]: *The Art of Computer Programming*, Vol. 1 and 2, Addison/Wesley, Reading, 1969.

Körner, F. [1980]: *Über die optimale Größe von Teiltableaus beim Simplexverfahren*, Technische Universität Dresden, 1980.

Krickeberg, K. [1963]: *Wahrscheinlichkeitstheorie*, Teubner Verlag, Stuttgart, 1963.

Kuhn, A. & Quandt, R. E. [1963]: *An Experimental Study of the Simplex Method*, Proc. Symp. Appl. Math. 15(1963), 107–124.

Lau, H. T. [1981]: *An Experimental Study of Some Pivoting Rules in Linear Programming*, Computer Science Department, Vanderbilt University, Nashville, Tennessee, June 1981.

Liebling, T. [1972]: *On the Number of Iterations of the Simplex Method*, Operations Research Verfahren 17 (1972), 248–264.

Lindberg, P. O. [1981]: *A Note on Random LP-Problems*, Department of Mathematics, Royal Institute of Technology, Stockholm, Sweden, 1981.

Lindberg, P. O. & Olafsson, S. [1980]: *On the Lengths of Simplex Paths: The Assignment Case*, Department of Mathematics, Royal Institute of Technology, Stockholm, Schweden, 1980.

Lovász, L. [1980]: *A New Linear Programming Algorithm — Better or Worse than the Simplex Method?*, The Mathematical Intelligencer 2 (1980), 141–146.

Mangasarian, O. [1969]: *Nonlinear Programming*, Mc Graw Hill, New York 1969.

May, J. H. & Smith, R. L. [1982]: *Random Polytopes: Their Definition, Generation and Aggregate Properties*, Mathematical Programming 24 (1982), 39–54.

McMullen P. [1970]: *The Maximum Numbers of Faces of a Convex Polytope*, Mathematika 17 (1970), Part 2, 179–184.

Megiddo, N. [1982]: *Solving Linear Programming in Linear Time when the Dimension is Fixed*, Statistics Department, Tel Aviv University, Israel, April 1982.

Megiddo, N. [1983]: *Improved Asymptotic Analysis of the Average Number of Steps Performed by the Self-Dual Simplex Algorithm*, Dept. of Computer Science, Stanford University, September 1983.

Megiddo, N. [1984]: *A Note on the Generality of the Self-Dual Algorithm with Various Starting Points*, Department of Computer Science, Stanford University, 1984.

Murty, K. G. [1978]: *Computational Complexity of Complementary Pivot Methods*, in: Balinski & Cottle (eds.), Complementarity and Fixed Point Problems, 1978, 61–73.

Murty, K. G. [1980]: *Computational Complexity of Parametric Linear Programming*, Mathematical Programming 19 (1980), 213–219.

Nachbin, L. [1965]: *The Haar Integral*, P. van Nostrand, Princeton, 1965.

Natanson, I. P. [1965]: *Theorie der Funktionen einer reellen Veränderlichen*, Harri Deutsch Verlag, Zürich, 1965.

Olafsson S. [1984]: *Studies of the Efficiency of the Stochastic Simplex Method*, Dissertation, Dept. of Mathematics, Royal Institute of Technology, Stockholm, Schweden, 1984.

Orden, A. [1974]: *Computational Investigation and Analysis of Probabilistic Parameters of Convergence of a Simplex Algorithm*, Colloquia Mathematica Socetatis János Bolyai, 12. Progress in Operations Research, Eger, Hungary, 1974, 705–715.

Orden, A. [1976]: *A Study of Pivot Probabilities in LP Tableaus*, in: A. Prekopa (ed.), Survey of Mathematical Programming, Publishing House of the Hungarian Academy of Sciences, 1976.

Philip, J. [1972]: *Plane Sections of Simplices*, Mathematical Programming 3 (1972), 312–325.

Prékopa, A. [1972]: *On the Number of Vertices of Random Convex Polyhedra*, Periodica Mathematica Hungarica 2 (1972), 259–282.

Raynaud, H. [1965]: *Sur le Comportement Asymptotique de l'Enveloppe Convexe d'un Nuage des Points Tirés au Hasard dans* $\mathbb{R}^n$, C. R. Acad. Sci. Paris 261 (1965), 627–629.

Raynaud, H. [1970]: *Sur l'Enveloppe Convexe des Nuages de Points Aléatoires dand* R^n, Journal of Applied Probability 7 (1970), 35–48.

Renyi, A. [1973]: *Wahrscheinlichkeitsrechnung*, VEB Deutscher Verlag der Wissenschaften, Berlin, 1973.

Renyi, A. & Sulanke, R. [1963]: *Über die konvexe Hülle von n zufällig gewählten Punkten I*, Zeitschrift für Warsch. Verw. Gebiete 2 (1963), 75–84.

Renyi, A. & Sulanke, R. [1964]: *Über die konvexe Hülle von n zufällig gewählten Punkten II*, Zeitschrift für Wahrscheinlichkeitstheorie 3 (1964), 138–147.

Renyi, A. & Sulanke, R. [1968]: *Zufällige konvexe Polygone in einem Ringgebiet*, Zeitschrift für Wahrscheinlichkeitstheorie 9 (1968), 146–157.

Rinnooy Kan, A. H. G. & Telgen, J. [1981]: *The Complexity of Linear Programming*, Statistica Neerlandica 2 (1981), 91–107.

Ruben, H. [1954]: *On the Moments of Order Statistics in Samples from Normal Populations*, Biometrica 41 (1954), 200–227.

Saigal, R. [1983a]: *On Some Average Results for Random Linear Complementarity Problems*, Dept. of Industrial Engineering, Northwestern University, Evanston, Illinois, June 1983.

Saigal, R. [1983b]: *An Addendum to: On Some Average Results for Random Linear Complementarity Problems*, Dept. of Industrial Engineering, Northwestern University, Evanston, Illinois, August 1983.

Saigal, R. [1984]: *A Variant that Solves Random Convex Quadratic Programs in Average Steps Bounded by a Quadratic Function of Size*, Department of Industrial Engineering, Northwestern University, Evanston, March 1984.

Santalo, L. A. [1954]: *Introduction to Integral Geometry*, Hermann Cie, Paris, 1954.

Schmidt, B. K. & Mattheiss, T. H. [1977]: *The Probability that a Random Polytope is Bounded*, Mathematics of Operations Research 2 (1977), 292–296.

Schmidt, W. M. [1968]: *Some Results in Probabilistic Geometry*, Zeitschrift für Warsch. Verw. Gebiete 9 (1968), 158–162.

Shamir, R. [1984]: *The Efficiency of the Simplex Method: A Survey*, Department of Industrial Engineering and Operations Research, University of California, Berkeley, May 1984.

Smale, S. [1982]: *The Problem of the Average Speed of the Simplex Method*, Proceedings of the 11th International Symposium on Mathematical Programming, Universität Bonn, August 1982, 530–539.

Smale, S. [1983]: *On the Average Speed of the Simplex Method*, Mathematical Programming 27 (1983), 241–262.

Stoer, J. & Witzgall, C. [1970]: *Convexity and Optimization in Finite Dimensions I*, Springer Verlag, Berlin, 1970.

Telgen, J. [1979]: *Redundancy and Linear Programs*, Dissertation, Erasmus Universität Rotterdam, 1979.

Todd, M. J. [1983]: *Polynomial Expected Behavior of a Pivoting Algorithm for Linear Complementarity and Linear Programming Problems*, Technical Report No. 595, School of Operations Research and Industrial Engineering, Cornell University, Ithaca, New York, November 1983.

Traub, J. & Wozniakowski, H. [1982]: *Complexity of Linear Programming*, Operations Research Letters 1 (1982), 59–62.

Tucker, A. [1963]: *Combinatorial Theory Underlying Linear Programs*, in: R. Graves & P. Wolfe (eds.) Recent Advances in Mathematical Programming, Mc Graw Hill, New York, 1963, 1–16.

Ulkucu, A. [1975]: *Efficiency Measures of Vertex Following Algorithms on Polyhedral Sets*, Operations Research Center, Report No. 75-9, College of Engineering, University of California, Berkeley, 1975.

Valentine, F. A. [1964]: *Convex Sets*, Mc Graw Hill, New York, 1964.

Van Dam, W. B. & Telgen, J. [1979]: *Randomly Generated Polytopes for Testing Mathematical Programming Algorithms*, Erasmus Universiteit Rotterdam, Report No. 7929/0, 1979.

Walkup, D. W. [1978]: *The Hirsch Conjecture Fails for Triangulated 27-Spheres*, Mathematics of Operations Research 3 (1978), 224–230.

Wan, Yieh-Hei [1983]: *On the Average Speed of the Lemke's Algorithm for Quadratic Programming*, Dept. of Mathematics, State University of New York at Buffalo, 1983.

Weyl, H. [1935]: *Elementare Theorie der konvexen Polyeder*, Comment. Math. Helv. 7 (1935), 290–306.

Wolfe, P. & Cutler, L. [1963]: *Experiments in Linear Programming*, in: R. Graves & P. Wolfe (eds.) Recent Advances in Mathematical Programming, Mc Graw Hill, New York, 1963, 177–200.

Zadeh, N. [1981]: *What is the Worst Case Behaviour of the Simplex Method?*, Technical Report No. 27, Dept. of Operations Research, Stanford University, 1981.

Ziezold, A. [1970]: *Über die Eckenanzahl zufälliger konvexer Polygone*, Dissertation, Heidelberg, 1970.

Subject Index

Algorithms and Combinatorics

Editors: R. L. Graham, B. Korte, L. Lovász

Combinatorial mathematics has substantially influenced recent trends and developments in the theory of algorithms and its applications. Conversely, research on algorithms and their complexity has established new perspectives in discrete mathematics. This new series is devoted to the mathematics of these rapidly growing fields with special emphasis on their mutual interactions.

The series will cover areas in pure and applied mathematics as well as computer science, including: combinatorial and discrete optimization, polyhedral combinatorics, graph theory and its algorithmic aspects, network flows, matroids and their applications, algorithms in number theory, group theory etc., coding theory, algorithmic complexity of combinatorial problems, and combinatorial methods in computer science and related areas.

The main body of this series will be monographs ranging in level from first-year graduate up to advanced state-of-the-art research. The books will be conventionally type-set and bound in hard covers. In new and rapidly growing areas, collections of carefully edited monographic articles are also appropriate for this series. Occasionally there will also be "lecture-notes-type" volumes within the series, published as *Study and Research Texts* in soft cover and camera-ready form. This will be mainly an outlet for seminar notes, drafts of textbooks with essential novelty in their presentation, and preliminary drafts of monographs.

Prospective readers of the series **Algorithms and Combinatorics** include scientists and graduate students working in discrete mathematics, operations reearch and computer science.

Forthcoming titles:

M. Grötschel, L. Lovász, H. Schrijver, **Geometric Algorithms and Combinatorial Optimization**

K. Murota, **Structural Solvability and Controllability of Systems**

Springer-Verlag
Berlin Heidelberg New York
London Paris Tokyo

A. Recski, **Matroid Theory and its Applications**

J. Nesétril, V. Rödl (eds.), **Mathematics of Ramsey Theory**

Springer